Withdrawn
University of Waterloo

*High-Pressure Shock Compression of Condensed Matter*

**Springer**
*New York*
*Berlin*
*Heidelberg*
*Barcelona*
*Hong Kong*
*London*
*Milan*
*Paris*
*Singapore*
*Tokyo*

**High-Pressure Shock Compression of Condensed Matter**

*L.L. Altgilbers, M.D.J. Brown, I. Grishnaev, B.M. Novac, I.R. Smith, I. Tkach,* and *Y. Tkach:* Magnetocumulative Generators

*J. Asay* and *M. Shahinpoor* (Eds.): High-Pressure Shock Compression of Solids

*A.A. Batsanov*: Effects of Explosion on Materials: Modification and Synthesis Under High-Pressure Shock Compression

*R. Cherét*: Detonation of Condensed Explosives

*L. Davison, D. Grady*, and *M. Shahinpoor* (Eds.): High-Pressure Shock Compression of Solids II

*L. Davison, Y. Horie*, and *M. Shahinpoor* (Eds.): High-Pressure Shock Compression of Solids IV

*L. Davison* and *M. Shahinpoor* (Eds.): High-Pressure Shock Compression of Solids III

*A.N. Dremin*: Toward Detonation Theory

*R. Graham*: Solids Under High-Pressure Shock Compression

*J.N. Johnson* and *R. Cherét* (Eds.): Classic Papers in Shock Compression Science

*M. Sueska*: Test Methods for Explosives

*J.A. Zukas* and *W.P. Walters* (Eds.): Explosive Effects and Applications

Larry L. Altgilbers
Igor Grishnaev
Ivor R. Smith
Yuriy Tkach
Mark D.J. Brown
Bucur M. Novac
Iaroslav Tkach

# Magnetocumulative Generators

With a Foreword by C.M. Fowler

With 203 Illustrations

Springer

Larry L. Altgilbers
Mark D.J. Brown
Advanced Technology Directorate
Missile Defense and Space
Technology Center
U.S. Army Space and Missile
Defense Command
Huntsville, AL 35807, USA

Igor Grishnaev
Peter the Great Military
Engineering Academy
Moscow K-47, Russia

Bucur M. Novac
Ivor R. Smith
Department of Electronic and Electrical
Engineering
Loughborough University
Loughborough, Leicestershire LE 11 3TU, U.K.

Yuriy Tkach
Iaroslav Tkach
Institute of Electromagnetic Research
Pravdi av.5
Kharkov 31022, Ukraine

*Editor-in-Chief*
Robert A. Graham
Director of Research
The Tomé Group
383 Entrada Road
Los Lunas, NM 87031, USA

Library of Congress Cataloging-in-Publication Data
Magnetocumulative generators / Larry L. Altgilbers . . . [et al.]
p. cm. — (High pressure shock compression of condensed matter)
Includes bibliographical references and index.
ISBN 0-387-98786-X (hardcover : alk. paper)
1. Pulsed power systems. 2. Magnetic flux compression.
I. Altgilbers. Larry L. II. Series.
TK2986.M34 1999
621.3—dc21 99-13257

Printed on acid-free paper.



Production managed by Terry Kornak; manufacturing supervised by Erica Bresler.
Photocomposed copy prepared by the authors.
Printed and bound by Sheridan Books, Ann Arbor, MI.
Printed in the United States of America.

9 8 7 6 5 4 3 2 1

ISBN 0-387-98786-X Springer-Verlag New York Berlin Heidelberg SPIN 10715275

# Foreword

Devices that convert explosive energy into electromagnetic energy are often called Flux Compression Generators (FCGs) in the United States, whereas the term Magnetocumulative Generators (MCGs) is more commonly used in Russia. Since the Russian literature is accessed more heavily in this book, the latter term is used here. In any event, the basic process involves using explosives to force an initial magnetic flux into a region of smaller inductance in such a manner that loss of flux is minimized. In the event that no flux is lost, the magnetic energy associated with the flux, inversely proportional to the inductance, must increase. Flux loss is minimized by confining it with good conductors which, in turn, are driven rapidly by the explosive to reduce the system inductance. The magnetic energy is increased by the work the conductors do as they are forcibly moved against the magnetic field, the energy, in turn, being supplied by the explosive driving them. As the reader may infer, there are different kinds of generators, some of which might be difficult to recognize as MCGs. Nonetheless, they all possess the features outlined above.

Explosives have some unique features as energy sources. They have very high available energy densities; they release energy rapidly, or at high power; they can develop very high pressures. Different exploitation of these features has led, historically, to MCG development and use along two different lines: as generators of ultrahigh magnetic fields and as compact energy sources for devices that require very high pulsed power levels. Some peak performance levels achieved with these devices include the following: magnetic fields in excess of two-thousand Teslas have been reported recently; pulsed power bursts of some hundred megajoules at multi-terawatt levels have been achieved; rocket launched experiments in the ionosphere have been performed that were only possible by using compact explosive powered electromagnetic sources.

The understanding, and thus the design and construction, of these devices requires knowledge in many disciplines such as equation of state of materials, frequently under extreme conditions of pressure, temperature, and magnetic field, shock-wave physics, many aspects of electromagnetic theory, including magnetic field diffusion, the behavior of explosives, and other topics. A number of monograms that unite some of these disciplines have been written over the years, but only one comprehensive treatment of the subject has been available—the well-known book by H. Knoepfel, *Pulsed High Magnetic Fields* (North-Holland, Amsterdam-London, 1970). The main thrust of this book was the production and use of high magnetic fields and the design of the systems that produce them. Some treatment was made of pulsed power supplies, but it was necessarily limited since contributions from the large government weapons laboratories, the source of most of these devices, were withheld.

In is probable that more than 80% of the publications in the field have been published since Knoepfel's book appeared nearly 30 years ago. Of the eight "Megagauss Conferences," the major forum for this field, several were held since the book appeared. The proceedings of the first conference contained only 26 papers, while well over 800 papers were given at the subsequent seven conferences. The present volume has successfully undated the field, particularly in the treatment of MCGs designed for pulsed power sources. Of necessity, in a field so broad, the authors have had to limit the scope of the book. They have omitted treatment of high magnetic fields, have restricted the treatment of pulsed power MCGs mainly to a class called helical generators, and have restricted discussion of applications to only a few devices. These restrictions, however, have allowed a more complete discussion of the topics treated.

The first two chapters of the book present a review of disciplines necessary for a good understanding of the subject. Included here are a fairly comprehensive background in the electrodynamics involved in the operation of MCGs as well as a brief survey of explosives, the properties, and effects. Several types of generators, among the many used in various applications, are next described, followed by a discussion of various techniques used to shape the generator output pulses so that they are suitable for various applications. Chapter 5 contains one of the most comprehensive collections of lumped parameter solutions to various generator-load configurations, a number of the solutions being reduced to closed form in terms of known functions. The preceding chapters are used somewhat as a prolog to now lead the reader, in Chapter 6, through a complete design of a helical generator, thus putting to use much of the preceding material. Various experimental techniques are discussed in the penultinale chapter, among other things, giving the reader insight into various diagnostics employed in the field. The final chapter discusses use of MCGs as power supplies for various devices, primarily laser and microwave sources. The book, enhanced by the generous use of figures, some 200 of them, is a useful and timely addition to the field.

*C.M. Fowler*
Los Alamos National Laboratory
March 20, 1999

# Preface

In the 1950s, several groups of researchers proposed the use of explosives to reduce rapidly the volume occupied by a magnetic field, thereby generating a very high magnetic field. Perhaps the most prominent among these was Andrei Sakharov, an original thinker widely known and well respected fro his many important contributions across a wide spectrum of theoretical work on atomic science and technology. Subsequently, experimentalists at both the All-Russian Institute of Experimental Physics at Sarov (Russia) and the Los Alamos National Laboratory (USA) very successfully exploited the basic concepts that had emerged from these initial studies, as they designed, built, and tested the first working forms of the devices widely known as either magnetocumulative generators (MCGs) or flux compression generators (FCGs).

Over the years, generators with many widely differenct and often highly ingenious geometrical forms have appeared, as they have found applications in increasing front-line areas of basic and applied research. Nowadays, they are primarily use for the production of either very strong magnetic fields or very high-energy and high-current electric pulses. Although the basic science involved of course remains the same, whatever the applicatio, those applications considered in this book describe mainly the use of MCGs as sources of pulsed power. Typical examples are found in areas such as high-power lasers and microwave generators; in detonator arrays and railguns; in ion, electron, and neutron radiation sources; and in many others besides.

In order to ensure a full appreciation of the science and technology that is involved, this book reviews the basic physics that is necessary, before considering in detail some of the basic generator designs that have been built and tested. Consideration is also given to the construction and design of a typical generator for specific applications and the important role played by the load in the overall design process.

It should be noted that most of the material presented in this book arises from the personal experiences, views, and contributions of the various authors and that MCG work is ongoing in many laboratories in a number of other countries. It was not the intention of the authors to overlook the valuable contributions of other researchers, but owing to limitations on the size of the volume, they decided to limit the material to that which they are most familiar.

To achieve the objective of this book and to present a coherent account of MCGs, Chapter 1 contains a thorough introduction to these devices. This chapter also provides the underlying electromagnetic and shock wave theory and the explosives technology that are required to appreciate the material of the later chapters. Chapter 2 describes the conditions that affect magnetic flux compression and the basic theory of how flux compression is achieved, and raises a number of important design issues. This is followed in Chapter 3 by a review of various basic MCG designs, including the cylindrical coaxial, conical, helical, plate, disk, and semiconductor generators. This chapter concludes by discussing cascaded MCGs, built up from two or more of the basic designs, and special purpose short-pulse designs.

Chapter 4 describes the various ancillary circuit components that ensure the MCG delivers its specified output pulse to the load, and includes novel switches, transformers, and transmission lines that have been developed. The precise form these take depends on the characteristics of the load, and Chapter 5 therefore discusses a number of possible loads and the corresponding impact they have on the operation of the MCG.

Chapter 6 is devoted to an in-depth account of the design, construction, and testing of a particular helical MCG developed by Loughborough University in the UK and termed the FLEXY I generator. This is followed in Chapter 7 by a description of the switches and conditioning circuits that were used with the FLEXY I, of the experimental techniques required to obtain its operating parameters, and of an explosive laboratory. This practical theme continues in Chapter 8, which describes the application of MCGs as power sources for high-power laser and microwave generators, topics selected because of the current high level of research and development interest in several countries.

As in any big undertaking, there are many people whom we need to acknowledge for their guidance and support provided over the three years or so that it has taken to produce this book. Particular thanks must be given to Dr. J. Richard Fisher (Director), Dr. Michael J. Lavan (Director, Advance Technology Directorate), and Dr. Ira Merritt (Chief, Concept Identification and Application Analysis Division), all of the US Army Missile Defense and Space Technology Center. We should also acknowledge Mr. Phillip Tracy (Teledyne Brown Engineering, USA), Dr. Saulius Balevicius (Semiconductor Physics Institute, Lithuania), Dr. Kris Kristiansen (Texas Tech University, USA), Dr. Alexander Prishchepenko (High Mountain Geophysical Institute, Russia), Drs. C.M. "Max" Fowler and Doug Tasker (Los Alamos National Laboratory, USA), Dr. Robert Hoeberling (Explosives Pulsed Power, Inc., USA), Dr. John M. Lyons (DERA, UK), and Prof. Michael J.Kearney (Loughborough University, UK). We would like to pay special attention to Dr. Lee Davison, Technical Editor for Springer-Verlag, who provided invaluable suggestions on the content and organization of this book.

# Contents

# 1
# Explosive-Driven Power Sources

## 1.1 Introduction

The generation of high-energy and high-current pulses is critical to the success of many scientific and engineering projects. High-energy, high-current pulses are needed to accelerate plasmas under laboratory conditions, power intense gas-discharge light sources, accelerate high-current electron beams, excite high-power lasers and microwave sources, accelerate projectiles in railguns, generate high-pressure pulses, and heat plasmas in thermonuclear fusion reactors [1.1,1.2]. These high-current pulses are used to simulate processes that occur in space, during powerful explosions, and in aerodynamics, as well as to generate powerful shock waves and plasma flows in industrial processes. The technology involved in the production of these pulses is referred to as *pulsed power technology*, and it deals with the generation of very-high-power electromagnetic pulses and the coupling of these pulses into loads [1.3].

Although numerous devices that produce high-energy, high-current pulses have been developed, the focus of this book is on pulsed power generators that convert the chemical energy of high explosives into electrical energy. Because explosives offer energy storage at the highest available density, explosive-driven generators can be much smaller and lighter than those employing other energy sources. It should be noted that explosive power sources are unsuited for some applications, because each device can be used only once: the high explosive destroys the power source itself and quite possibly the load.

Explosive-driven generators have been designed to produce high current and energy pulses, but ancillary equipment—switches, transformers, and transmission lines— is required to deliver properly shaped pulses to a load. The characteristics of the load profoundly affect system operation and careful, coordinated design of both the generator and the ancillary equipment is required for pulse shaping and the efficient transfer of energy.

Following a brief introduction to several types of explosive power sources, attention is focused on magnetocumulative generators (MCGs), the most versatile and efficient of these devices, and their physics. This is followed by a description of various types of MCGs in Chapter 3, their ancillary equipment in Chapter 4, and their relationship with various types of loads in Chapter 5. In Chapters 6 and 7, the design, construction, and testing of one type of MCG and its ancillary equipment is presented in detail. Finally, in Chapter 8, two specific examples—excitation of high-power lasers and microwave sources—are discussed.

## 1.2 Overview of Explosive-Driven Power Sources

Among those power sources that generate high power pulses of electromagnetic energy, only a few are capable of storing, converting, and releasing this energy over the wide range of time domains from tens of nanoseconds to 1 second. The characteristics of several of these devices are presented in Table 1.1. Of these, only those that use chemical explosives and rechargeable batteries are self-contained; capacitors, inductive stores, and rotating machines must be connected to an external energy source.

Stored electromagnetic energy must be concentrated spatially; that is, within the boundaries of a capacitor, inductor, or other storage device. Usually the wave nature of an electromagnetic field can be neglected, permitting a description of the sources in terms of quasistatic theory. In the case of inductive stores, the transfer of energy can be described solely in terms of the electric current within the conductor. This is because the magnetic field is proportional to the current and the magnetic energy is a function of the magnetic field. In the case of a long, straight wire, for example, the magnetic field strength, $H$, at a radial distance $r$ (greater than the wire diameter, $r_c$) is given by

$$H = \frac{I}{2\pi r_c}, \tag{1.1}$$

where $I$ is the current flowing through the wire, and the energy store per unit volume is $E = BH/2 = \mu_0 H^2/2 = B^2/2\mu_0$, where $B$ is the magnetic flux density and $\mu_0$ is the permeability of free space.

| Energy Source | Energy Density ($MJ/m^3$) | Pulse Length ($\mu s$) | Conversion Efficiency (%) |
|---|---|---|---|
| Chemical Explosives | 8000 | 0.1 | 2 |
| Capacitors | 0.01–0.1 | 0.01 | 80 |
| Rotating Machines | 100–500 | 100 | 30 |
| Inductive Stores | 10 | – | – |
| Rechargable Battery | 500 | 1000 | 0.5 |

TABLE 1.1. Prime energy and energy storage systems.

The devices in Table 1.1 rely on the storage of energy in electric fields, magnetic fields, and/or chemical bonds. In Table 1.1, the energy densities are the maximum amounts of energy stored per unit volume and the pulse lengths and conversion efficiencies are those that have been achieved with power conditioning. The primary focus of this book is on those devices that store energy in molecular bonds of an explosive, but it will be shown that as this energy is released from the explosive, it must be stored in a magnetic field before it can be delivered as electric current to the load.

The first experiments to generate superstrong magnetic fields were conducted in the 1914 by H. Deslandres and A. Perot and in the 1920s by P.L. Kapitsa [1.4]. In the experiments performed by Deslanders and Perot, a rechargeable battery was connected to a specially designed inductive storage device, in which 50 kOe (1 Oe = $10^3/4\pi$ A/m) fields were generated in a very small volume. By using the kinetic energy of the rotor of an AC generator connected to a multicoil solenoid, Kapitsa generated a 320 kOe pulsed magnetic field in a volume of 2 x $10^{-6}$ $m^3$ for a period of 0.01 s. In the early 1950s, experiments were carried out in which the energy of high explosives was applied to a solenoid. In these and subsequent experiments, pulsed magnetic fields in the megagauss regime were achieved. It is these devices that are the basis of the *magnetocumulative generator* (MCG), also referred to as *magnetic flux compression generators* (MFCG) or *flux compression generators* (FCG).

High-energy physics laboratories have usually used high-voltage capacitors as the basis of their energy storage systems. After extensive research, magnetic fields in excess of 1 MOe were achieved by discharging these sources through single-coil solenoids. However, the dimensions of the solenoids were very large, making it impossible to reduce their overall inductance to low levels. The density at which energy can be stored in capacitors is 80 $kJ/m^3$ (40 J/g), a limit that is set by dielectric breakdown (attributable to autoelectron emission) at field strengths of $10^9$ V/m. For example, to store 100 MJ of energy in a dielectric with an energy density of $10^5$ $J/m^3$ would require 1000 $m^3$ of dielectric weighing 25 x $10^5$ kg. Of course, the resulting capacitor bank would be very large, inefficient, and expensive.

The need to circumvent problems associated with large capacitor banks has motivated extensive investigation of devices that convert the chemical energy of high explosives into electromagnetic energy for charging inductive stores. This work has been conducted in the United States, Russia, China, Japan, United Kingdom, and other countries. These explosive-driven energy sources are collectively called *explosive magnetic generators* (EMGs). They are divided into families according to how they convert chemical energy into electromagnetic energy. The five families of EMGs are:

- *Ferromagnetic generators* (FMG), which generate an electromotive force (EMF) by the demagnetization of magnetic materials with explosive generated shock waves.

- *Ferroelectric (Segnetoelectric) generators*[1] (SEG), which generate an EMF by the depolarization of ferroelectric (segnetoelectric) materials.

- *Magnetohydrodynamic generators* (MHDG), which generate an EMF when a plasma is moved through a magnetic field.

- *Magnetoelectric generators* (MEG), which generate an EMF by the motion of an explosively propelled magnetic body through a magnetic field, thus changing the structure of the field.

- *Magnetocumulative generators* (MCG), which generate an EMF by the compression of a magnetic field.

Of these families of devices, the most advanced is the MCG. MCGs are essentially pulsed current sources that generate high energy densities and very high powers. During their final stage of operation, the magnetic field compression process has been completed and the device becomes an inductive energy store. The maximum density of the energy stored in the generator may reach $10^9$ J/m$^3$; that is, the energy stored in a volume of 0.1 m$^3$ would be $10^8$ J. This illustrates well one advantage of MCGs, since the stored energy density is very much greater than the sources in Table 1.1. Another advantage of the MCG is its lower cost in comparison to these sources.

---

[1] In the 1930s, I.V. Kurchatov, while working at the Ioffe Physical Technical Institute in Leningrad, was studying the properties of dielectric materials. During the course of his studies, he examined anomalies in the dielectric properties of Rochelle salts, which led to the discovery of a special class of crystals that behave in electric fields exactly like ferromagnets in magnetic fields. These crystals are known as ferroelectrics, but Kurchatov called them *segnetoelectrics* after the French chemist Segnier.

## 1.3 Magnetocumulative Generator History

Development of magnetic flux compression techniques began in the early 1950s in Russia and the United States, followed by the United Kingdom, France, Italy, and China in the 1960s. Research groups in Japan, Poland, Germany, and Romania have since joined what has been referred to as the *Megagauss Club*. These countries and their achievements are summarized in Table 1.2.

According to an unpublished report [1.4] in 1944 by J.L. Fowler, magnetic flux compression was first used during World War II at Los Alamos National Laboratory to measure the implosion of explosive-driven liner systems. In 1954, W.B. Garn, at the suggestion of E. Teller and F.J. Willig, conducted flux compression experiments to generate high magnetic fields. In 1957, C.M. "Max" Fowler resumed this work [1.1]. In 1951, A.D. Sakharov at the Kurchatov Institute in the Former Soviet Union (FSU) first proposed using imploding devices to compress magnetic fields [1.2]. In 1952, R.Z. Lyudaev and others conducted the first implosion experiments to generate ultrahigh magnetic fields. The first journal article proposing the magnetic field compression method was published in 1957 by Ia.P. Terletskii [1.5] in which he suggested using this method for particle acceleration. Other prominent scientists from Russia who have further developed magnetic field compression technology include A.I. Pavlovskii and V.K. Chernyshev. Other institutes involved in the development of MCGs in the FSU are the Institute of Nuclear Physics, the Efremov Research Scientific Institute of Electrophysical Devices, and Arzamas-16, with the latter taking the lead in the development and application of MCGs. Based upon publications in the 1980s, it appears that 13 scientific and research institutes in the United States were at one time engaged in the development and application of magnetic compression techniques. The greatest number of publications; however, belonged to F. Herlach, G. Knoepfel, and U. Linhart at the Laboratorio Gas Ionizzati (Euroatom-CNEN), Frascati, who were applying magnetic compression to the problem of controlled thermal fusion. These teams have, with comparative ease, been able to generate magnetic fields of $10^3$ Tesla (T), currents of in excess of $10^8$ A, voltages of 1 MV, and magnetic energies of tens of megajoules. At the Megagauss VIII conference in 1998, Arzamas-16 reported achieving a record magnetic field in an MCG of 28 MG (2800 T). The limiting values of output energy and power for certain types of MCGs have been estimated as 1 GJ and $10^{14}$ W, respectively.

In the late 1990s, the number of laboratories in the United States and Russia working on MCGs has decreased. In the US, Los Alamos National Laboratory, the United States Air Force's Phillips Laboratory, Texas Tech University, and the private company Explosive Pulsed Power, Inc., are con-

| | Magnetic Field Density MG($10^2$ T) | Energy MJ | Current MA | Program Start |
|---|---|---|---|---|
| Russia (USSR) | 17/28 | 100 | >300 | 1952 |
| United States | 10/14 | 50 | 320 | 1950 |
| France | 10/11.7 | 8.5 | 24 | 1961 |
| Japan | 5.5 | – | – | 1970 |
| EURATOM (Italy) | 5.4/7 | 2 | 16 | 1961 |
| United Kingdom | 5 | 10 | 20 | 1956 |
| Romania | 5/7.5 | 0.5 | 12 | 1979 |
| Poland | 3.5 | – | 0.8 | 1973 |
| P.R. China | – | – | 2(?) | 1967 |
| Germany | – | – | 1.2(?) | 1975 |

TABLE 1.2. Countries that had started MCG programs in the 1950s through the 1970s. In the column labeled magnetic field density, the second number reflects values that are probably rarely obtained, the boxes with the symbel (–) means that data are not available, and the boxes with a question mark imply that much higher values were probably obtained.

ducting research on and with MCGs. In Russia, the laboratories continuing to do research are Arzamas-16, the Institute of High Temperatures, and the Institute of Chemical Physics in Chernogolovka. However, the number of laboratories conducting research on MCGs in other countries has increased to include Loughborough University in England, Forsvarets Forskningsanstack (FOA) in Sweden, the Gramat Research Center in France, and the Institute of Fluid Physics in China. Other countries conducting research on MCGs are Germany and Japan.

From the 1970s onward, the study of magnetic compression began to move in two different directions; that is, the development of power sources based on magnetic compression and that of devices to generate ultrahigh magnetic fields [1.2]. Systems in the first category are *current* or *energy generators*, in which the flux compression occurs in a region separated from the inductive load, and they are designed to generate very high currents. Systems in the second category are called *energy density generators*, and they are designed to generate ultrahigh magnetic fields by the radial implosion of a hollow, conducting cylinder containing a magnetic field. Each type of device has its own set of characteristics. In the case of the power sources, the aim is to improve the parameters and to enhance the stability of the MCG and thereby increase the current (MA) and energy (MJ) delivered to an external load. In the case of the intense magnetic field sources, the goal is to develop methods that accelerate metal liners by pulsed magnetic fields or by compressed gases to concentrate magnetic field energy at the center of the generator to create ultrahigh magnetic fields (MG). Linerless generators in which magnetic fields are compressed by shock-ionized gases have also been investigated.

Initially, MCGs were used only to create ultrahigh magnetic fields. However, as the physics of these devices became understood through experimentation, it became evident that MCGs could also be used as power sources to generate pulses of electromagnetic energy with pulse lengths ranging from $10^{-6}$ to $10^{-4}$ s. To date, experiments have been conducted in which pulses with currents of 300 MA, powers of $10^{13}$ W, and energies of tens to hundreds of megajoules have been delivered to loads with pulse durations in the microsecond regime.

There are a large number of MCG designs, and they are categorized by whether they are implosive or explosive and by their geometrical configuration. Explosive generators (Fig. 1.1) consist of a seed source (capacitor, battery) that delivers current to a solenoid, a hollow conductor, called a *liner,* filled with high explosives, and a load. When peak current is flowing through the solenoid, the explosives are detonated and the liner expands, compressing the magnetic field in the annular region between the solenoid and the liner. Chemical energy from the high explosives is converted into electromagnetic energy. Implosive generators consist of the same components, but the liner is now imploded to compress the magnetic field.

MCGs take on different geometries, including planar, conical, helical, and disk shapes. Several different designs are shown in Fig. 1.2. However, they all work in basically the same way; that is, explosive-driven conductors are used to compress magnetic fields. Each type of MCG is discussed in Chapter 3.

In the remainder of this chapter, a brief introduction to the principles of electromagnetic theory, shock wave physics, and explosives technology is presented. This material is essential to the proper understanding of the operation of MCGs, developed in later chapters.

## 1.4 Electromagnetic Theory

### *1.4.1 Field Theory: Maxwell's Equations*

In order to understand the relationship between electric current and magnetic fields, let us consider Maxwell's equations [1.6]:

$$\nabla \times \mathbf{H} = \mathbf{j} + \frac{\partial \mathbf{D}}{\partial t}, \tag{1.2}$$

$$\nabla \times \mathbf{E} = -\frac{\partial \mathbf{B}}{\partial t}, \tag{1.3}$$

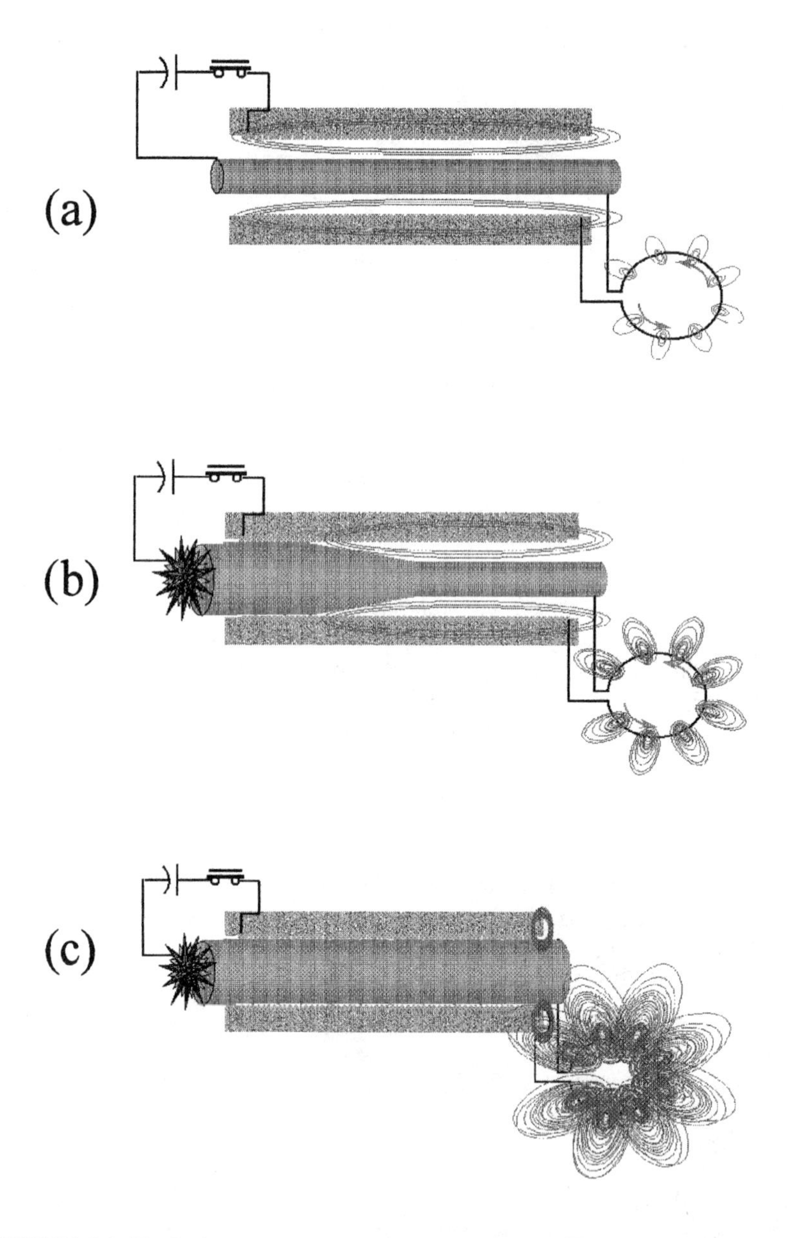


FIGURE 1.1. Explosive magnetocumulative generator. When the detonator initiates the explosives, the liner expands outward at one end, making contact with the first coil of the helix. Protusions, called crowbars, from the first coil are used to trap the magnetic flux within the generator. As the explosive continues to detonate, the liner continues to expand and short out succeeding coils of the helix thereby converting chemical energy into electrical energy. a – shows the MCG prior to detonation, b – during detonation, and c – the end of operation.

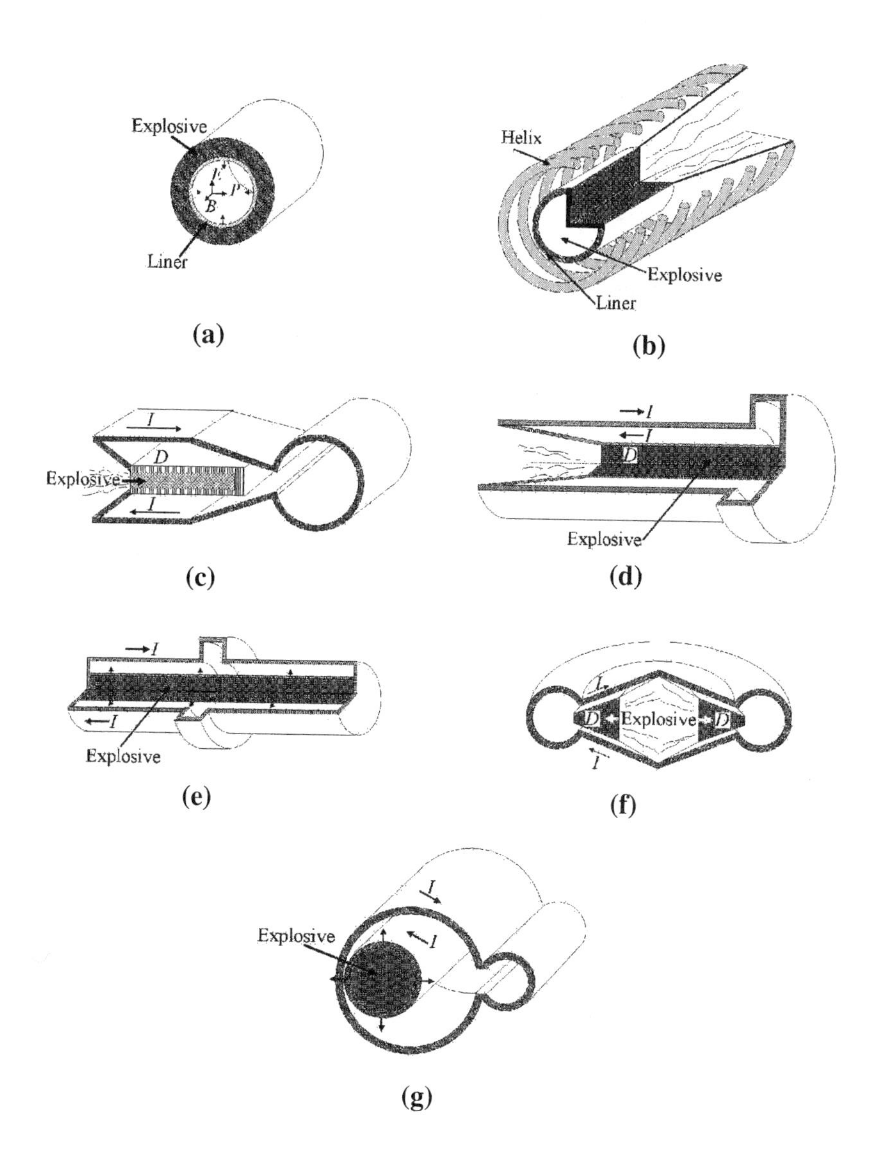


FIGURE 1.2. Different MCG geometric configurations: a – cylindrical field generator, b – helical MCG, c – plate MCG, d – coaxial MCG, e – coaxial MCG with axial initiation, f – disk MCG, and g – loop MCG.

$$\nabla \cdot \mathbf{B} = 0, \tag{1.4}$$

$$\nabla \cdot \mathbf{D} = \rho_e, \tag{1.5}$$

where $\mathbf{H}$ is the *magnetic field strength* or *magnetic intensity* (A/m), $\mathbf{j}$ is the *free current density* (A/m$^2$), $\mathbf{D}$ is the *electric flux density* or *electric displacement* (coulombs/m$^2$), $\mathbf{E}$ is the *electric field strength* or *electric intensity* (V/m), $\mathbf{B}$ is the *magnetic flux density* or *magnetic induction* (webers/m$^2$), and $\rho_e$ is the *density of free electric charges* (coulombs/m$^3$). The constitutive relations that specify the properties of materials (or free space) must also be specified. In isotropic media, these are:

$$\mathbf{B} = \mu \mathbf{H}, \tag{1.6}$$

$$\mathbf{D} = \varepsilon \mathbf{E}, \tag{1.7}$$

in which $\mu = \mu_0 \mu_R$ is the *magnetic permeability* with $\mu_0 = 4\pi \times 10^{-7}$ H/m and $\varepsilon = \varepsilon_0 \cdot \varepsilon_R$ is the *dielectric constant* with $\varepsilon_0 = 8.854 \times 10^{-12}$ F/m. The relative quantities $\mu_R$ and $\varepsilon_R$ can be complicated functions of space, time, fields, material properties, and field variables. An additional constitutive relation is *Ohm's law*:

$$\mathbf{j} = \sigma \mathbf{E}, \tag{1.8}$$

in which the *conductivity,* $\sigma$, is a material coefficient that is again a complicated function of temperature and magnetic field. The *constitutive parameters* $(\varepsilon, \mu, \sigma)$ are used to characterize the electrical properties of a material. In general, materials are described as dielectrics (insulators), magnetics, and conductors depending on whether polarization, magnetization, or conduction, respectively, is the predominant response to electromagnetic stimuli.

Throughout this book, it is assumed that the magnetic fields are quasistationary; i.e., the displacement term $\partial \mathbf{D}/\partial t$ can be neglected in Eq. 1.2. In *free space*, where $\mathbf{j} \equiv \mathbf{0}$, the magnetic field varies with time but is independent of position. Therefore, the quasistationary field approximation is valid if $\lambda >> l$, where $\lambda$ is the characteristic wavelength of any field variation and $l$ is a characteristic length of a magnetic coil or other spatial configuration under consideration. In *conducting media*, the quasistationary field approximation is valid if $1/\omega >> \varepsilon/\sigma$, where $\omega$ is the frequency at which the magnetic field varies. For good conductors such as metals, this condition is valid up to frequencies far in excess of those encountered in the operation of MCGs [1.4].

The integral form of Maxwell's equations describes the relationship between field vectors, charge densities, and current densities over extended regions of space. They are usually used to solve boundary-value problems that possess complete symmetry. However, the fields and their derivatives do not need to be continuous functions. Applying *Stoke's theorem*

$$\int_{\mathrm{S}} (\nabla \times \mathbf{H}) \cdot ds = \oint_{\mathrm{C}} \mathbf{H} \cdot d\mathbf{l} \tag{1.9}$$

to an open surface S bounded by a closed curve C, one obtains *Ampere's Law* (which states that the line integral of the magnetic field over a closed path is equal to the current enclosed):

$$\oint_{\mathrm{C}} \mathbf{H} \cdot d\mathbf{l} = I_{\mathrm{S}}, \tag{1.10}$$

where $I_{\mathrm{S}} = \int \mathbf{j} \cdot d\mathbf{s}$ is the total current flowing through S. In a similar manner, one obtains *Faraday's law* (which states that the electromotive force (EMF) or electric field appearing at the open-circuited terminals of a loop is equal to the time rate of change of magnetic flux linking the loop):

$$\oint_{\mathrm{C}} \mathbf{E} \cdot d\mathbf{l} = -\int_{\mathrm{S}} \frac{\partial \mathbf{B}}{\partial t} \cdot ds = -\frac{d}{dt} \int_{\mathrm{S}} \mathbf{B} \cdot ds. \tag{1.11}$$

The magnetic induction can also be expressed in terms of the number of magnetic lines of force that pass through a given surface, which is called the *magnetic flux*, $\Phi$ (T·m$^2$ or webers) and is defined by

$$\Phi = \oint \mathbf{B} \cdot d\mathbf{S}. \tag{1.12}$$

When $\Phi = 0$, this expression is known as *Gauss's Law*, which asserts that the magnetic flux through any closed Gaussian surface must be zero.

The power density of varying electromagnetic waves is given by

$$\mathbf{S} = \mathbf{E} \times \mathbf{H}^*, \tag{1.13}$$

where $\mathbf{S}$ (W/m$^2$) is the Poynting vector and $\mathbf{H}^*$ is the complex magnetic field. The equation governing the time-domain Poynting vector can be derived by subtracting the dot product of Eq. 1.2 with $\mathbf{E}$ from the dot product of Eq. 1.3 with $\mathbf{H}$. Using the vector identity:

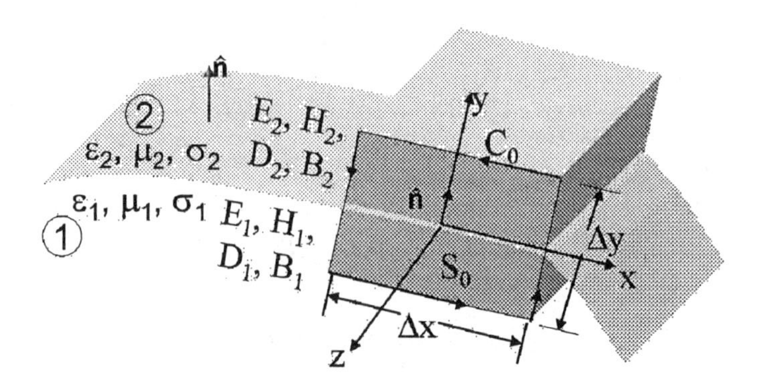


FIGURE 1.3. Geometry for boundary conditions of (1) tangential and (2) normal components of the electric and magnetic fields.

$$\mathbf{H} \cdot \nabla \times \mathbf{E} - \mathbf{E} \cdot \nabla \times \mathbf{H} = \nabla \cdot (\mathbf{E} \times \mathbf{H}), \tag{1.14}$$

it can be shown that:

$$\nabla \cdot (\mathbf{E} \times \mathbf{H}) = -\mathbf{H} \cdot \frac{\partial \mathbf{B}}{\partial t} - \mathbf{E} \cdot \frac{\partial \mathbf{D}}{\partial t} - \mathbf{J} \cdot \mathbf{E}. \tag{1.15}$$

Assuming that $\mathbf{J} = 0$ and using the constitutive relations of Eqs. 1.6 and 1.7, Eq. 1.15 can be rewritten as

$$\nabla \cdot (\mathbf{E} \times \mathbf{H}) = -\frac{\partial}{\partial t}\left(\frac{B^2}{2\mu}\right) - \frac{\partial}{\partial t}\left(\frac{\varepsilon E^2}{2}\right). \tag{1.16}$$

The quantity in the first set of parentheses is the energy stored in the magnetic field and the quantity in the second set is the energy stored in the electric field.

In general, most problems of interest involve free space with conducting boundaries. The electromagnetic equations apply in both regions; i.e., free space and the conductor, and they are coupled together by their *boundary conditions*. Identifying the two media as 1 and 2, and assuming that the conductor has finite conductivity (Fig. 1.3), Maxwell's equations can be used to derive the following boundary conditions [1.6]:

- The tangential component of the electric field is continuous across an interface between two media, with no impressed magnetic current densities along the interface:

$$\widehat{n} \times (\mathbf{E}_2 - \mathbf{E}_1) = 0, \tag{1.17}$$

where $\widehat{n}$ is a unit vector normal to the interface between the two media.

- The tangential component of the magnetic field across an interface is discontinuous by an amount equal to the electric current density, $\mathbf{j}$ (A/m), flowing on the surface:

$$\widehat{n} \times (\mathbf{H}_2 - \mathbf{H}_1) = \mathbf{j}. \tag{1.18}$$

Note that the tangential component of $\mathbf{H}$ is continuous at the interface if $\mathbf{j} = \mathbf{0}$.

- The normal components of both the electric flux density and the electric field on an interface on which a surface charge density resides are discontinuous, by an amount equal to the surface charge density:

$$\widehat{n} \cdot (\mathbf{D}_2 - \mathbf{D}_1) = \rho_e, \tag{1.19}$$

$$\widehat{n} \cdot (\mathbf{E}_2 - \mathbf{E}_2) = \rho_e. \tag{1.20}$$

Nothe that when $\rho_e = 0$, these fields are continuous at the interface.

- The normal components of the electric field intensity across an interface are discontinuous:

$$\widehat{n} \cdot (\varepsilon_2 \mathbf{E}_2 - \varepsilon_1 \mathbf{E}_1) = 0. \tag{1.21}$$

- The normal components of the magnetic flux density across an interface between two media where there are no sources are continuous:

$$\widehat{n} \cdot (\mathbf{B}_2 - \mathbf{B}_1) = 0. \tag{1.22}$$

- The normal components of the magnetic field intensity across an interface are discontinuous:

$$\widehat{n} \cdot (\mu_2 \mathbf{H}_2 - \mu_1 \mathbf{H}_1) = 0. \tag{1.23}$$

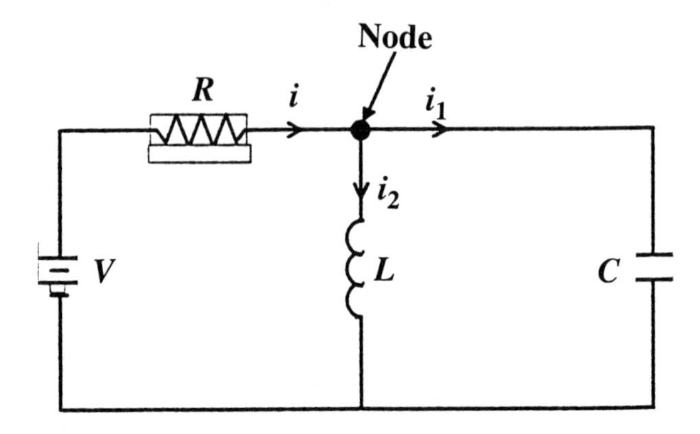


FIGURE 1.4. A node is where three or more branches of an electrical circuit are connected.

### 1.4.2 *Circuit Equations: Kirchhoff's Equations*

Since the equivalent circuit diagrams of MCGs and their loads are treated by lumped-element circuit analysis, Kirchhoff's circuit equations are now introduced. *Kirchhoff's voltage law* states that the algebraic sum of the potential differences around any complete loop within a network is zero, or

$$\sum V = 0. \tag{1.24}$$

*Kirchhoff's current law* states that the algebraic sum of the currents at any node within the network is zero [a node is a point at which three or more conductors are joined (Fig. 1.4)]:

$$\sum i = 0. \tag{1.25}$$

The major elements of most MCG circuits are resistances ($R$), capacitances ($C$), and inductances ($L$). The potential difference across these elements are defined by

$$V = iR \tag{1.26}$$

for resistors (Ohm's law),

$$V = \frac{Q}{C} = \frac{1}{C}\int idt \tag{1.27}$$

for capacitors, and

$$V = L\frac{di}{dt} \tag{1.28}$$

for inductors. In these equations, $V$ is potential difference (V), $i$ is current (A), $Q$ is charge (A·s), and $t$ is time (s).

The differential and integral forms of Maxwell's equations are usually referred to as *field equations*, since the quantities appearing in them are *field quantities*. However, Maxwell's equations can also be written in terms of the *circuit quantities* presented above.

By definition, the work done by an electric field to move a unit charge through a displacement $d\mathbf{l}$ is

$$W = -q\int \mathbf{E}\cdot d\mathbf{l}. \tag{1.29}$$

Since potential difference is also defined to be the work required to move the charge, it can be related to the field quantity $\mathbf{E}$ by

$$V = -\int \mathbf{E}\cdot d\mathbf{l}. \tag{1.30}$$

Substituting this expression into Ohm's law,

$$V = -\int \mathbf{E}\cdot d\mathbf{l} = iR, \tag{1.31}$$

it can be seen that current is generated by either a potential difference or an electric field gradient.

Substituting Eq. 1.30 into the definition of capacitance, $C = QV$,

$$C = -Q\int \mathbf{E}\cdot d\mathbf{l}, \tag{1.32}$$

it can be seen that the capacitance is related to a potential difference or an electric field gradient.

Using Eq. 1.11, the voltage can be related to changes in magnetic flux,

$$\oint_{\mathrm{C}} \mathbf{E}\cdot d\mathbf{l} = -\frac{\partial}{\partial t}\int_{\mathrm{S}} \mathbf{B}\cdot d\mathbf{s} = -\frac{\partial \Phi}{\partial t}, \tag{1.33}$$

which can be rewritten in terms of the circuit quantities,

$$V = -\frac{\partial \Phi}{\partial t} = -L\frac{\partial i}{\partial t}, \tag{1.34}$$

where, according to Faraday's law, a change in magnetic flux due to a change in current causes a voltage drop.

# 1.5 Electromagnetic Phenomena

## 1.5.1 Magnetic Pressure and Diffusion

If a magnetic field is localized within a certain spatial domain, then, according to Maxwell's equations, the current that maintains this field must circulate on the boundaries of the domain. If the boundaries are ideal conductors, this current is concentrated in a thin surface layer and does not penetrate into the material. If the conductor possesses a certain resistance, the charge carriers are scattered in the lattice and the thickness of the current layer increases with time. The current layer is called the *skin layer* and its thickness the *skin depth.*

To study the penetration of magnetic fields into *incompressible, electrically isotropic, conducting media,* Maxwell's equations and Ohm's law are used. Assuming the electrical conductivity is a material constant and independent of space and time, then $\sigma = \sigma_0$. Substituting Eqs. 1.7 and 1.8 into the curl of Eq. 1.2 yields

$$\nabla \times (\nabla \times \mathbf{H}) = \sigma_0 \nabla \times \mathbf{E} + \varepsilon \frac{\partial}{\partial t} (\nabla \times \mathbf{E}) . \tag{1.35}$$

By using the identity

$$\nabla \times \nabla \times \mathbf{A} = \nabla (\nabla \cdot \mathbf{A}) - \nabla^2 \mathbf{A} \tag{1.36}$$

and Eqs. 1.3 and 1.4, Eq. 1.35 is transformed into the *wave equation*:

$$\nabla^2 \mathbf{H} = \sigma_0 \frac{\partial \mathbf{B}}{\partial t} + \varepsilon \frac{\partial^2 \mathbf{B}}{\partial t^2} . \tag{1.37}$$

In a conducting medium, the displacement currents can be neglected in comparison to the conduction currents and the last term of Eq. 1.37 can be omitted. This reduces the equation to the *magnetic diffusion equation:*

$$\nabla^2 \mathbf{H} - \frac{1}{\kappa_0} \frac{\partial \mathbf{H}}{\partial t} = 0, \tag{1.38}$$

where $\kappa_0 = 1/\sigma_0 \mu$ is the *magnetic diffusivity.*

The diffusion of magnetic energy into incompressible conductors is accompanied by an inflow of energy, which appears in the form of magnetic energy and Joule heat generated as a result of eddy currents. Complication arises if the conductivity $\sigma$ is not constant when the diffusion becomes nonlinear. If the conductor is a metal, there are further complications including melting of its surface layer and the formation of instabilities at the metal–field interface. An additional complication arises if the compressibility of the material is taken into account. In fact, compressibility of the conductor sets the most stringent limitations on the generation of ultrahigh magnetic fields [1.4]. Compression of the conductor by the magnetic pressure results in its inner layer moving at a lower velocity than would be the case if it were incompressible.

The depth of the skin layer in a conductor with constant electrical conductivity can be estimated from the equation

$$\delta = \sqrt{\frac{2}{\omega\mu\sigma}}, \tag{1.39}$$

where $\mu$ is the magnetic permeability and $\sigma$ is the conductivity of the conducting material and $\omega$ is the frequency of the pulsed field. The time

$$\tau_S = \frac{1}{2}\pi\mu\sigma\delta^2 \tag{1.40}$$

required for the magnetic field to penetrate to a depth equal to that of the skin layer, which is called the *skin time*, can be determined from Eq. 1.39. Magnetic compression is accompanied by an increase in magnetic field strength and magnetic energy, provided that the time interval during which the cavity compresses the magnetic field is much less than $\tau_S$.

### *1.5.2 Magnetic Force*

A conductor carrying a current density $\mathbf{j}$ in a magnetic field $\mathbf{B} = \mu\mathbf{H}$ experiences a force per unit volume called the *magnetic stress*:

$$\mathbf{F}_M = \mathbf{j} \times \mathbf{B}. \tag{1.41}$$

Neglecting the displacement term in Eq. 1.2 and substituting $\nabla \times \mathbf{H}$ for $\mathbf{j}$ in the above equation yields

$$\mathbf{F}_M = (\nabla \times \mathbf{H}) \times \mathbf{B} = \mu\left[(\mathbf{H}\cdot\nabla)\,\mathbf{H} - \frac{1}{2}\nabla H^2\right]. \tag{1.42}$$

It can be shown that the magnetic force acting on a conductor carrying a current can be expressed in terms of the surface field alone [1.4].

Likewise, a conductor moving through a magnetic field experiences a similar force, which is the basic operating principle of the MCG. To illustrate this, consider Fig. 1.5. As the rectangular loop is pulled out of the magnetic field with velocity $\mathbf{v}$, a force is exerted on the electrons. This force is the Lorentz force

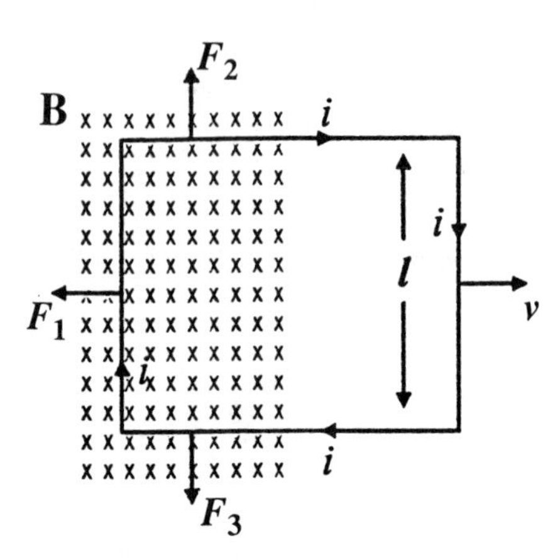


FIGURE 1.5. Forces are exerted on a rectangular loop moving through a magnetic field.

$$\mathbf{F}' = q\mathbf{v}_d \times \mathbf{B}, \tag{1.43}$$

where $q$ is the electric charge and $\mathbf{v}_d$ is the drift velocity of the charge carriers in the wire. The drift velocity is related to the current density by $v_d = j/nq$, where $n$ is the charged particle density. If $l$ is the length of a section of wire and $A$ is its cross-sectional area, then the total force on this section is

$$F = nlAF' = ilB, \tag{1.44}$$

where $i = jA$. Making the appropriate substitutions, Eq. 1.44 becomes

$$\mathbf{F} = i\mathbf{l} \times \mathbf{B}. \tag{1.45}$$

According to Faraday's law, a change in magnetic flux induces an EMF in the conductor:

$$V = -\frac{d\Phi}{dt} = -\frac{d}{dt}(Blx) = -Bl\frac{dx}{dt} = -Blv, \tag{1.46}$$

where $x$ is the distance the rectangular loop is displaced and $v$ is the speed at which it is moved. This EMF sets up a current in the loop:

$$i = \frac{V}{R} = \frac{Blx}{R}. \tag{1.47}$$

Substituting this expression into Eq. 1.44 yields an expression for the force on the wire:

$$F = \frac{B^2 l^2 v}{R}, \tag{1.48}$$

if it is assumed that the $\mathbf{l}$ and $\mathbf{B}$ are perpendicular. It will be noted that the magnetic force on the conductor is directly proportional to the square of the magnetic field.

### *1.5.3 Magnetic Pressure*

Consider a conductor whose surface is subjected to a magnetic pulse $\mathbf{H}(t)$. For simplicity, it is assumed that the conductor is infinitely thick and planar and that the magnetic field lines are parallel to its surface. Neglecting the displacement current, Eq. 1.2 for this case becomes

$$j_y = -\frac{\partial H}{\partial x}. \tag{1.49}$$

Substituting this expression into Eq. 1.41, and integrating along the $x$ axis, gives the hydrodynamic pressure at a depth $x$:

$$p(x,t) = -\int_0^x \frac{\partial H_z}{\partial x} B_z dx = p_H - \frac{1}{2}\mu H^2(x,t), \tag{1.50}$$

where

$$p_H = \frac{1}{2}\mu H^2(0,t) \tag{1.51}$$

is the *magnetic pressure* due to the magnetic field at the surface of the conductor.

### *1.5.4 Electric Fields*

Faraday's law of induction states that a changing magnetic field produces an electric field. Such is the case in the MCG, which is a closed circuit. As the magnetic field is compressed, an EMF, $V = d\Phi/dt$, is induced around the circuit. The work done by the EMF on a unit charge is $qV$ and, from another point of view, the work done by the electric force, $qE$, to move the charge through a distance $l$ is $qEl$. Equating these two expressions for work yields

$$V = El. \tag{1.52}$$

Generalizing this expression gives

$$V = \oint \mathbf{E} \cdot d\mathbf{l} = -\frac{d\Phi}{dt}. \tag{1.53}$$

Thus, as magnetic flux is compressed, electric fields are induced in the MCG circuit, and these, in turn, circulate a current.

Another way of looking at the relationship between magnetic and electric fields is to consider the force acting on the conductor moving through a magnetic field by using the Lorentz force:

$$\mathbf{F} = q(\mathbf{E} + \mathbf{v} \times \mathbf{B}). \tag{1.54}$$

Dividing both sides by $q$, yields an expression for the electric field, $\mathbf{E}^*$, detected by an observer moving with the conductor:

$$\mathbf{E}^* = \mathbf{E} + \mathbf{v} \times \mathbf{B}. \tag{1.55}$$

If the external electric field is $\mathbf{E}^* = 0$, the electric field generated by the motion of the conductor is:

$$\mathbf{E} = -\mathbf{v} \times \mathbf{B}. \tag{1.56}$$

The electric fields generated within the MCG can greatly effect the efficiency of the generator through such energy loss mechanisms as electrical breakdown in the insulator or between the liner and the solenoid.

## 1.6 Shock and Detonation Waves

The surfaces of conductors in MCGs are subject to very strong transient forces produced by detonating explosives and by strong magnetic fields. Since the pressure applied to the surface of a conductor is transmitted to its interior by a shock wave, some knowledge of wave propagation is required for analysis of generator performance.

A *shock wave* is a discontinuous high-pressure disturbance moving through matter [1.7–1.10]. The pressures in strong shock waves stress solid materials far beyond their elastic limits. Since differences among the stress components are limited by the stress yield, the principal stress components are

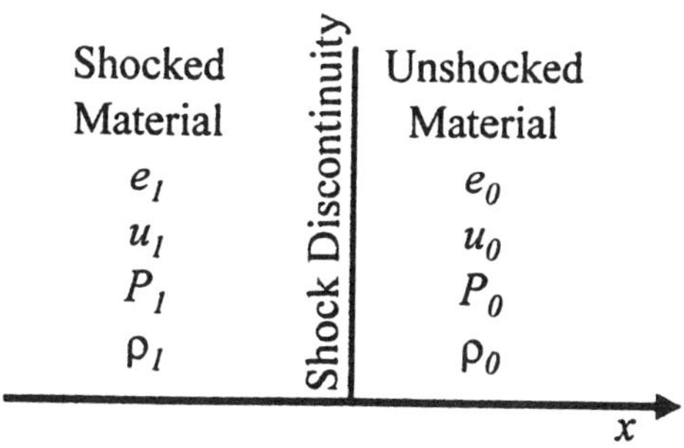

FIGURE 1.6. Shock discontinuity propagating at a velocity $U$ and producing a transition from material in the state $(P_0, \rho_0, e_0, u_0)$ to the state $(P_1, \rho_1, e_1, u_1)$.

almost equal and the stress can be approximated by its average value, the pressure. The fields on either side of a shock discontinuity propagating at a velocity $U$ are as shown in Fig. 1.6.

Applying the principles of conservation of mass, momentum, and energy to the fields produced when a shock wave passes through a material gives rise to the equations

$$\frac{\rho_1}{\rho_0} = \frac{U - u_0}{U - u_1} \tag{1.57}$$

representing the conservation of mass,

$$P_1 - P_0 = \rho_0(u_1 - u_0)(U - u_0) \tag{1.58}$$

representing the conservation of momentum, and

$$e_1 - e_2 = \frac{P_1 u_1 - P_0 u_0}{\rho_0(U - u_0)} - \frac{1}{2}(u_1^2 - u_0^2), \tag{1.59}$$

representing the conservation of energy. In these equations, $P$ is the shock pressure (Pa), $U$ is the shock velocity (m/s), $u$ is the particle velocity (m/s), $\rho$ is the density (kg/m$^3$), and $e$ is the internal energy (J/kg). The subscripts 0 and 1 refer to the state of the unshocked and shock material, respectively. Usually, pressure is reported in GPa (1 GPa $= 10^9$ Pa $= 10^9$ N/m$^2$) and the velocity in km/s. Conveniently, 1 km/s $=$ 1 mm/$\mu$s, the latter units comparable to the dimensions and time intervals arising in the analysis of MCGs.

If the state of the material into which a shock is propagating is known, Eqs. 1.57–1.59 provide three relationships with the five unknown quantities $P_1$, $u_1$, $\rho_1$, $e_1$, and $U$. One of these quantities, e.g., the pressure or particle velocity in the material behind the shock, represents the strength of the shock and must be specified as a boundary condition. The additional relationship needed to completely characterize the shock arises as a description of the material behavior. In the context of shock physics, this relation is a *Hugoniot curve* (or simply a *Hugoniot*), the locus of states achievable by passing a shock into material in a given initial state. A Hugoniot curve can relate any two of the variables in question and a Hugoniot relationship between any two of the variables can be transformed into a relationship between any other two of these variables.

Experimental measurements of the response of materials (in a given initial state) to shock compression can often be represented by the Hugoniot relation [1.7]

$$U = c_0 + u_0 + s(u - u_0), \tag{1.60}$$

in which $c_0$ and $s$ are measured coefficients that depend on the initial thermodynamic state. This equation is broadly applicable to describing the response of materials to compression by strong shocks, but requires refinement when phase transitions are encountered and at low stresses where elastoplastic phenomena often dominate observed responses. As an example of other Hugoniot curves that can be derived from the linear $U$–$u$ Hugoniot, the $P$–$\rho$ Hugoniot obtained by combining Eqs. 1.57, 1.58, and 1.60 is

$$P_1 - P_0 = \rho_0 \left(1 - \frac{\rho_1}{\rho_0}\right) \left[1 - s\left(1 - \frac{\rho_1}{\rho_0}\right)\right]^{-2}. \tag{1.61}$$

When the boundary forces imposed to produce a shock are removed, a decompression wave originates at the boundary and propagates into the interior of the material. When heat conduction is neglected, the decompression wave is a smooth wave rather than the shock that is observed on compression. Each decompression wavelet advances at the rate $c(\rho) + u$, where $c$ is the speed of sound at the point in the wave where the density is $\rho$. The speed of sound at the point immediately behind the shock exceeds $U$, but decreases with the decrease of $\rho$ that occurs as the wave passes. Because of this decrease in the rate of advance of the wavelets, the decompression wave spreads as it propagates. The speed of sound in the material immediately behind the shock exceeds the shock velocity, so that the decompression overtakes the shock. This causes attenuation, decreasing both the energy and pressure changes produced by the shock. When the rarefaction wave overtakes the shock, the pressure pulse becomes sawtooth-shaped, decreasing in amplitude and increasing in length as the attenuation

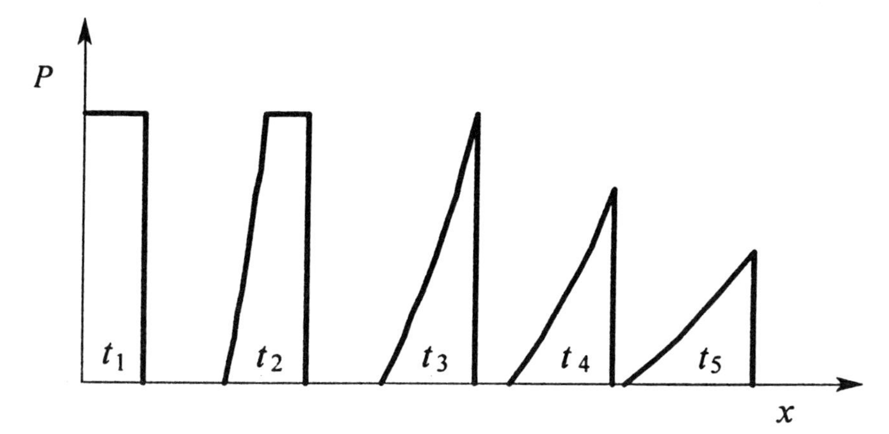


FIGURE 1.7. This sketch of an attenuating pressure pulse is drawn on the premise that the pressure applied to the boundary to produce the shock is sustained until $t = t_1$, at which time it is removed. The decompression wave introduced into the material at this time overtakes the shock at $t = t_3$ and the attenuation process begins.

process proceeds, as shown in Fig. 1.7. Eventually, the shock wave becomes a sound wave.

A *detonation wave* is a shock wave that is supported by an exothermic chemical reaction. This reaction continuously adds energy to the flow, maintaining the shock strength in much the same way as work done on the boundary in the case of a shock in a nonreactive medium. When the chemical energy liberated at a shock front exceeds that required to sustain the shock, it will grow in strength. When it is too little, the detonation wave will be attenuated by an overtaking decompression wave, as in the case of a nonreactive shock. In the case of a detonation, there exists an equilibrium amplitude for which the chemical energy is just sufficient to maintain a steady shock. In this case, the shock propagates at a constant velocity called the *detonation velocity* and has a constant pressure and particle velocity called the *Chapman–Jouguet pressure* and *velocity.*

In a nonreactive medium, the structure of the detonation wave is more complicated than a pressure pulse, because of the exothermic reaction occurring within it (Fig. 1.8). The wave front moves at the detonation velocity and initiates the explosive reaction as it passes through the material. The reaction takes place in a *reaction zone* immediately behind shock. The end of the reaction zone is called the *Chapman–Jouguet (C–J) plane* and its state is characteristic of the explosive. The hot, high pressure gas formed by reaction couples energy into other materials as it expands.

The shock equations 1.57–1.59 apply to a detonation wave just as they do to a nonreactive shock, except that the chemical energy released in the reaction must be taken into account. In this case, in which the pressure

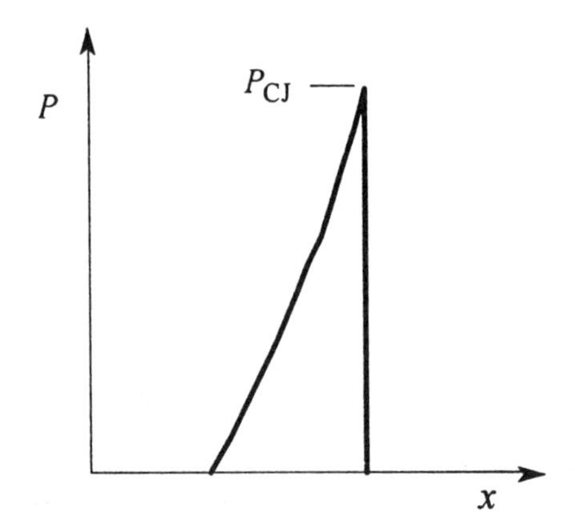


FIGURE 1.8. Schematic illustration of a Chapman–Jouguet detonation wave. This waveform consists of a detonation shock followed by a centered simple decompression wave called a Taylor wave. This waveform is very similar to the attenuating waveform shown in Fig. 1.7, but it differs from the nonreactive wave in that it is supported by the exothermic chemical reaction occurring in the shock and does not attenuate.

and particle velocity are zero in the material ahead of the detonation shock, these equations can be written as

$$\frac{\rho_{cj}}{\rho_0} = \frac{D}{D - u_{cj}} \tag{1.62}$$

and

$$P_{cj} = \rho_0 u_{cj} D, \tag{1.63}$$

where $P_{cj}$ is the pressure, $\rho_{cj}$ is the density, $u_{cj}$ is the particle velocity at the C–J plane. The density of the unreacted material is $\rho_0$, and the detonation velocity is $D$.

The detonation velocity depends on the density of the high explosive. For most explosives, the relationship between the detonation velocity and the density of the unreacted explosives is linear:

$$D = a + b\rho_0, \tag{1.64}$$

where $a$ and $b$ are empirical constants that depend on the type of explosives used. How the explosive is packed into the MCG, and therefore its density, is critical, relative to the overall efficiency of the MCG.

## 1.7 Explosives and Explosive Components

The science and the technology of explosives and components for initiating and controlling detonation of explosives is highly developed [1.8,1.9,1.11, 1.12]. There are many explosive materials, although only a few have sufficient power and detonate with the reproducibility required for use in high-performance power sources. Only a few of these materials satisfy the additional requirements of safety and stability that may be associated with specific applications. The explosives used to drive power sources are almost entirely those in the class call *secondary high explosives*, a category comprising materials of sensitivity between the easily detonable primary explosives and the very insensitive tertiary explosives. Primary explosives are used in blasting caps and other initiating devices, but are unsuitable for use as main-charge explosives. Tertiary explosives are the safest materials to use, but often prove impracticably difficult to initiate, do not function well in small charges, and are less powerful than the best available secondary explosives. Materials such as pyrotechnics and propellants (e.g., gun powder) are used to accelerate projectiles in some types of generators, but are not suited for driving magnetocumulative generators.

### 1.7.1 *Categories of Explosives*

Almost all explosives used in MCGs are organic compounds formed from the elements C, H, N, and O and called CHNO explosives. Their molecular structure is usually quite complicated (see Fig. 1.9) and the explosive materials are often mixtures of explosive compounds and include polymeric binders that effect the mechanical properties, safety, etc., but are not intended to alter the detonation performance. Explosives of this class liberate their detonation energy by molecular decomposition and reformation into detonation products such as $CO_2$, $N_2$, and $H_2$. The detonation process does not require atmospheric oxygen. A traditional secondary high explosive is trinitrotoluene (TNT), which is easily melted and cast into the shapes required for power supply and other applications. More powerful materials in the same class are hexhydrotrinitrotriazine (called RDX, cylonite, or hexogen) and cyclotetramethylenetetranitamine (HMX). These cannot be cast to shape, but granules of the compounds can be mixed into molten TNT and the mixture cast in much the same way as TNT. More powerful (but more expensive) explosives can be made by coating granules of RDX and HMX with polymeric binders and pressing the resulting material into high-

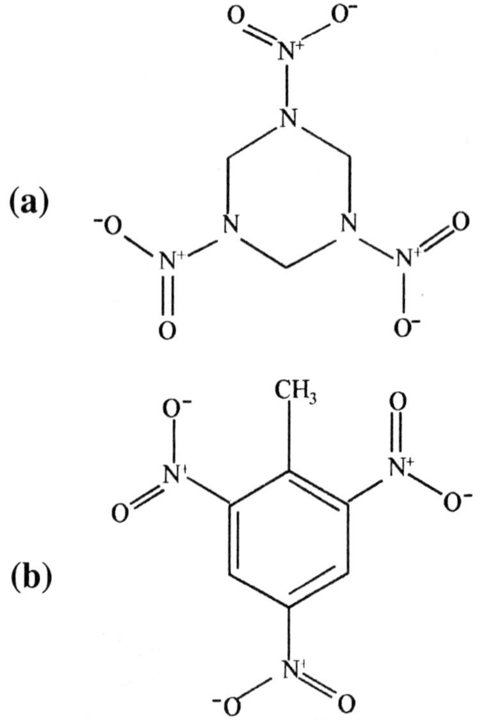

FIGURE 1.9. Chemical structure of (a) RDX and (b) TNT, explosives commonly used in MCGs.

| Material | Chemical Formula | Theoretical Density (kg/m³) | Detonation Density (km/s) | Estimated CJ Pressure (GPa) |
|---|---|---|---|---|
| PETN | $C_3H_5N_3O_8$ | 1780 | 8.26 (1770) | 30 (1670) |
| RDX | $C_3H_6N_6O_6$ | 1810 | 8.7 (1770) | 34 (1770) |
| HMX | $C_4H_8N_8O_8$ | 1900 | 9.11 (1890) | 39 (1890) |
| TNT | $C_7H_5N_3O_6$ | 1650 | 6.93 (1640) | 21 (1630) |
| TATB | $C_6H_6N_6O_6$ | 1940 | 7.76 (1880) | 29 (1880) |
| NM | $CH_3NO_2$ | 1130 | 6.35 (1130) | 13 (1130) |

TABLE 1.3. Properties of selected secondary high explosives.

density billets, which are then machined to the shape required. Properties of several secondary high explosives are listed in Table 1.3[2].

At Los Alamos National Laboratory, one of the explosives used in MCGs is composition B, which is a mixture containing 40% TNT and 60% RDX. In Russia, an explosive, called hexogen, is used , which is a 50:50 mixture of TNT and RDX. Throughout the rest of this book, the latter definition of hexogen is used.

### 1.7.2 *Explosive Components*

Most explosive systems use a number of explosive components in addition to the main charge. These components include *detonators* (devices to initiate detonation of the main charge) and *lenses* or other configurations used to control the shape of a detonation wave. Explosive systems may also include *mild-detonating fuses*, a slender column of high explosive in a metal sheath, and *explosive delay lines*.

Detonation of the explosive used in MCGs is usually initiated by one or more exploding bridgewire (EBW) detonators. These initiate a reaction by electrically vaporizing (exploding) a thin gold or gold/platinum alloy wire in contact with a pellet of a granular secondary explosive such as pentaerythritoltetranitrate (PETN). Additional explosives may be included in the detonator, to ensure that it provide a sufficiently strong stimulus to initiate the main explosive charge. The advantage of EBW detonators, in addition to greater safety than is characteristic of blasting caps, is their short initiation delay and uniformity of performance. These latter characteristics facilitate their use in systems where precisely timed initiating stimuli must be provided at several different places in the system.

---

[2]Numbers in parenthesis are densities in kg/m$^3$. Data from Ref. 1.10. The "TATB" of density 1880 kg/m$^3$ is the plastic–bonded explosive PBX 9502, which is 95/5 wt % TATB/Kel-F and is one of the most powerful tertiary explosives.

Explosive lenses allow the production of a plane wave from tens to hundreds of millimeters in diameter from a single initiation point. Convergent cylindrical detonation waves and other configurations can also be produced.

Mild-detonating fuses can be used to propagate a detonation to an explosively driven switch or other device that may be located away from the main charge. Mild-detonating fuses are usually made with a small amount of explosive material and are confined in such a way that other materials and components can be protected from any damage that the explosion might produce.

Explosive delay lines are used to initiate a series of detonators at different times. By changing the length of the delay line to a given detonator, the time it initiates can be controlled. Explosive delay lines are usually made from strips of Deta sheet, which is a plastic explosive.

## 1.8 Introduction to MCGs

The magnetocumulative generator (MCG) is a device in which chemical energy released by a detonating explosive is converted into the kinetic energy of a conductor, which, in turn, is converted into magnetic energy as the moving conductor is used to compress a magnetic field [1.13,1.14]. This conversion process is very efficient (losses are less than 50%) provided there are not high flux losses. MCGs may be divided into two classes:

- *Generators of high-energy densities* or *field generators*, in which superstrong magnetic fields are created during the radial collapse of a cylindrical hollow conductor.
- *Generators of current* or *energy* or *energy generators*, in which the volume that is compressed and the inductive load are separated.

A characteristic feature of the first type of generator is that both the magnetic energy and its density increase significantly during the deformation of the magnetic field—a process called *energy accumulation*. Generators of either of these classes can be analyzed by field theories, although the second type may be described more simply using the electrical circuit shown in Fig. 1.10. Because only power sources are considered in this book, only generators of the second type will be considered.

### 1.8.1 Circuit Equations

The time-dependent inductance in Fig. 1.10 is described by the expression

$$L(t) = L_r(t) + L_L, \tag{1.65}$$

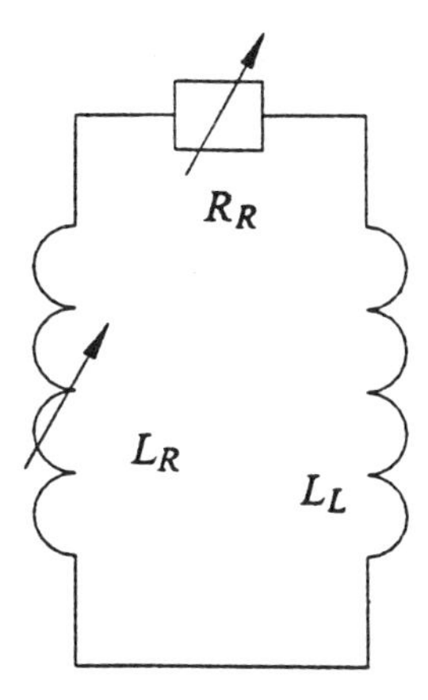


FIGURE 1.10. Equivalent circuit diagram for an MCG, where $L_r(t)$ is the inductance of the generator, which changes in time, and $L_L$ is the inductance of the load.

which includes the constant load inductance $L_L$ and the "compressed" time-varying inductance $L_r(t)$, that is, the inductance of the generator. The resistance $R_r(t)$ includes all forms of magnetic flux losses. The current $I$ flowing through the circuit can be found from the differential equation

$$\frac{d(LI)}{dt} + R_r I = 0. \tag{1.66}$$

The solution of this equation, at any instant in time, is

$$LI = L_0 I_0 \exp\left\{-\int_0^t \left(\frac{R_r}{L}\right) dt\right\}, \tag{1.67}$$

where $I_0$ is the initial current and $L_0 = L_{r0} + L_L$ is the initial $(t = 0)$ inductance. Introducing the *flux compression coefficient*

$$\lambda_k(t) = \frac{LI}{L_0 I_0} = \exp\left\{-\int_0^t \left(\frac{R_r}{L}\right) dt\right\} \tag{1.68}$$

and the *inductance coefficient* or *inductive compression ratio*

$$\gamma_L(t) = \frac{L_0}{L}, \tag{1.69}$$

Eq. 1.67 can be rewritten as

$$I(t) = I_0 \gamma_I(t) = I_0 \gamma_L(t) \lambda_k(t), \tag{1.70}$$

where $\gamma_I$ is the *current gain coefficient* or *current multiplication ratio.* Because the magnetic field is proportional to the current,

$$\frac{I}{I_0} = \frac{H}{H_0}, \tag{1.71}$$

which describes, to a first approximation, the increase in the magnetic field and the current in the generator resulting from compression of the magnetic flux.

The energy equation can be derived by multiplying Eq. 1.66 by $I$ and integrating:

$$\frac{W(t)}{W_{\mathrm{M0}}} = \frac{LI^2}{L_0 I_0^2}, \tag{1.72}$$

where

$$W_{M_o} = \frac{1}{2} I_o^2 L_o \tag{1.73}$$

is the initial energy. Introducing the coefficients defined by Eqs. 1.68 and 1.69 into Eq. 1.72, yields the basic energy equation

$$\frac{W(t)}{W_{\mathrm{M0}}} = \gamma_L(t)\lambda_k^2(t) \tag{1.74}$$

for the MCG.

### 1.8.2 Field Equations

The effects of magnetic compression can clearly be observed in those types of MCGs that are designed to create superstrong magnetic fields. In these generators, an axial magnetic field is created inside a cylindrical shell, usually called the *liner* or *armature*, by an external current source. The liner is surrounded by high explosives, which, when detonated, creates a converging cylindrical detonation wave. The moment at which the explosive is initiated is chosen such that the liner begins to move when the magnetic field created by the external source is at its peak. When the liner is accelerated inward by the high explosives, it collapses coaxially, compressing the magnetic field with a current-carrying cylindrical surface (called the *stator*) of constant diameter. In the limiting case of infinitely fast field compression or of a liner with ideal conductivity, the enveloped magnetic field is conserved; i.e.,

$$\int H\, ds = const, \tag{1.75}$$

showing that $H \rightarrow \infty$, as $s \rightarrow 0$. The strength of the magnetic field, $H(t)$, increases until the area bounded by the contour $s$ stops decreasing. In the ideal case in which flux is conserved, the magnitude of the magnetic field depends on the radius, $r$, of the liner cavity according to the relation

$$H(t) = H_0 \frac{r_0^2}{r(t)^2}. \tag{1.76}$$

The initial kinetic energy of the conductor, $W_{k0}$, is completely converted into magnetic energy. The radius of the cavity at the time it stops decreasing is equal to

$$r_k = \frac{r_0}{\sqrt{\frac{W_{k0}}{W_{M0}} + 1}}, \tag{1.77}$$

where $W_{M0}$ is the initial magnetic field energy and $r_0$ is the initial radius of the liner. The magnetic field strength at this instant is

$$H(t) = H_0 \left( \frac{W_{k0}}{W_{M0}} + 1 \right). \tag{1.78}$$

Both the magnetic pressure and the magnetic energy density increase as $r$ decreases and are proportional to $r^{-4}$.

### 1.8.3 *Magnetocumulative Generator Performance*

Pulsed magnetic fields of $\sim 10^7$ Oe have been achieved in MCGs designed specifically to create superhigh magnetic fields. Magnetic energy densities have reached values of $4 \times 10^{11}$ J/m$^3$, which is 40 times greater than the chemical energy density of the explosive. The specific power in the superstrong fields of MCGs is $\sim 10^{17}$ W/m$^3$.

In MCGs designed as power sources, magnetic energy densities of $10^9$ J/m$^3$ and specific powers of $10^{13}$–$10^{14}$ W/m$^3$ have been achieved. When the dimensions of the former are compared with the latter type of MCG, each having the same set of characteristics, it is that the latter is nearly one order of magnitude smaller. The amplitude of the pulsed currents from the latter MCGs has reached values of 300 MA. Energies of 10–100 MJ have been generated at instantaneous powers of $10^{13}$ W. Based on the present state-of-the-art, it is believed that MCGs may be developed as power sources with even higher values of these parameters.

The significant advantage of using MCGs in experiments, as well as in various technological applications, is that their output characteristics may be varied over a wide range. For example, the characteristic time in which energy is generated by MCGs is 10–100 $\mu s$. By applying special devices, such as switches and transformers, the spatial–temporal output characteristics of the generator can be changed to match those of the load. These techniques allow researchers to change the shape of the output current pulse delivered to the load, and to vary its pulse length from 0.2–0.4 $\mu s$ to 1–10 ms. The output power and energy may also be changed, even when the length of the pulse remains unchanged. The high-energy density of MCGs facilitates the delivery of high energies to the load. Models have been developed that allow researchers to predict the expected characteristics of MCGs with satisfactory precision [1.1,1.2,1.13].

Based on the value of 1991 Rubles in the Former Soviet Union, without taking into account inflation, the cost of energy produced by MCG's was $\sim 10^{-3}$ Rublc/J (18 cents/J), and could possibly be decreased to $\sim 10^{-4}$ Ruble/J (1.8 cents/J). The cost of the hardware required to deliver a pulse having the required shape and length to a load is roughly equivalcnt to that of the MCG itself. This is why it seems to be economically feasible to use MCGs not only in expendable installations, but also in experimental facilities where one needs to carry out $\sim$ 100 experiments that require energies of 10–100 MJ. In this case, the energy from the MCG can be transmitted a safe distance from an explosive containment vessel to the installation where the experiments are to be carried out. For example, MCGs are currently being considered to drive imploding liners at the National Ignition Facility at Lawrence Livermore National Laboratory to achieve controlled magnetized target fusion.

# References

[1.1] *Megagauss Technology and Pulsed Power Applications* (eds. C.M. Fowler, R.S. Caird, and D.J. Erickson), Plenum Press, New York (1987).

[1.2] *Megagauss Magnetic Field Generation and Pulsed Power Applications* (eds. M. Cowan and R.B. Spielman), Nova Science Pub., New York (1994).

[1.3] I. Vitkovitsky, *High Power Switching*, Van Nostrand Reinhold Company, New York (1987).

[1.4] G. Knoepfel, *Pulsed High Magnetic Fields*, North-Holland Publishing Company, Amsterdam (1970).

[1.5] Ia.P. Terletskii, "Production of Very Strong Magnetic Fields by Rapid Compression of a Conducting Shell," Sov. Phys. JETP, **32**, p. 301 (1957).

[1.6] C.A. Balanis, *Advanced Engineering Electromagnetics*, John Wiley & Sons, New York (1989).

[1.7] M.B. Boslough and J.R. Assay, *High Pressure Shock Compression of Solids* (eds. J.R. Asay and M. Shahinpoor), Springer-Verlag, New York, pp. 7–42 (1993).

[1.8] D.S. Drumheller, *Introduction to Wave Propagation in Nonlinear Fluids and Solids*, Cambridge University Press, Cambridge (1998).

[1.9] R. Cheret, *Detonation of Condensed Explosives*, Springer-Verlag, New York (1993).

[1.10] R. Engleke and S.A. Sheffield, *High-Pressure Shock Compression of Solids III* (eds. L. Davison and M. Shahinpoor), Springer-Verlag, New York, pp. 173–239 (1998).

[1.11] P.W. Cooper and S.T. Kurowski, *Introduction to the Technology of Explosives*, Wiley–VCH, New York (1996).

[1.12] M.A. Meyers, *Dynamic Behavior of Materials*, Wiley–Interscience, New York (1994).

[1.13] *Superstrong Magnetic Fields: Physics, Techniques, and Applications* (eds. V.M. Titov and G.A. Shvetsov), Nauka, Moscow (1984).

[1.14] R.S. Caird and C.M. Fowler, "Conceptual Design for a Short-Pulse Explosive-Driven Generator," *Megagauss Technology and Pulsed Power Applications* (eds. C.M. Fowler, R.S. Caird, and D.J. Erickson), Plenum Press, New York, pp. 425–432 (1987).

# 2

# Magnetocumulative Generator Physics and Design

## 2.1 Conditions That Affect Magnetic Field Compression

The conditions required to sustain compression of the magnetic field in an MCG depend on the magnitude of the magnetic field to be generated, but are characterized by high pressure, high current density, and high temperature. The deformation of a magnetic field by an explosive-driven conductor is described by the *magnetohydrodynamic equations*, which can only be solved by numerical methods. Modern computers allow these calculations to be performed for specific MCG designs, but analytical methods can also be used if certain simplifying assumptions are made [2.1–2.3].

In the case of ideal compression, there is only one restriction on device performance—the *energy condition*. To investigate the implications of this condition in the simplest context, the effect of the explosive on the driven conductor is modeled as an impulse. That is, it is supposed that the explosive imparts an initial kinetic energy to the conductor, but provides no sustained driving pressure. In this case, magnetic compression will proceed until the initial kinetic energy of the collapsing conductor is completely converted into magnetic energy. At this point, called the *turnaround point*, the conductor begins to move outward and the field strength decreases.

Consider a device that produces high magnetic energy densities by compressing an initial field within an imploding cylinder. Let the initial cross-sectional area of the cavity be $Z_0$ and the initial axial field be $H_0$. Because the flux in the cavity is conserved, $H(t)Z(t) = H_0 Z_0$, where $H(t)$ and $Z(t)$

are the field and cavity cross-sectional area, respectively, at a time $t$ during the compression process. The magnetic energy (per unit length of the cylinder) is given by $W_M = (\mu/2)\,H^2 Z$. The *magnetic energy gain coefficient* is defined as $\gamma_M = W_M/W_{M_0}$, where $W_M$ is the magnetic energy (per unit of cylinder length) and $W_{M_0}$ is its initial value. Combining these equations yields the relation

$$\gamma_M(t) = \frac{W_M(t)}{W_{M_0}} = \frac{H(t)}{H_0} = \frac{Z_0}{Z(t)}. \tag{2.1}$$

At the turnaround point, designated by the subscript $k$, the energy conservation equation is $W_{M_k} = W_{K_0} + W_{M_0}$, where $W_{K_0}$ is the initial kinetic energy of the collapsing conductor. Evaluation of Eq. 2.1 at the turnaround point yields

$$\gamma_{M_k} = \frac{W_{M_k}}{W_{M_0}} = \frac{W_{K_0}}{W_{M_0}} + 1 = \frac{H_k}{H_0} = \frac{Z_0}{Z_k}. \tag{2.2}$$

In real devices, the energy condition is affected by diffusion of the field into the conductors and by compressibility, loss of conductivity, and surface instability of the conductors. If not controlled, loss of conductivity and surface instability can cause catastrophic failure of a generator.

### 2.1.1 *Field Diffusion*

Consider the diffusion of a uniform magnetic field from a cylindrical cavity of cross-sectional area $Z_0$ into the surrounding infinite conductor shown in Fig. 2.1 [2.4]. If an initial magnetic field, $B_0$, is created in the cavity during an excitation time $t_0$, then an initial skin-layer depth, $\delta_0$, is created around the perimeter of the cavity, $P_0$. The magnetic flux $\Phi_{01} = B_0 Z_0$ is concentrated in the cavity and the flux $\Phi_{02} = B_0 \delta_0 P_0$ has penetrated into the skin layer. Since the field in the conductor decreases exponentially with distance from the surface of the conductor, then, from Faraday's law

$$\oint \mathbf{E}\cdot d\mathbf{l} = \frac{d\Phi}{dt}, \tag{2.3}$$

it can be shown that the integral approaches zero as the path recedes to infinity. This implies that the total flux inside this circuit is conserved; that is,

$$\Phi_1(t) + \Phi_2(t) = \Phi_{01} + \Phi_{02} = \Phi_0. \tag{2.4}$$

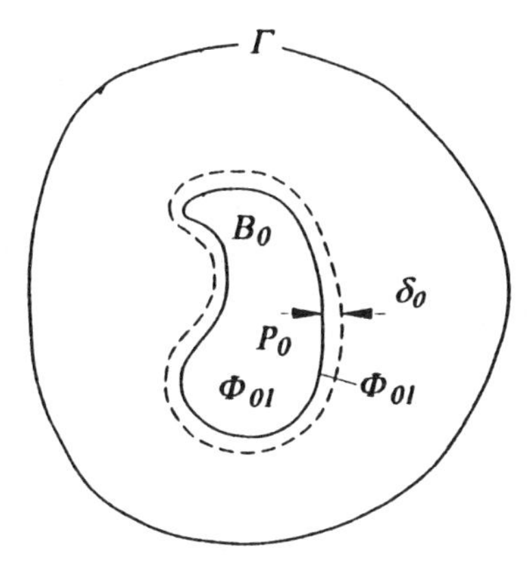


FIGURE 2.1. Diffusion of magnetic field into an infinite conductor, when a magnetic field is introduced into a cavity having a cross-sectional area $Z_0$.

The result that the total flux is conserved can be applied to conductors of finite size, provided that the thickness of the conductor is significantly greater than the skin depth. If the field inside the closed cavity is uniform, Eq. 2.4 can be written as

$$B(t) = B_0 \frac{Z_0 + P_0\delta_0}{Z(t) + P_S(t)\delta(t)}, \tag{2.5}$$

where $P_S(t)$ is the perimeter of the diffusion region, $\delta(t)$ is the skin layer, and $Z(t)$ is the cross-sectional area at time $t$.

The field inside the cavity increases when the rate of decrease of the cross-sectional area of the cavity, $Z(t)$, exceeds the rate of increase of the skin layer cross section, $P_S(t)\delta(t)$. The compression velocity required to cause the magnetic field to increase can be found, if the initial magnetic flux inside the cavity is sufficiently low and the skin-layer depth, $\delta$, is comparable to the diameter of the cavity:

$$\delta = \frac{2Z}{P_S}. \tag{2.6}$$

The time

$$\tau_S = 0.5\pi\mu\sigma\delta^2 \tag{2.7}$$

required for the magnetic field to penetrate into the conductor is called the *skin time*, and can be determined using Eq. 1.40. The magnetic compression is accompanied by an increase in the magnetic field strength and energy,

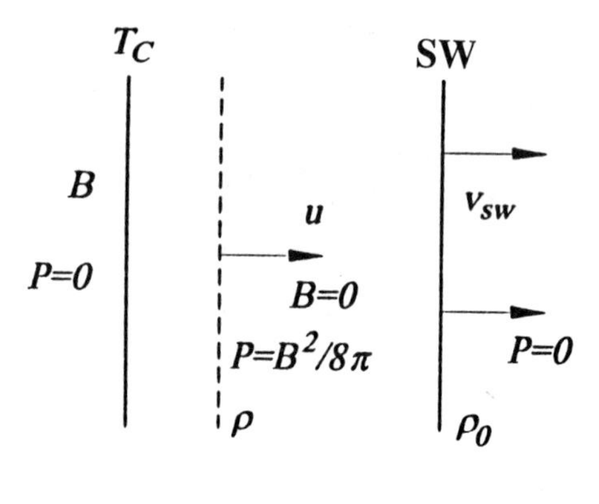


FIGURE 2.2. When a conductive material moves toward a magnetic field with a velocity $v$, a simple compression wave forms at its surface. As it propagates into the conductor, the compression wave becomes increasingly steep and evolves into a shock wave.

provided the time interval during which the cavity compresses the magnetic field is much less than the time $\tau_S$.

The efficiency with which the electromagnetic energy accumulates increases with the magnetic Reynold's number:

$$N_P = \frac{\tau_S}{\tau_k} = \frac{\pi\mu\sigma\delta^2}{2\tau_k}, \tag{2.8}$$

which is the skin time divided by the time from the beginning of the compression process to the turnaround time. The condition that must be satisfied if there is to be magnetic accumulation is $N_P >> 1$. If the conductor moves with constant a velocity, $v$, that causes it to displace one skin thickness in a time $\tau_k$, this conditions is satisfied when

$$v >> \frac{2}{\pi\mu\sigma\delta}. \tag{2.9}$$

Using this expression, it can be shown that, for conductors with $\sigma \sim 36/(\text{M}\Omega\text{·m})$, which is a typical conductivity for metals, $v >> 0.0716/\delta$ m/s. Therefore, to attain efficient magnetic compression in a cavity having a diameter of 10 mm, the compression velocity must be hundreds of meters per second.

### *2.1.2 Liner Compressibility*

The magnitude of the magnetic field strength that can be attained in an MCG is limited by the compressibility of real conductors, because the conductor that is compressing the magnetic field is also being compressed. Even in the case of an ideal conductor, kinetic energy is used not only to compress the magnetic field, but also to increase the internal energy of the conductor itself. To estimate the effects of the compressibility of the conductor, the wave propagation problem illustrated in Fig. 2.2 will be examined. When a conductive material of specific density $\rho_0$ moves toward the field with a velocity $v$, a simple compression wave forms at the surface of the conductor. As it propagates into the interior of the conductor, the compression wave becomes increasingly steep, evolving into a shock. The shock compresses the conductor material, decelerates it, and increases its internal energy. A pressure, $P$, equal to the magnetic pressure is established behind the shock front, and the velocity of the shock-compressed material decreases to zero at the turnaround point. A current layer that exists at the interface of the liner with the magnetic field acts like a piston to transfer pressure to the conductor.

If follows from Eqs. 1.57 and 1.60 that

$$P = \rho_0 v(c_0 + sv), \tag{2.10}$$

with a peak magnetic pressure, obtained from Eq. 1. 51, of

$$P_{M_k} = \frac{\mu}{2} H_{\text{max}}^2. \tag{2.11}$$

Equating these results shows that the magnetic field strength (and, therefore, the energy density), which does not depend on the thickness of the walls of the liner, cannot exceed the value

$$H_{\text{max}} = \sqrt{\frac{2}{\mu} \rho_0 v(c_0 + sv)}. \tag{2.12}$$

Estimates based on Eq. 2.12 show that $H_{\text{max}} = 20$–$30$ MOe when the velocity of the conductor is 10 km/s.

During the final stages of magnetic field compression, the kinetic energy of the conductor is not converted into magnetic energy, but rather is used to increase the elastic (when $H_{\text{max}}$ approaches 10 MOe) and thermal (when $H_{\text{max}} \approx 10$–$100$ MOe) energies of the conductor, to ionize its atoms, and to generate electromagnetic waves. Each of these processes causes a loss of efficiency in the conversion of kinetic energy to magnetic energy. In the event that the field is compressed by a planar conductor, the plot of the peak magnetic field strength as a function of the initial velocity of the conductor is shown in Fig. 2.3.

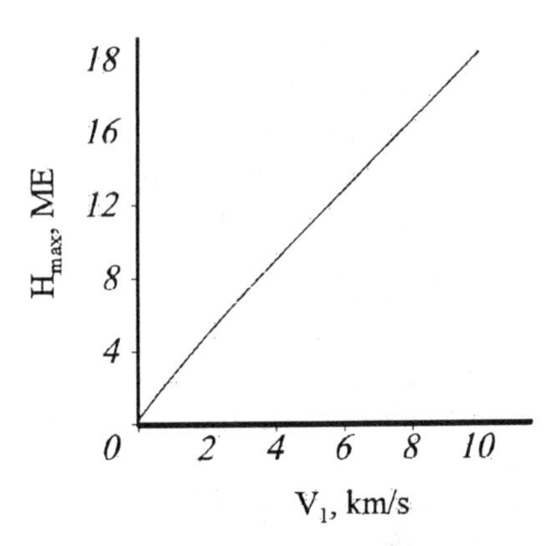


FIGURE 2.3. The maximum magnetic field that can be achieved within a magnetic flux compressor depends on the value of the seed magnetic field and the initial velocity of the conductor. However, this relationship is not linear owing to various losses that occur during compression process.

For a substance with a third-order equation of state in the elastic compression domain; that is, when $H \approx 10^6$ Oe,

$$p = \frac{\rho_0 C^2}{3\left[\left(\frac{\rho}{\rho_0}\right)^3 - 1\right]}, \tag{2.13}$$

the maximum field strength is equal to

$$H_{\text{max}} = \left\{\frac{8\pi\rho_0 C^2}{3}\left[\left(\frac{v_1}{C} + 1\right)^3 - 1\right]\right\}^{\frac{1}{2}}, \tag{2.14}$$

where $C$ is the velocity of sound. If the thermal pressure is much greater than the elastic pressure (i.e., when $H \approx 10^8$ Oe), the maximum magnetic field is

$$H_{\text{max}} = \sqrt{\frac{32\pi\rho_0}{3}} v_1. \tag{2.15}$$

When the radiation pressure is much greater than the pressure in the conductor, it has been shown that the value of $H_{\text{max}}$ is proportional to $v^{1/4}$.

The problem of determining the limiting value of $H_{\text{max}}$ for an ideal conductor reduces to that of finding the restrictions on the kinetic energy density acquired during the acceleration of the liner by the explosive products. Magnetic compression will occur if

$$\frac{\rho v_1^2}{2} > \frac{1}{2}\mu H^2; \tag{2.16}$$

i.e., when the hydrodynamic pressure is greater than the magnetic pressure.

The compressibility of the liner material noticeably limits the compression of the magnetic field. A high velocity is imparted to the liner by the explosive, and the velocity of its inner surface tends to increase further owing to the convergence of the motion. During the collapse of an incompressible cylindrically shaped liner, the velocity of its inner surface approaches infinity according to the expression $\left(r^2 \ln r\right)^{-1/2}$ as $r \to 0$; that is, the kinetic energy is concentrated at the inner surface of the liner [2.1].

### *2.1.3 Conductivity Change*

Diffusion of the magnetic field into the conductor leads to losses of both magnetic flux and magnetic energy. The magnetic field does not remain localized in the cavity enclosed by the conductor, because part of the magnetic energy diffuses into the conductor before it is released in the from of Joule heat. This part of the flux is lost from further compression. The current density flowing through the MCG may be estimated from Ampere's equation (cf. Eq. 1.2):

$$j = -\frac{dH}{dx} \approx \frac{H}{z}, \tag{2.17}$$

where $z$ is a characteristic length of the generator. If a magnetic induction of $B = \mu H = 100$ T were created within 10 $\mu$s, then the skin current density, according to Eq. 2.12, would be $j = 2.5 \times 10^9$ A/m$^2$.

High current densities lead to strong Joule heating of the conductor, which may cause increasing vaporization of the material from the surface of the liner as the magnetic field increases. This results in a hydrodynamic explosion, and when this occurs, the normal thermal conductivity of the material may be neglected. During the explosion process, a compression wave, followed by a rarefaction wave, passes into the material. The rarefaction wave converts the material from a conducting medium into a nonconducting medium. Inside the rarefaction wave, the pressure decreases from the value inside the medium down to zero at the boundary with the field, i.e., the surface of the liner. The front of the wave at which the conductivity loss occurs separates by a certain distance from the rarefaction wave front. Magnetic compression will take place as the front propagates into the conductor, provided that the velocity of the conductor exceeds the velocity of the front at which conductivity is lost. This is because the magnetic field is compressed by the deeper layers of the medium that remain conductive. If this last condition is met, the hydrodynamic explosive process will not be catastrophic, even with the unfavorable conductivity of the plasma layer. When the field increases, it can be expected that heating the vapors to high temperatures will yield high conductivities. The plasma layer is formed between the solid surface of the liner and the magnetic field in the cavity. The conductivity of the plasma layer increases as the magnetic field is compressed, but is limited by radiative cooling at fields greater than $10^8$ Oe. Therefore, the finite value of the actual conductivity does not restrain magnetic compression. The decrease in conductivity may, in principle, be compensated by an increase in the compression velocity and by selectively increasing the dimensions of the region occupied by the compressed magnetic field. Thus, when the conductivity is finite, it is still possible to generate the same magnitudes of magnetic field strength, as those obtained in the ideal case of infinite conductivity.

### *2.1.4 Surface Instability*

Another phenomenon limiting magnetic compression is instability of the surface of the conductor that is in contact with the magnetic field [2.1]. When the field strength is high, the inner surface layer is heated to its melting temperature during the magnetic field compression process. When this occurs, the solid matter is no longer in contact with the magnetic lines of force. In principle, the liquid metal that separates the solid part of the liner from the magnetic field may experience various magnetohydrodynamic instabilities that prevent it from maintaining the smooth cylindrical shape required to compress the magnetic field effectively. The main type of instability is the Raleigh–Taylor instability.

Consider two ideal liquids having mass densities of $\rho_1$ and $\rho_2$, which are separated by a boundary that has a shape at time $t = 0$ defined by the equation (see Fig. 2.4a)

$$\eta(t = 0) = \eta_0 \cos(kx), \tag{2.18}$$

where

$$|\mathbf{k}| = \frac{2\pi}{\lambda} \tag{2.19}$$

is the wave number. It follows from the theory of *weak disturbances* that the surface tension, $T_S$, at the boundary between the two liquids, which reside in the gravitational field, $g$, determines the temporal behavior of the displacement of the boundary. This behavior can be expressed by the equation

$$\eta = \eta_0 \cosh \omega t \cos(kx), \tag{2.20}$$

where the frequency, $\omega$, is determined from the dispersion equation:

$$-\omega^2 = gk\frac{\rho_2 - \rho_1}{\rho_2 + \rho_1} + \frac{k^3 T_S}{\rho_2 + \rho_1}. \tag{2.21}$$

If $\rho_2 > \rho_1$, then the amplitude of the interfacial disturbance does not exceed its initial value, $\eta_0$. In other words, the interaction is stable for the case of small disturbances (i.e., $\eta_0 < \lambda$). If $\rho_2 < \rho_1$, then stable behavior will only occur if the restoring force caused by the surface tension is much greater than the inertial force, i.e., $k^2 T_S > g(\rho_2 - \rho_1)$; otherwise, the amplitude of the perturbation will increase exponentially in time.

These results are still valid for the hydromagnetic problems shown in Figs. 2.4b, 2.4c, and 2.4d, if the problem is interpreted correctly. For example, if the density of one of the fluids is set equal to zero and if the magnetic field lies only along the axis of the cylinder, $H_z$, then one obtains

$$\omega^2 = \pm gk, \tag{2.22}$$

where the plus sign corresponds to the unstable condition depicted in Fig. 2.4b and the minus sign to the stable condition depicted in Fig. 2.4c.

For the case shown in Fig. 2.4d, it is expected that a term containing the magnetic surface tension must be added to Eq. 2.21, since it is well known that a tension $\mu H^2/2$ exists along the magnetic lines of force. For the case of an ideally conducting fluid, analysis shows that the magnetic surface tension is equal to

$$T_M = \frac{\mu_0 H_\infty^2 \cos^2\varphi}{k}, \tag{2.23}$$

where $\varphi$ is the angle between $\mathbf{H}_\infty$ and $\mathbf{k}$. Thus, in place of Eq. 2.21, the following hydromagnetic equation may be derived by using Eqs. 2.22 and 2.23:

$$\omega^2 = \pm gk - \frac{k^3}{\rho}(T_M + T_S). \tag{2.24}$$

The sign of the first term is selected in the same way as it is for Eq. 2.22; that is, the plus sign corresponds to unstable conditions and the minus sign to stable conditions. In the majority of the practical cases, the $T_S$ term may be neglected.

For the case of an unstable condition, the second term may stabilize the disturbances even if $T_S \approx 0$, provided that the wavelengths are sufficiently short. This occurs when

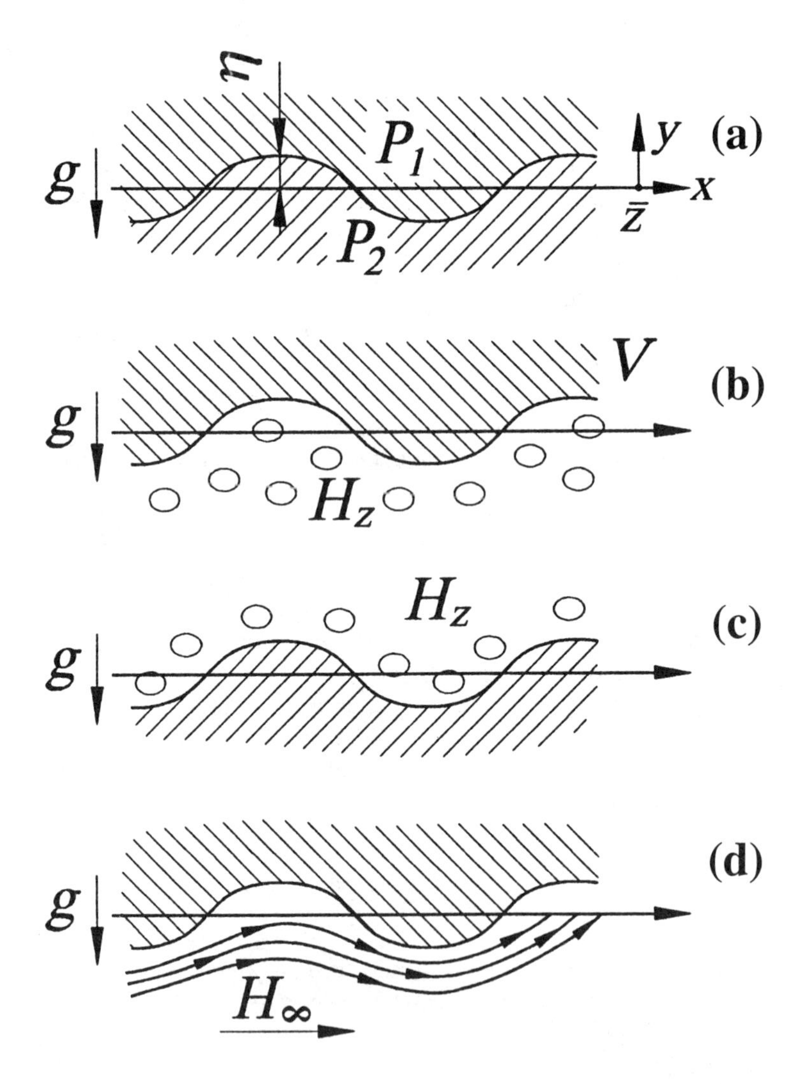


FIGURE 2.4. During magnetic compression, heating causes the surface of the conductor to melt and, depending on the level of disturbances that form, can lead to magnetohydrodynamic instabilities. The four cases depicted in this figure are: a – weak disturbance at the boundary separating two liquids, b – unstable disturbance at the boundary separating a liquid and magnetic field, c – stable disturbance at the boundary separating a magnetic field and liquid, and d – effect of magnetically induced surface tension. When the acceleration vector is perpendicular to the boundary, the surface is stable for small disturbances as shown in a and b. Otherwise, the system is unstable as shown in c.

$$\lambda < \frac{2\pi}{\rho g} \mu_0 H_\infty^2 \cos^2 \varphi. \tag{2.25}$$

The above expressions are useful in more general circumstances than are suggested by the derivation. For example, the stability criterion does not change, if instead of an infinitely thick layer of fluid, a layer with thickness $h_L$, supported by the magnetic pressure

$$\frac{1}{2} \mu_0 H_\infty^2 = h_L \rho g \tag{2.26}$$

is considered. The stability criterion in this case takes on the simpler form:

$$\lambda < 4\pi h_L \cos^2 \varphi. \tag{2.27}$$

The above results can be used to analyze the stability of the moving liner in a MCG by replacing the gravity forces with inertial forces and the direction in which these forces act by changing the sign of the acceleration of the fluid. Therefore, when the acceleration vector at the boundary separating the magnetic field and the fluid is directed into the fluid (see Figs. 2.4a and 2.4b), this surface will be stable for small disturbances with wavelengths that satisfy the condition in Eq. 2.27. Otherwise, as is shown in Fig. 2.4c, the system is stable for any wavelength.

Excitation of Raleigh–Taylor instabilities as the liner moves into the volume where superstrong magnetic fields are being created strongly limits the capabilities of the generator. This occurs when the main conductive portions of the liner are being slowed by the retarding pressure created by the magnetic field. This is the final, and most critical stage of the compression process.

To decrease hydrodynamic instabilities in coaxial MCGs, it has been proposed that several coaxial stages, each acting as a liner, be used to successively compress the magnetic field. This process increases magnetic field accumulation and generates magnetic fields having strengths in the thousands of Tesla.

## 2.2 Theory of Magnetocumulative Current Generators

The discussion of the previous section has been couched in terms of devices that compress flux to produce superhigh magnetic fields. All the issues addressed are also relevant to generators that are intended to produce large currents in a load. Some additional matters arise when this latter class of devices is considered, and the relative importance of the various phenomena is altered [2.5–2.7].

Magnetocumulative generators that operate in the energy-generation mode produce magnetic fields no higher than 0.5 MOe. In this case, the conductor is a solid and has a smooth, well-defined boundary. The conductor motion is much more affected by the accelerating effect of the explosive driver than it is by the retarding effect of the magnetic pressure, so that liner motion can often be analyzed without regard to the electromagnetic behavior of the device. In this case, the electromagnetic analysis can be conducted using *a priori* knowledge of the conductor motion. Because flux diffusion into the conductors is also less important than for superhigh field devices, the laws that describe the change in the magnetic field (or electric current) are also simplified. Capitalizing on these observations permits the creation of simplified models that can be solved analytically.

Consider a MCG analysis based on a specific assumption regarding the change in the magnetic field, $H(t)$. In this case, the problem can be solved with relative ease and with a precision comparable to that of available experimental measurements.

When taking into account the leakage of magnetic flux in a planar MCG, it is sufficient to use the field diffusion equation for incompressible conductors:

$$\frac{dH}{dt} = \frac{d}{dx}\left(\chi \frac{dH}{dx}\right), \tag{2.28}$$

where

$$\chi = \frac{1}{\mu_0 \sigma} \tag{2.29}$$

and $\sigma$ is the electrical conductivity (Eq. 2.28 specializes Eq. 1.40 to plane fields but generalizes it to allow variable conductivity). It is more convenient to rewrite this equation in the form

$$\delta^2 = \chi^{\tau}(1 + \delta') - \chi'^{\tau}\delta, \tag{2.30}$$

where

$$\delta = -\frac{H}{\frac{\partial H}{\partial x}}, \quad \tau = \frac{H}{\frac{\partial H}{\partial t}}, \quad \delta' = \frac{\partial \delta}{\partial x}, \chi' = \frac{\partial \chi}{\partial x}. \tag{2.31}$$

All these quantities are functions of time, $t$, and the coordinate, $x$, which is the distance from the surface of the conductor into its medium. If the condition

$$-\frac{\chi(1+\delta')}{\chi'} >> \delta, \tag{2.32}$$

is satisfied, as it is for various types of MCGs, when $\sigma \approx$ const, then the diffusion equation is

$$\delta = \pm\sqrt{\chi^{\tau(1+\delta)}}. \tag{2.33}$$

For most MCGs, any current-carrying path formed on the conducting boundary of the surface of the MCG cavity encloses the magnetic flux $\Phi = \int_S H\, dS$, where $S$ is the area bounded by the path. Applying Faraday's and Ohm's laws to this circuit yields Kirchhoff's equation:

$$R'I + (dL'/dt)I + L\dot{I} = 0, \tag{2.34}$$

where $I$ and $\dot{I}$ are the current and its derivative with respect to time; $R'$ is the resistance, if it is assumed that $\delta$ is the skin depth at the surface of the conductor, $L = \Phi/I$ is the inductance of the MCG cavity; $\Phi$ is the magnetic flux in the cavity (which does not include the flux that lies beyond the boundaries of the cavity), and $dL'/dt$ is the rate at which the inductance changes, which depends on the motion of the conductor as the magnetic field is deformed. Using these definitions of $R$, $L$, $dL'/dt$, Kirchhoff's equation for the MCG is as precise as Maxwell's equations. A portion of the magnetic flux may not participate in the magnetic compression process if it is separated from the magnetic flux in the main cavity of the MCG. For example, some of the flux trapped in the solid dielectric or in slots between the conductors in the outside coil of the MCG will not be compressed.

If the flux cut-off occurs continuously (or almost continuously), then Kirchhoff's equation may be rewritten to have the same form as Eq. 1.66. This equation determines the flux and energy losses. During operation of the MCG, the lost flux is that which leaks through the internal surface of the conductor and is equal to

$$\int_0^t R'I\, dt = L_0 I_0 (1 - \lambda_k), \tag{2.35}$$

while the lost energy is equal to

$$W_{\text{Loss}} = \int_0^t I^2 R'\, dt = I_0^2 L_0^2 \int_1^{\lambda_k} \frac{\lambda_k}{L}\, d\lambda_k, \tag{2.36}$$

which is that part of the magnetic energy that passes through the boundary of the conductor and is converted into Joule heat during the time interval, $t$. The energy loss, $W_{\text{Loss}}$, tends to zero as $\lambda_k \to 1$.

After field compression in the MCG has been completed, the flux and energy losses in the conductor may also stop; i.e., the value of $R$ may become negative: $\partial H/\partial x > 0, \delta < 0$. The value of $\delta(0,t)$ required to calculate $R$ can be easily determined, if it does not depend on time. For the case under consideration, it is useful to assume that the magnetic field strength at the conducting boundary is given by the equation

$$H(0,t) = H(0,0)\left(1 - \frac{t}{n\tau_0}\right)^{-n}, \tag{2.37}$$

where $\tau_0 = \tau(0,0)$ and $n$ is some number. If $n \geq 0$, then the field $H(0,t)$ increases hyperbolically. The magnetic field in MCGs with ideal conductors ($\sigma = \infty$) also changes hyperbolically. If $n > 1$, where the value of $n$ depends on the type of MCG being considered, $\delta' = 1/4$. As $n \to \infty$, $H(0,t) \to H(0,0)\exp(t/\tau_0)$, $\tau \to$ const, $\delta' = 1/4n \to 0$, and $\delta \to \sqrt{\chi^{\tau_0}} =$ const (which is the well-known exponential approximation).

For the case of a uniform magnetic field, the value of $R'$ is proportional to the perimeter $p_c$ of the cross-sectional area of the cavity, and the value of $L$ is proportional to the cross-sectional area, $S$, of the cavity. The geometry of the field compression region is related to the ratio $p_c = p_0\,(S/S_0)^{\alpha}$, where $p_0$ and $S_0$ are the initial values of $p$ and $S$, $\alpha = 1 - 1/m$, and $m$ is an arbitrary number. If it is assumed that the kinematics of the compression process are described by the hyperbolic equation, $H(t)$, i.e., $n > 1$, then from Kirchhoff's equation the *magnetic flux conservation coefficient* is

$$\lambda_k = \left[1 - \frac{2}{\gamma_m m\sqrt{(4n+1)R_m}}\,(h^{\gamma} - 1)\right]^{m}, \tag{2.38}$$

where $m \neq \infty$, $\gamma_m = (2n - m)/2nm$, $R_m = S_0^2/p_0^2\chi^{n\tau_0}$, and $\gamma_H = H(0,t)/H(0,0)$. In most typical cases, $n = m = 1, 2$, or $\infty$.

The electrical current $I$ is equal to $\Phi/L$, and it is amplified if $(-\dot{L}) > R$. For the case of a uniform magnetic field, the value of $I$ is proportional to $H$. The magnetic field may also be amplified if the thickness of the surrounding conductor is $\delta > -2\chi p_c/\dot{S}$. When $(-\dot{L}) >> R$, then $\gamma_I = L_0/L$, which implies an ideal amplifier. The magnetic energy is $W = I^2L/2 = \Phi^2/2L$, and it is amplified (i.e., $\dot{W} > 0$), if $(-\dot{L}) > 2R$.

If Kirchhoff's equation is applied to a uniform magnetic field, then the cross-sectional area of the cavity, where the energy amplification is critical (i.e., $\dot{W} = 0$), can be found:

$$S_* = \frac{2\chi p_c^2}{(-\dot{S}_*)(1+\delta'_*)} = \frac{\delta_* p_{c*}}{(1+\delta'_*)}. \tag{2.39}$$

When $\delta'_* \to 1$, $s_* \to \infty$, and when $\delta'_* \to 0$, $s_* \to \delta_* p_{c*}$.

## 2.3 Current Generator Design Issues

### *2.3.1 Eliminating Electric Breakdown*

At the surface of an ideal conductor moving with a velocity of $\mathbf{v}$ through a magnetic field, the electric field strength is given by the equation

$$\mathbf{E} = -\mathbf{v} \times \mathbf{B}. \tag{2.40}$$

If the magnetic field is parallel to the surface of the conductor, then the magnitude of the electric field strength depends only on the normal component of the velocity, $v_n$, i.e.,

$$E = v_n B. \tag{2.41}$$

When the magnetic induction is $B = 100$ T and the velocity of the conductor is $v_n = 1$ km/s, the electric field strength is $E = 100$ kV/m. Thus, one may conclude that MCGs can be distinguished as devices that create strong magnetic fields and very high electric currents but comparatively low electric fields.

However, in some types of MCGs, an angle forms between the collapsing liner and the current-carrying coil, which leads to the formation of strong electric fields, as shown in Fig. 2.5. In this case, the strength of the electric field is determined from the expression

$$E = \frac{vB}{\sin\alpha}, \tag{2.42}$$

where $\alpha$ is the angle formed by the collapsing shell.

Breakdown occurs ahead of the colliding surfaces at small angles, $\alpha$, owing to the formation of high electric fields, $E$. The short-circuiting of the cavity compressing the magnetic field leads to high losses and decreases their capability to compress the magnetic fields. One way to increase the efficiency of MCGs is to fill the compression region with gases having higher breakdown thresholds than air. In experiments that have been done using gases with high breakdown thresholds, the current-bearing surfaces have

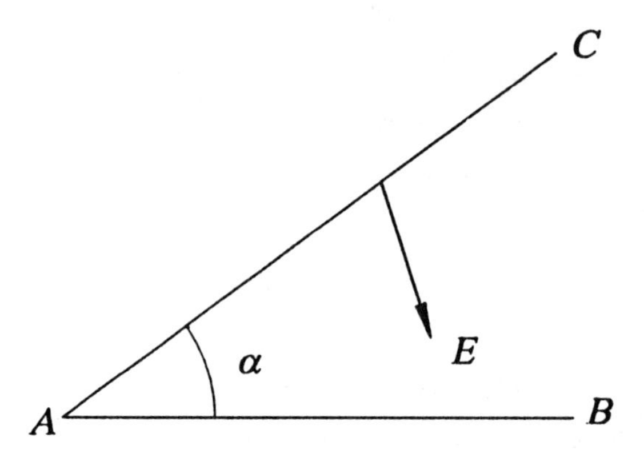


FIGURE 2.5. The angle between collapsing liner and current carrying-conductor can lead to the formation of strong electric fields, which can cause breakdown and reduce the efficiency of the flux compressor.

made contact at angles of $\alpha = 0.02$ to 0.06 radian with no noticeable loss of flux due to breakdown.

On the basis of the material provided above, it may be concluded that magnetic field compression can generate high magnetic fields, high currents, high temperatures, and high electric fields, depending on the particular design. The factors that limit the capabilities of MCGs are finite conductivity, compressibility of the liner material, hydrodynamic instabilities, and arc formation.

### 2.3.2 *Increasing the Energy Amplification Factor*

Helical MCGs, which are discussed in Section 3.4, are designed to amplify a small input current. The amplification factor can be increased without increasing the cavity volume by artificially increasing its length. This is done by adding an infinite number of turns with infinitesimally decreasing step sizes, i.e., decreasing the distance between each succeeding turn without changing the length of the cavity. For a helix of specified length, the magnitude of the energy amplification factor $\gamma_M(\lambda_c)$, where $\lambda_c$ is the length of the selected section of the MCG spiral in which the inductance has decreased by a factor of $e$, reaches a maximum when the length of the section is chosen to be equal to $\lambda_c \approx \rho_R D/u^2 \times 10^{10}$. In this expression, $D$ is the detonation velocity, $u$ is the mass velocity, and $\rho_R = 1/\sigma$ $\Omega$·cm. Over the whole length, $l$, of the preamplifier, it is found that

$$\gamma_M = \left(\frac{l}{2}\right)^{\frac{l}{\lambda_c}}, \ \lambda_c \approx \left(\frac{1}{2}\right)^{\frac{l}{2\lambda_c}}. \tag{2.43}$$

The value of $\gamma_M$ increases rather strongly with an increase in the velocity $u$, and the value of $D/u$ is in the range 3–5 in most MCGs. A generator that has an optimized gain must have a high magnetic flux conservation coefficient, not a high efficiency. However, this is not required for the case where the input energy to the amplifier is low. During the initial stages of magnetic field compression, the magnetic energy density may be several orders of magnitude less than the energy density of the high explosives.

### 2.3.3 Delivering the Maximum Possible Energy to the Load

When a high value of the energy gain, $\gamma_M$, has been specified and when the inductance of the load is low, it is possible to select generator parameters that ensure delivery of the maximum possible energy to the load. For given values of $\gamma_M$ and fixed initial inductance and when the condition $\delta_\triangle << \delta/2$, where $\delta_\triangle$ is the thickness of the helical turns, is satisfied, the maximum possible energy is delivered to the load when the Q-factor of the MCG is at its peak value. The *quality- factor* (or Q-factor) is defined to be ratio of the energy stored in the MCG to the energy lost. The Q-factor is:

$$\alpha = \frac{L}{2TR}, \tag{2.44}$$

where $T$ is the characteristic MCG coupling time and $L$ is the total inductance of the MCG. Likewise, the Q-factor of the load is

$$\alpha_L = \frac{L_L}{2TR_L}. \tag{2.45}$$

The amount of energy delivered to the load is given by the expression

$$W_L = \frac{\varepsilon_0^2 \lambda_{c1}}{2L_L D^2 \gamma_M^{\frac{1}{\alpha_k - 1}}}, \tag{2.46}$$

where $\lambda_{c1}$ is the length of the coupling section.

### 2.3.4 Attaining the Maximum Possible Gain

When a specified amount of energy is to be delivered to a load having a high inductance, $L_L$, it is possible to select generator parameters to achieve the maximum possible gain. The critical limitation on generators having a high load inductance is the high electric field strength, which depends on the EMF and the geometry of the MCG. The magnitude of the EMF, $E = -\, L\, I$, will be maintained at its maximum allowed value, $E = E_*$, if

$$L = L_0 \exp[-2(\alpha - \alpha_0)] = L_0 \exp(-t/\tau), \tag{2.47}$$

where $\alpha = \alpha_0 \exp(t/\tau_*)$, $\tau = t/2\alpha_0(e^{t/\tau_*} - 1) \neq$ const, and $\tau_* = L/R =$ const.

If the energy delivered to the load is $W_L = E_*^2\tau_*^2/8\alpha^2 L_L$, then the maximum value of $\gamma_M$ is achieved at the initial value of $\alpha = \alpha_0 = 1$. In this case, the energy delivered to the load is $\gamma_M^{1/(\alpha-1)}$ times greater than the energy that would be delivered if the inductance of the generator varies exponentially for the same $L_L$; however, the energy amplification factor is limited by the value of $\alpha_*$.

### 2.3.5 *Unconstrained Energy Amplification*

When a given value of energy, $W_L$, is delivered to the load, then to achieve an infinite amplification factor, $\gamma_M$, for an MCG having a constant cross-sectional area that does not tend to zero, an ideal dielectric insulator, having infinitesimal conductivity, must be used. In this case, the EMF and $\gamma_M$ could approach infinity and the value of $W_L$ remains the same for a limited value of the Q-factor, $\alpha$. The use of a single ideal conductor ($\sigma = \infty$) appears to be insufficient. The magnetic flux needs to be completely conserved. No flux cut-off is possible, including that related to the irregular motion of the conductor. Ideally, a compressed dielectric is needed, for only in this case will $\alpha$ and $\tau$ tend to infinity.

The finite conductivity of real conductors allows an unconstrained increase in $\gamma_M$ at given values of $W_L$, but only when the dimensions of the generator increase in all three dimensions. This is referred to as the *quasi-time-independent* model of the MCG. The presence of a conductive medium (a gas) inside the generator may limit the parameters of the generator.

According to the quasi-time-independent model, infinitely large values of $\gamma_M$ can be achieved through an increase in the length of the MCG and when $W_L$ goes to zero. However, the finite value of the velocity of the electromagnetic wave constrains this possibility. For example, in the spiral MCG, the step size of the spiral must be made infinitely small, so that the phase velocity with which the liner makes contact with the turns of the spiral does not exceed the speed of light. The upper limit of $\gamma_M$ for a spiral with a diameter of 1 cm appears to be on the order of $10^3$. When the diameter increases, i.e., tends to infinity, the upper limit of $\gamma_M$ tends to $10^8$.

In MCG amplifiers, the conversion of chemical energy from high explosives into magnetic energy does not need to be highly efficient. To achieve the highest values of $\gamma_M$ from an MCG with an optimized design and a coupling inductance that obeys the relation $L = L_0 \exp(t/\tau)$, only high explosive charges having a length of $\lambda = \pi D \leq \lambda_1$, where $D = const$, a diameter of $d_1 = const$, and that releases its energy linearly, can be utilized efficiently.

Energy amplification can be achieved with high efficiency, if the equation that describes the release of energy by the high explosives ($W_{HE}$) is in good agreement with the relation that describes the optimal generation of magnetic energy. This may be accomplished in two ways:

- Properly shaping the explosive charge so as to increase the effectiveness of the explosive.
- Distributing the high-explosive detonators that drive the detonation process.

The second method can be accomplished by using a large number of volumetrically distributed detonators, which are detonated according to a specified scheme. It is also possible to use a wire along which are located a chain of detonators to initiate the explosives. This last method is shown in Fig. 3.4.

Using the expression $\dot{L}I^2/2 = k_{HE}W_{HE}$, which was derived for helical MCGs by assuming that the inductance varies exponentially according to the expression $L = L_0 \exp(t/\tau)$ and that the diameter of the high explosives is constant, the expression

$$D \approx \frac{W_{M0}}{\pi k_{HE} r_c^2 q \tau} \exp\left[\frac{t}{\tau}\left(1 - \frac{1}{\alpha}\right)\right] \tag{2.48}$$

can be derived for the detonation velocity. In this expression, $k_{HE} = \text{const}$ is the efficiency with which magnetic energy is generated, $r_c$ is the radius of the explosive charge, and $q$ is the energy density of the high explosive.

Replacing a single-point initiation of the high explosives with volumetric, surface, or axial detonation can lead to an increase in the rate of magnetic field compression during the final stages of the compression process. Higher efficiencies and specific energies can also be achieved by using these initiation systems.

One of the most efficient MCGs is the coaxial MCG with axial initiation, which is shown in Fig. 3.4. The equations that describe the operation of this generator, which do not take into consideration the compressibility or strength of the walls, consists of Kirchhoff's equation and the equations that describe the motion of the conductor and field diffusion. The boundary condition is

$$\dot{r}_1 = \frac{(p_{HE} - p_M)}{M_1}, \tag{2.49}$$

where $p_{HE}$ is the pressure created by the detonation of the explosives, $M_1$ is the mass of a unit area of the tube, and $r_1$ is the radius of the liner wall.

An analytical expression for the initial velocity imparted to the liner wall by an axial detonation is

$$u = \frac{D}{4}\sqrt{\frac{2\gamma_u K_m}{(\gamma_u + K_m)}}, \tag{2.50}$$

where $K_m = m/M$, $m$ is the mass of the high explosive, $M$ is the mass of the tube, and $\gamma_u = 0.5\log_2 K_m \sim 1.5 + 0.75\ln K_m$. The value of $K_m$ lies in the range $0.25 < K_m < 4$. The efficiency with which the chemical energy of the high explosives is converted into magnetic energy is given by

$$\eta_{HE} = \frac{\gamma_u}{\gamma_u + K_m}. \tag{2.51}$$

For the case in which the maximum energy of the explosion is converted into magnetic energy, there is an optimal value for $H(t)$. In the case of coaxial MCGs, which are designed to optimize the specific energy, the magnetic field strength is $H \sim 0.7$ MOe. When $r_1 \geq 100$ mm, this value is much less than the maximum possible achieved values. The restrictions on performance are due to field diffusion and irregular motion of the liner walls and are most significant in MCGs with small dimensions.

For the case of axial detonation, the coaxial MCG may achieve rather high powers, because the liner has a large surface area, $s$, and higher velocities, $u$, which reach $4 \times 10^3$ m/s. The magnetic energy flux (power density) is $P_W = 2\omega_W$, where $\omega_W = H^2/8\pi$. The total energy flux depends on the surface area of the liner, $s$, i.e., $P = P_W s$. For example, if $u \sim 10^3$–$10^4$ m/s, $H \sim 1$ MOe, and $E \sim 10^5$–$10^6$ V/m, then $P_W \sim 10^{13}$–$10^{14}$ W/m$^2$.

When coupling the energy from the MCG into the load, the derivatives of the power and current are limited by the electrical strength of the conductors that make up the transmission line from the MCG to the load. These values may be increased by increasing the radius of the transmission line.

# References

[2.1] G. Knoepfel, *Pulsed High Magnetic Fields*, North-Holland Publishing Company, Amsterdam (1970).

[2.2] A.I. Pavlovskii and R.Z. Lyudaev, in *Aspects of Modern Theoretical and Experimental Physics*, Nauka, Leningrad, pp. 206–270 (1984).

[2.3] *Superstrong Magnetic Fields: Physics, Techniques, and Applications* (eds. V.M. Titov and G.A. Shvetsov), Nauka, Moscow (1984).

[2.4] R.E. Kidder, "Compression of Magnetic Field Inside a Hollow Explosive-Driven Cylindrical Conductor," *Proceedings of the Conference on Megagauss Magnetic Field Generation by Explosives and Related Experiments*" (eds. H. Knoepfel and F. Herlach), European Atomic Energy Community, Brussels, pp. 37–54 (1966).

[2.5] M. Cowan, "Energy Conversion Efficiency in Flux Compression," *Proceedings of the Conference on Megagauss Magnetic Field Generation by Explosives and Related Experiments* (eds. H. Knoepfel and F. Herlach), European Atomic Energy Community, Brussels, pp. 167–181 (1966).

[2.6] G. Lehner, "Maximum Magnetic Fields Obtainable by Flux Compression: Limitation Due to the Diffusion of Magnetic Field," *Proceedings of the Conference on Megagauss Magnetic Field Generation by Explosives and Related Experiments* (eds. H. Knoepfel and F. Herlach), European Atomic Energy Community, Brussels, pp. 55–66 (1966).

[2.7] J.P. Somon, "Magnetic Fields Obtained by Flux Compression: Limitation Due to the Dynamics of the Cylindrical Liner," *Proceedings of the Conference on Megagauss Magnetic Field Generation by Explosives and Related Experiments* (eds. H. Knoepfel and F. Herlach), European Atomic Energy Committee, Brussels, p. 67–99 (1966).

# 3
# Magnetocumulative Generators

## 3.1 Introduction

As was discussed in the previous chapter, MCGs convert chemical energy from high explosives into electromagnetic energy. A highly conductive body, accelerated as the result of an explosion, moves through a magnetic field, which, in turn, does work against magnetic forces and induces an electromotive force in a second conductor that surrounds the magnetic field. If the deformation of the magnetic field is sufficiently fast, the magnetic flux will not penetrate through the walls of the moving conductor and the energy in the magnetic field increases. This energy is accumulated in an inductive coil and delivered to a load.

The energy transformation process consists of three stages, that in reality occur simultaneously:

- transfer of explosive energy to the moving conductor in the form of kinetic energy,
- conversion of the kinetic energy into magnetic energy, and
- conversion of magnetic energy into electrical energy.

The conversion of explosive energy into *kinetic energy* is optimal when the high explosive completely covers the surface of the moving conductor (the liner) and when the mass of the explosive is approximately equal to that of the liner. Exact calculations and experiments show that more than half of the explosive energy can be transformed into kinetic energy. However, the velocity of the liner is not very high in this case, being only about 1–2 km/s. Effective conversion of explosive energy into *magnetic energy* depends on the velocity of the liner and on the initial magnetic energy density. The optimal condition occurs when the initial magnetic energy density has the same order of magnitude as the energy density stored in the high explosive. This condition is satisfied when the magnetic field strength has a value in the range of 0.25–1 MOe. This field is optimal because it provides the most effective force for stopping the liner. Moreover, there is no benefit in increasing the initial magnetic field beyond 1 MOe, because to do so disproportionately increases the Joule heating in the walls of the liner, thus decreasing the efficiency of the MCG.

Another factor that affects the output of the MCG is its specific design. Each design has its own unique set of parameters that makes certain designs better for one application and others better for other applications. That is, the characteristics of the load dictate which type of MCG is best for use with that load. In this chapter, various MCG designs are examined.

## 3.2 Classifications of MCGs

Since 1952, two general classes of MCGs, referred to as MK-1 and MK-2 [3.1], have evolved. The MK-1 class (Fig 3.1a) is an "explosive-driven" or "implosive-driven" flux type of generator in which the detonation wave covers a broad area and accelerates the entire surface of the liner or outer conductor, respectively. The MK-2 class (Fig. 3.1b) is an "explosive-driven flux" type of generator, in which the detonation wave moves along its axis and accelerates the liner continually outward to form a flange that moves from one end to the other end.

There are several MCG designs based on the MK-1 and MK-2 classes of generators, or some combination of the two. The MK-1 class of generators is more effective, because the explosives fully surround the liner and the conversion of explosive energy into magnetic energy can be preformed under optimal conditions. However, this design requires rather a high initial magnetic field, which increases the mass of the generator if the coil used to create the initial magnetic field is powered either by capacitors or batteries. The MK-2 class of generators is less effective, but requires lower initial magnetic fields, thus making it more portable. Which class of generator is optimal for a particular application will be determined by the nature of the electrical load and the mass and volume constraints imposed on the system. In some applications, the MK-2 is used to create the initial magnetic field in an MK-1 generator, and in other cases the two designs have been integrated into a multistage generator.

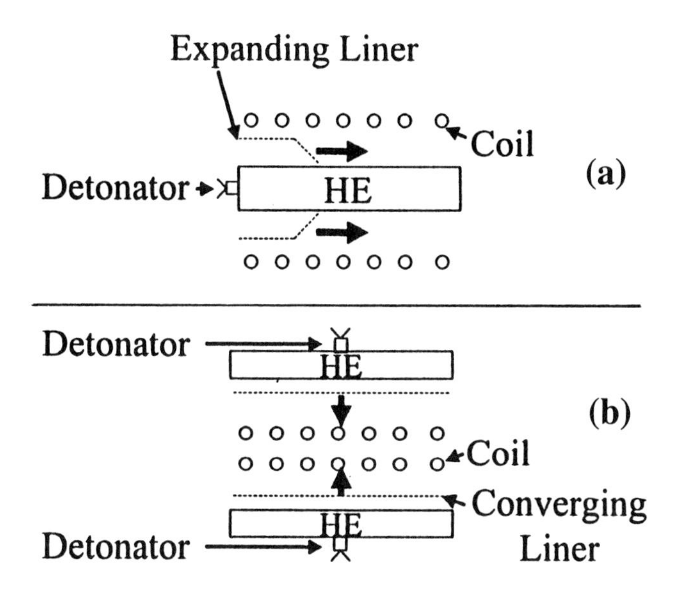


FIGURE 3.1. Schematic drawings of (a) MK-1 and (b) MK-2 type MCGs. In the MK-1 generator, the implosive waves are directed towards the axis of the MCG, while in the MK-2 the explosive wave is directed along the axis.

There are several ways of initiating MK-1 and MK-2 generators. Typically MK-2 generators are end-initiated; i.e., the detonator is placed at the end of a column of explosives, while MK-1 generators are detonated by an array of detonators, which results in either

- surface initiation,
- axial initiation,
- combined surface and axial initiation.

MCGs can also be classified according to whether their rate-of-change in inductance is *time constant* or *time variable.* Time-constant inductance rate of change is observed in plate, bellows, cylindrical-coaxial, and single-pitch helical generators, and time-variable inductance rate of change is observed in all generators including all the above (or multisection and/or variable-pitch helical generators). In addition, MCGs can be classified as either *slow* (tens to hundreds of microseconds) or *fast* (less than ten microseconds). The characteristic values of parameters MK-1 and MK-2 generators is presented in Table 3.1 [3.2].

| Class | Final Characteristics | Driving Force | | |
|---|---|---|---|---|
| | | Explosives | | Electromagnetic |
| | | Metallic Liners | Dielectric-Metallic Shock-Wave Transition | Metallic Liners |
| MK-1 | $B_f = B_0\left(\lambda \frac{S_0}{S_f}\right)$ | $\lambda \approx 0.8$ | $\lambda \approx 0.056$ | $\lambda \approx 0.6$ |
| | | $\frac{S_0}{S_f} \approx 200$ | $\frac{S_0}{S_f} \approx 1600$ | $\frac{S_0}{S_f} \approx 440$ |
| | | $B_{max} \approx 1700$ T | $B_{max} \approx 1600$ | $B_{max} \approx 550$ T |
| MK-2 | $I_f = I_0\left(\lambda \frac{L_0}{L_f}\right)$ | $\lambda \approx 0.2 - 0.8$ | ? | – |
| | | $\frac{L_0}{L_f} \approx 5 - 16000$ | | |
| | $W_f = W_0\left(\lambda \frac{L_0}{L_f}\right)$ | $I_{max} \approx 320$ MA | $I_{max} \approx 0.027$ MA | |
| | | $W_{max} \approx 100$ MJ | $W_{max} \approx 0.20$ MJ | |

TABLE 3.1. Characteristic parameter values of the MK-1 and MK-2 generators: $B$ – magnetic induction, $I$ – electrical current, $W$ – magnetic energy, $L$ – inductance, $S$ – cross-sectional area, $\lambda$ – magentic flux conservation factor, and the subscripts 0 and f are the initial and final values respectively.

In this chapter, the following designs are examined:

- coaxial MCGs (CMCG)
- spiral (helical) MCGs (SMCG)
- plate MCGs (PMCG)
- loop MCGs (LMCG)
- disk MCGs (DMCG)
- shock wave or semiconductor MCGs (SWMCG).

Of these generators, the CMCG, SMCG, PMCG, and SWMCG can be either MK-1 or MK-2. The class and design of MCG used in a particular application depends on the requirements of the load, that is, pulse length, rise time, impedance, and/or voltage. The reasons for selecting a particular class and design of MCG is discussed in Chapter 5.

In addition, multistage MCGs and multi-MCG systems called "batteries" are discussed. Multistage MCGs are where different classes and designs of MCGs are integrated into the same system, such as when an MK-2 spiral generator is used to drive an MK-1 spiral generator through flux trapping. Multistage MCGs are two or more MCGs connected in series, such as when an MK-2 spiral generator is used to drive another MK-2 spiral generator through a transformer.

## 3.3 Coaxial MCGs

One of the simplest MCG designs is the coaxial MCG (CMCG) [3.3–3.5], which consists of two coaxial cylinders, where the inner cylinder has an

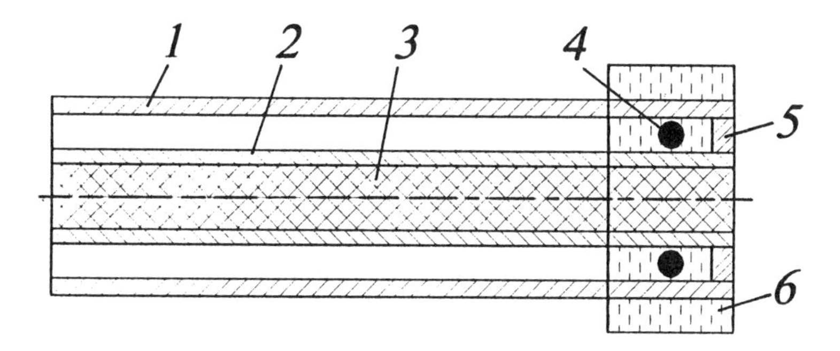


FIGURE 3.2. General view of a coaxial MCG: 1 – external conductor, 2 – internal conductor (liner), 3 – explosive, 4 – measuring coil, 5 – switch, and 6 – Plexiglas ring. The explosive is detonated at one end of the generator. The liner expands toward the outer conductor and makes contact with the outer conductor. This expansion process moves along the length of the generator.

outside radius of $r$ and the outer an inside radius of $R$. Flux compression takes place in the annular space between the two coaxial cylinders by explosively expanding the inner cylinder, called the *liner*. The inner cylinder begins to expand toward the outer cylinder at one end. This expansion process moves down the cylinder reducing the length of the annular region, thus compressing the magnetic field.

A general view of the CMCG is presented in Fig. 3.2. It consists of two tubes, portions of which are pressed into a strong turned barrel. The portions of the coaxial cylinders placed in the barrel act like a load for the generator. A measuring coil is embedded in the cylindrical Plexiglas ring.

Flux compression may be achieved by either explosively expanding the inner conductor or explosively collapsing the outer cylinder. In some designs, both processes have been carried out simultaneously. The first method is the more interesting, since it is energetically efficient, easy to design, and more effectively used with other MCGs in cascaded systems.

Let us now derive the basic expressions that describe magnetic compression in the CMCG. Analysis of experimental data shows that the output current of CMCGs using low-energy explosives is limited by the maximum work that the liner does against the magnetic field during flux compression. When the high explosives are detonated, the amount of power generated is proportional to the expression $qS_\Phi D$, where $q$ is the density of the high explosive, $S_\Phi$ is the area of the detonation front, and $D$ is the detonation velocity. The fraction of the explosive energy that is converted into kinetic energy of the liner is denoted by $k_{\mathrm{HE}}$.

It is worth noting that the pressure of the explosion products decreases rapidly, so that the acceleration of the liner is completed after the explosive products have expanded to 1.5–2 times the initial diameter of the high explosive. Experimental studies of the acceleration of copper tubes by high explosives show that the tubes acquire 70–85% of their maximum kinetic energy within the first third of their final increase in diameter. When the tube has expanded to more than 2.5 times its original diameter, longitudinal fractures develop, which implies that the circumferential conductivity goes to zero.

The efficiency with which the high-explosive energy is converted into kinetic energy is determined mainly by the ratio of the mass of the high explosive, $m$, to that of the liner, $M$:

$$k_m = \frac{m}{M}. \tag{3.1}$$

Calculations show that the smaller the value of $k_m$, the weaker the dependence of the acceleration efficiency is on both the type of high explosives used and the materials from which the liner is constructed. The maximum velocity of the liner is [3.4]

$$u = D\frac{\sqrt{1+\frac{32}{27}k_m}-1}{\sqrt{1+\frac{32}{27}k_m}+1}. \tag{3.2}$$

The optimal value of $k_m$ is 81/32, which is explained by the fact that the liner is efficiently accelerated by only the comparatively small layer of high explosive that is in direct contact with the surface of the liner. Increasing the mass of the high explosive, such that it increases the value of $k_m$ beyond the optimal value, only decreases $k_{\mathrm{HE}}$ and does not increase the velocity of the liner. It should be noted that Eq. 3.2 has to do only with explosive performance and can be applied to other types of MCGs.

If it is assumed that the velocity of the explosion products is linearly distributed, then the acceleration efficiency can be estimated from the following expression:

$$k_{\mathrm{HE}} = \frac{2}{2+k_m}. \tag{3.3}$$

Analysis of this expression shows that as $k_{\mathrm{HE}} \to 1$, so $k_m \to 0$. Experiments confirm that $k_m$ does decrease as $k_{\mathrm{HE}}$ increases; however, $k_{\mathrm{HE}}$ does not increase to 1, but rather to nearly 0.8, before decreasing to $0.5-0.6$. This is because, along with the decrease in $k_m$, there is an increase in the thickness of the liner walls in comparison to the thickness of the high explosives and in the pulse length of the compression wave created by the high explosives. This length is proportional to the thickness of the explosive charge, and so the wave processes play an important role in the energy distribution.

In designing MCGs, it is frequently necessary to know the angle formed between the liner and the center line of the cylinder. If the thicknesses of the liner and outside cylinder are 0.4–2 mm and the distance between the two cylinders is 12–15 times this thickness, then a specific angle, $\beta_p$, is formed, which is time independent. This angle is

$$\beta_p = \sqrt{\frac{\eta_A - 1}{\eta_A + 1} \frac{\pi\sqrt{r_m}}{2} \left(\sqrt{r_m} + 1.86 + \frac{0.1}{\xi}\right)}, \qquad (3.4)$$

where $r_m = \rho_0 r_1^2 / \rho_1 (R^2 - r^2)$ is the normalized distance moved by a mass per unit area of high explosive and of liner, $\xi = (r_1 - r)/r$ is an increment of radius, $r$ is the outer radius of the liner, $r_1$ is the radius of the liner during its motion, $R$ is the inner radius of the outer conductor, $\rho$ is the density of the high explosive, and $\rho_0$ is the density of the liner material.

After being accelerated, the moving liner compresses the magnetic field, thus performing work against the magnetic field defined by the expression $-I^2 \Delta L/2$. Part of the kinetic energy of the liner is converted into magnetic energy when the magnetic field is compressed, which means that a fraction, $\eta_{\mathrm{HE}}$, of the explosive energy is converted into magnetic energy. The energy balance for this process is

$$-\frac{1}{2} I^2 R_* = \eta_{\mathrm{HE}} q S_\Phi D, \qquad (3.5)$$

where $R_* = dL/dt$ is the resistance due to the change in the inductance.

From Eq. 3.5, an expression for the maximum output current of the MCG can be derived as

$$I_{\max} = \sqrt{\frac{2\eta_{\mathrm{HE}} q S_\Phi D}{-R_*}}. \qquad (3.6)$$

Because the skin depth is usually much smaller than the radius $r$, its curvature may be neglected, and it can be assumed that the CMCG has a planar skin-layer. However, the fact that the fields at the surfaces of the two conducting cylinders are different must be taken into account.

If it is assumed that the magnetic field at the inner surface of the outer cylinder is $H$, then the field at the outer surface of the liner is equal to $HR/2$ and the current in the circuit is

$$I = 2\pi HR. \qquad (3.7)$$

The inductance of the generator during the longitudinal compression of the magnetic field is

$$L(t) = \frac{\mu_0}{2\pi}(l_0 - Dt)\ \ln\frac{R}{r}, \tag{3.8}$$

where $l_0$ is the length of the CMCG. The magnetic flux in the CMCG can be derived from Eqs. 3.7 and 3.8 as

$$\Phi = LI = \mu_0 HR(l_0 - Dt)\ \ln\frac{R}{r}. \tag{3.9}$$

The flux losses in the CMCG are determined from the formula

$$R\ln\frac{R}{r}\frac{d}{dt}\left[\left(1-\frac{D}{l_0}\right)\mu H(t)\right] \tag{3.10}$$

$$= \left[\mu\delta\left(1+\frac{r}{R}\right)\left(1-\frac{D}{l_0}t\right)\frac{dH}{dx}\right]_R,$$

or, in normalized form, where time is related to the compression time $\tau_c = l_0/D$ and the distance $x$ to a characteristic linear dimension,

$$\frac{d}{dt}(1-t)\mu H(t) = \left[\delta\left(\frac{1}{R}+\frac{1}{r}\right)\frac{l_0}{D}\frac{1}{\ln\frac{R}{r}}\frac{1-t}{a}\mu\frac{dH}{dx}\right]_R. \tag{3.11}$$

The magnetic Reynolds number depends on the diameter, $a$, and length, $l_0$, of the cavity:

$$N_p = \frac{a^2}{\delta}\frac{D}{l_0}. \tag{3.12}$$

If the coefficient on the right-hand side of Eq. 3.11 is equal to $2/\mu$ and if $a$ is

$$a = \frac{2Rr}{r+R}\ln\frac{R}{r}, \tag{3.13}$$

so that

$$N_p = \frac{1}{\delta^2}\frac{D}{l_0}\left(\frac{2Rr}{r+R}\right)\ln^2\frac{R}{r}, \tag{3.14}$$

then the flux compression is reduced to that of longitudinal compression along the axis of the generator.

To derive the basic parameters of the CMCG, it is useful to express all the formulae in terms of the dimensions of the generator. With this purpose in mind, and taking into account the assumptions that the compression process is time-independent, and that $R_* = dL/dt = -DdL/dx$, substituting the expression for the inductance per unit length of the generator into Eq. 3.6

$$L = \frac{\mu_0}{2\pi} \ln \frac{R}{r} \tag{3.15}$$

yields the maximum current as

$$I_{\text{max}} = \pi \sqrt{\frac{\eta_{\text{HE}} q}{\mu_0 \ln \frac{r}{R}}} \left(R - \delta_c\right), \tag{3.16}$$

where $\delta_c$ is the thickness of the liner wall.

The maximum field strength, generated by the field compression, can be derived by substituting Eq. 3.16 into Eq. 3.7:

$$H_{\text{max}} = \frac{1}{2} \sqrt{\frac{\eta_{\text{HE}} q}{\mu_0 \ln \frac{r}{R}}} \frac{R - \delta_c}{R}. \tag{3.17}$$

The energy density of modern high explosives is about $8 \times 10^9$ J/m$^3$. If it is assumed that $\eta_{\text{HE}} = 1$ and that $\ln(r/R) = 1$, then Eqs. 3.16 and 3.17 become

$$I_{\text{max}} = 2.5 \times 10^5 (R - \delta_c) \tag{3.18}$$

and

$$H_{\text{max}} = \frac{R - \delta_c}{2\mu_0 R} 10^2, \tag{3.19}$$

respectively. It can seen from these expressions, that the maximum currents and fields do not increase with the length of the CMCG. They can only be increased by increasing the radius of the generator.

Analysis of Eqs. 3.18 and 3.19 reveals that the CMCG can create azimuthal magnetic fields on the order of 100 T and currents of up to several megaamperes.

The basic disadvantages that limit these generators as energy sources are their comparatively low output currents and energy gains. These limitations

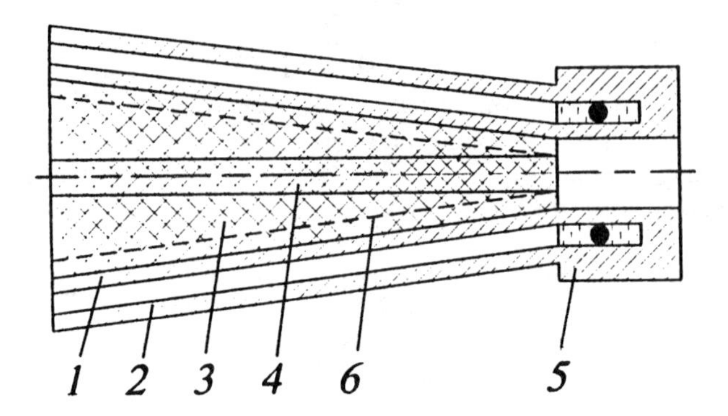


FIGURE 3.3. Cone-shaped MCG: 1 – internal conductor (liner), 2 – external conductor, 3 – explosives, 4 – detonation cord, 5 – load, and 6 – detonation wave. By constructing the inner and outer conductors as truncated cones, the compression time is decreased, which reduces magnetic flux losses.

are due to several reasons, with the most important of these being the high losses of magnetic flux and the small changes in inductance.

Because the high magnetic flux losses are caused by the comparatively long compression time, magnetic flux may be conserved by reducing the compression time. One possible MCG design that has a short compression time is shown in Fig. 3.3, and it will be seen that both the inner and outer conductors are constructed as truncated cones. The high explosives are placed inside the liner, and a detonation cord is placed along the central axis of the generator. The ratio of the explosive and detonation cord velocities depends on the sine of the angle between the liner and the axis of the generator:

$$Q = \arcsin \frac{D_1}{D_2}, \tag{3.20}$$

where $D_1$ is the detonation velocity of the high explosive and $D_2$ is the detonation velocity of the detonation cord. This particular design causes the entire length of the liner wall to move, since, as the detonation wave moves along the cord, it detonates the high explosive that burns radially outwards, but because of the decreasing amount of explosive in the direction of the load, the detonation wave of the high explosives reaches the liner simultaneously along the entire length of the generator, thus causing the entire liner to move at the same time. The compression time now becomes

$$\tau_c = \frac{h_w}{u_l}, \tag{3.21}$$

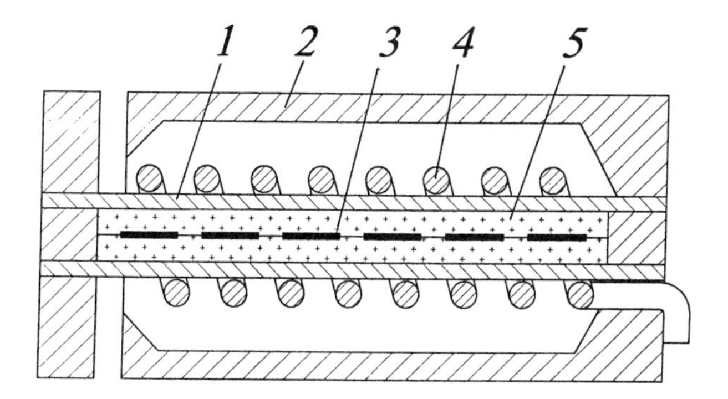


FIGURE 3.4. Coaxial MCG with simultaneous HE initiation: 1 – liner, 2 – case, 3 – exploding conductor, 4 – spiral feed, and 5 – gas like exploding mixture. Simultaneous axial initiation of the explosive can reduce the compression time by a factor of 10 compared to that of other CMCG designs.

where $h_w$ is the radial distance between the inner and outer conductors and $u_l$ is the liner velocity. This particular generator design decreases the compression time by a factor of 10 over that of typical CMCG designs.

The compression time can also be shortened by simultaneously initiating high explosives along the entire length of the generator, as shown in Fig. 3.4. The detonator consists of a conductor with a cross section that varies along its length and located on the center line of the generator. When a pulse of current flows through this conductor, it is explosively vaporized at those points along the length of the wire where the cross section changes sharply. The high explosives are detonated simultaneously at these points, and they can also be detonated along the entire length of the CMCG by uniformly distributing a series of electric detonators along its axis.

The output current of a generator that uses the simultaneous initiation of detonators placed along the axis of the generator is

$$I(r) = I_0 \frac{L_0}{L(r)} = I_0 \frac{A_k}{B_k - \ln\ r}, \tag{3.22}$$

where

$$A_k = \frac{L_L}{2l_c} + \ln \frac{R_0 + 2\delta_c}{r_0}, \tag{3.23}$$

$$B_k = \frac{L_L}{2I_c} + \ln\left(R_0 + 2\delta_c\right), \tag{3.24}$$

$l_c$ is the length of the coaxial generator, $\delta_c$ is the thickness of the outer wall, $R_0$ is the inner radius of the outer conductor, and $L_L$ is the load inductance.

Taking into account magnetic field diffusion, the inductance of the generator, which depends on the radius of the inner conductor during the time it is operating, is

$$L(r) = L_L + 2l_c \ln \frac{R_0 + 2\delta_c}{r}. \tag{3.25}$$

The design shown in Fig. 3.4 uses a spiral coil, which is positioned concentric to the liner, to increase the initial inductance. When the liner makes contact with the spiral, the magnetic energy in the CMCG remains the same, but the inductance decreases rapidly. The current in the generator changes according to the expression

$$I_2 = I_1 \sqrt{\frac{L_1}{L_2}}, \tag{3.26}$$

where $I_1$ and $L_1$ are the current and inductance before the liner makes contact with the spiral coil and $I_2$ and $L_2$ are the current and inductance after the liner makes contact with the spiral coil. The current flowing through the generator will increase only after the liner makes contact with the coil, and this particular design allows the current to be amplified by a factor of 10.

Therefore, to obtain maximum current and energy gains, it is necessary to use either simultaneous initiation of the high explosive over the entire length of the generator or a cone-shaped design that uses two kinds of explosives having different detonation velocities. To increase the energy initially stored in the generator, the outside conductor needs to be placed near to and concentric with the liner.

## 3.4 Spiral (Helical) MCGs

The most advanced and easy to design generators, which can amplify the output energy by a 100-fold, are the highly inductive multisectioned spiral MCGs. The *spiral* or *helical MCG* (SMCG) [3.3,3.5] is a modification of the coaxial generator, in which the solid outside conductor is replaced by a spiral winding, which gives the generator its comparatively high specific inductance. The magnetic flux is compressed by the expanding inner conductor (the liner). A drawing of a typical spiral generator is presented in Fig. 3.5. Depending on the specific application, numerous variations of this basic design have been produced. A summary of these important variations and their modifications are

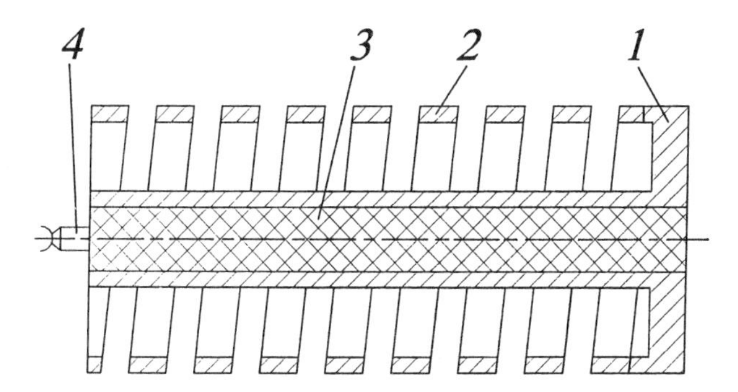


FIGURE 3.5. Sprial MCG: 1 – liner, 2 – spiral coil, 3 – explosives, and 4 – detonator. The spiral MCG is a modification of coaxial MCG, where the outside solid conductor is replaced by a spiral winding, which gives the generator a higher inductance than the corresponding coaxial MCG.

- Fast pulse — two-end initiation and/or inductive coupling plus late crowbarring.
- High efficiency — $dL/dt$ optimization.
- High voltage — inductive feed plus axial initiation, shaping, and/or nesting.
- High current — shaping of helical coil and/or armature, tilted turns, parallel "battery" coupling, and low value inductive/resistive load.
- High energy — $dL/dt$ optimization, high value inductive/resistive load.

The operating principles of the spiral MCG are similar to those of the CMCG. Magnetic flux compression occurs as described in Section 2.2. Because the magnetic flux diffuses into the conductive elements of the circuit, the diameter of the wires of the solenoid and the thickness of the walls of the liner must be selected accordingly. It is necessary that the diameter of the wire and the thickness of the liner walls are greater than the skin depth, as described in Section 2.1.1.

The inductance of the SMCG is

$$L = \mu_0 \int_z^l \pi \left(R^2 - r_1^2\right) n^2(z) dz, \tag{3.27}$$

where $\mu_0$ is the magnetic permeability of vacuum, $n(z)$ is the number of turns per unit length, $R$ is the inner diameter of the spiral coil, and $r_1$ is the outer diameter of the liner. This equation may be used to calculate the inductance of other generator designs. However, when designing a SMCG, the flux losses that occur during magnetic flux compression must be taken into account. These losses can be accounted for by deriving an expression that describes the corresponding changes in the inductance of the generator.

It is well known that, owing to the rapid increase in magnetic flux during magnetic field compression, electric fields are formed in the active volume of the generator. These fields can lead to electric breakdown, resulting in less energy being delivered to the load. In the limiting case, the maximum voltage in spiral generators must satisfy the condition: $LdI/dt = IdL/dt = (\Phi/L)\, dL/dt$. Especially high voltages are formed in the highly inductive spiral coils, since these coils are a fed high magnetic flux; i.e., for a specified flux in the load, the value of the initial flux, $\Phi_0$, must be lower, so that the ratio $L_0/L_L$ is higher. Voltages in the SMCG may reach tens, or even hundreds of kilovolts, depending on the initial energy in the generator, how the inductance changes, and its dimensions. The operating voltage will remain at a minimum value for specified generator parameters, if, during the entire compression process, the value of $LdI/dt$ remains constant.

Experiments show that for highly inductive spiral coils operating with no significant flux losses, i.e., without breakdown, the ratio $\alpha' = R/L$ remains practically constant during the entire time that the liner is coupling with the spiral turns. If it is assumed that $\alpha'$ is constant, then Eq. 1.57 can be rewritten as

$$I(t) = \frac{\Phi_0 e^{-\alpha' t}}{L(t)}. \tag{3.28}$$

Taking the derivative of Eq. 3.28 with respect to time,

$$\frac{dI}{dt} = \frac{\Phi_0 e^{-\alpha' t}\left[\frac{dL}{dt} + \alpha' L(t)\right]}{L^2(t)}, \tag{3.29}$$

multiplying both sides of this expression by $L$, and assuming that $LdI/dt = E = \text{const}$, it follows that

$$\frac{dL}{dt} + \left(\frac{E}{\Phi_0} e^{\alpha' t} + \alpha'\right) L = 0, \tag{3.30}$$

which can be solved to give,

$$L(t) = L_0 \exp\left[\left(1 - e^{\alpha' t}\right)\frac{E}{\Phi_0 \alpha'} - \alpha' t\right]. \tag{3.31}$$

Eq. 3.30 describes the coupling inductance of the MCG, where the maximum voltage between the liner and the spiral coils during the operating time of the generator has been assumed to be constant.

In order to achieve maximum conversion of chemical energy from the high explosive into magnetic energy, the distance between the spiral turns must be increased along the length of the generator. This will conserve the current density. The requisite distance between the spiral turns can be derived from the law of the conservation of energy, which yields

$$\Lambda_c = \frac{1}{n(z)} = \sqrt{\mu_0 \pi \frac{(R^2 - r_1^2) W_{M0}}{\beta' L_0 W_K}} \left( \frac{\alpha' W_K}{W_{M0}} z + 1 \right), \tag{3.32}$$

where $W_K$ is the kinetic energy per unit length of the liner, $W_{M0}$ is the initial magnetic energy, $L_0$ is the initial inductance of the generator, and the coefficient $\beta'$ is less than 1.

The energy coupled into the load is

$$W_L = \lambda_k^2(t) W_0 \left( \frac{L_0 + L_L}{L_L} \right)^2 \left( \frac{L_L}{L_L + L_B + L_G} \right), \tag{3.33}$$

where $W_0$ is the initial energy in the system, which includes the capacitive stores (provided it is used to create the initial magnetic field in the SMCG), generator, and load; $W_L$ is the energy coupled into the load; and $L_B$, $L_G$, and $L_L$ are the inductances of the capacitive store, generator, and load, respectively. From this equation, the ratio of the energy coupled into the load to the initial energy can be found:

$$\frac{W_L}{W_0} = \lambda_k^2(t) \frac{\left( 1 + \frac{L_G}{L_L} \right)^2}{1 - \frac{L_G + L_B}{L_L}}. \tag{3.34}$$

The efficiency with which the energy of the generator is coupled into the load depends on the conservation of magnetic flux and on the ratio of the load inductance to the inductance of the generator.

Figure 3.6 presents plots of the normalized energy versus the normalized inductance for different values of $\lambda_k$. From these curves, the inductance of the generator can be found for given values of $W_L/W_0$, $L_L$, and $\lambda_k$.

Taking into account the magnetic field losses, an expression for the current gain can be derived. The magnetic flux losses are due to the diffusion of magnetic field into the liner at the rate

$$\frac{d\Phi}{dt} = -\mu_0 \frac{2\pi R D}{\Lambda_c} \left[ \delta_1 B_1(0) + \delta_2 B \right], \tag{3.35}$$

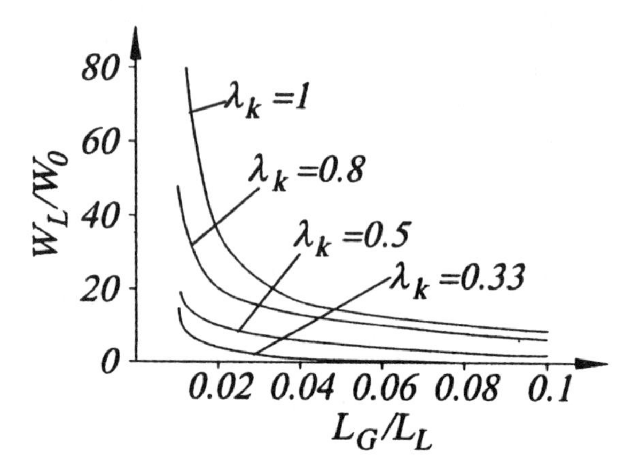


FIGURE 3.6. Normalized load energy and inductance for different values of the coupling factor. From these curves, the inductance of the generator can by found for different values of the ratio of load energy to initial energy, $W_L/W_0$, load inductance, $L_L$, and magnetic flux conservation coefficient, $\lambda_k$.

where $R$ is the inner radius of the spiral; $D$ is the detonation velocity; $\delta_1$ is the thickness of the skin layer, which depends on the time of flight of the liner, which is the time for the liner to almost make contact with the spiral coils; and $\delta_2$ is the thickness of the skin layer after the liner and spiral coils have made contact, which depends on the total operating time of the generator. The total flux losses due to diffusion of magnetic flux into the metal and due to the eccentricity of the liner and spiral coil are

$$\frac{B}{B_0} = \left[\frac{L_0}{\frac{(l-z)L_0}{l} + L_L}\right]^F, \tag{3.36}$$

where

$$F = 1 - \left[\frac{2\delta_2 R}{R^2 - r_1^2} + \frac{2\delta_1 \tan(\beta\Lambda_c)}{(R - r_1)(\Lambda_0 - d)} + \frac{\delta_L \tan\beta}{\pi R(R - r_1)}\right] \tag{3.37}$$

is the *quality* or *perfectness factor*. In this case, the current gain is

$$\gamma_I = \frac{I}{I_0} = \left[\frac{L_0}{\frac{(l-z)L_0}{l} + L_L}\right]^F, \tag{3.38}$$

where $z$ represents the distance along the generator axis that the liner has expanded and $d$ is the diameter of the wire used to make the spiral coil.

An empirical formula for the quality factor is

$$F_E = \frac{\ln(\frac{I}{I_0})}{\ln(\frac{L_0}{L_L})}, \tag{3.39}$$

which has been found to have values that lie in the range 0.5 to 0.7. From Eq. 3.37, it can be seen that the magnitude of the quality factor can be increased (i.e., the magnetic flux losses decreased) by increasing the difference in the radii, $R - r_1$, but an increase by a factor greater than 2.5 makes no sense due to the destruction of the liner.

To increase the energy and current gains, the expression that describes the conversion of chemical energy from the high explosives must be adjusted by using the optimal expression, which is exponential, for the generation of magnetic energy. This may be achieved by two methods:

- Properly shaping the explosive charge and varying its diameter.
- Properly driving the detonation of the high explosive, for example, by distributing a large number of detonators along the whole length of the high explosive or by exponentially releasing the energy of the high explosive, assuming that it has a constant cross sectional area.

In Fig. 3.7, which shows an SMCG constructed according to the second of these methods, ED represents a chain of simultaneously initiated electric detonators. The high explosive in a section is detonated when the walls of the liner simultaneously impact with the coils in the preceding section. To adjust the conversion of chemical energy into magnetic energy, the amount of high explosives in each section must exceed the amount in its subsequent section.

SMCGs that use simultaneous initiation of the high explosive may be used as high voltage generators [3.6]. This is due to the fact that these generators have the high inductances that are characteristic of spiral generators and shortened generation times. As a result, the total resistance of the generator increases, and, therefore, high voltages are created. For these generators, the parameter $dL/dt$ is much greater than it is for other types of generators, and they operate more efficiently when connected to high resistance loads. In the schematic diagram of Fig. 3.8, the thickness of the liner walls varies along the length of the generator in such a way that at the moment the liner makes contact with the spiral coils, the input to the load is opened. This is designed in such a way that it will survive the high voltages formed in the load.

Let us now determine how the inductance of a simultaneously initiated spiral generator changes. Assuming that the axial field is uniform, that is, neglecting boundary effects and the existence of any small radial components of the field, the effective cross-sectional area of the spiral coils can

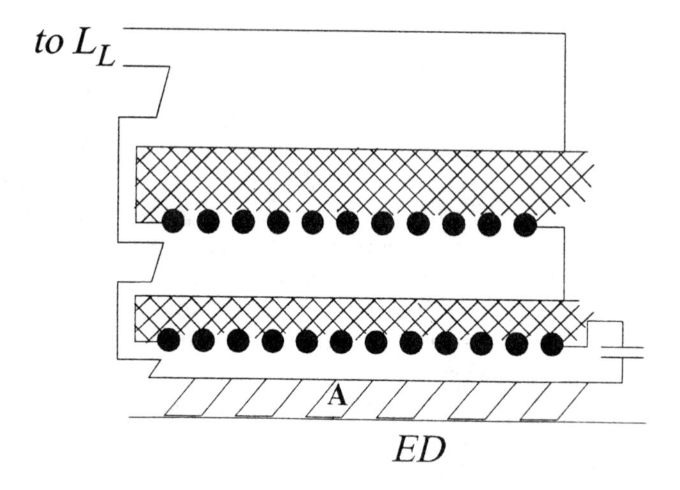


FIGURE 3.7. "Matrioshka" type of multisection spiral MCG has a chain of electric detonators, which are simultaneously initiated. This is one method for increasing the energy and current gains of the generator.

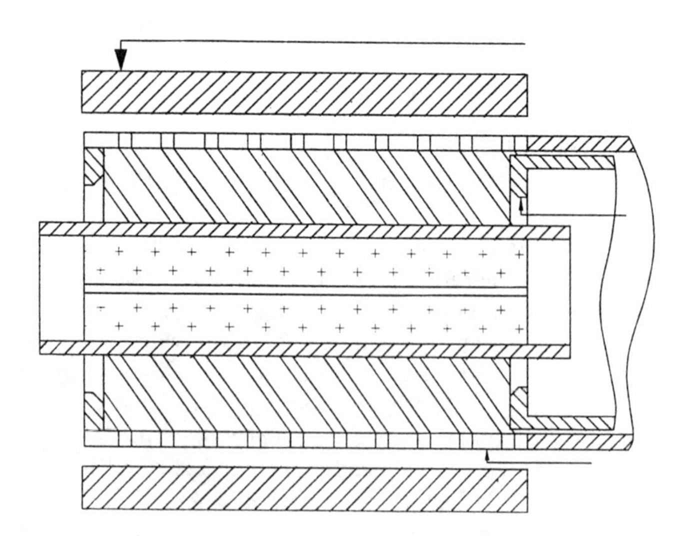

FIGURE 3.8. The short-pulse spiral MCG generate high voltages as a result of its high inductance. These generators operate more efficiently when connected to high-resistance loads.

be calculated. Denoting the radius of the $i$-th coil by $R_i(t)$, its effective cross-sectional area is

$$A_{ef} = \frac{1}{m} \sum_{i=1}^{N} \pi \left[ R^2 - R_i^2(t) \right], \tag{3.40}$$

where $m$ is the number of conductors in a section, $R$ is the radius of the spiral coils in a section, and $N$ is the number of sections that make up the spiral coil. Since $Nm$ is a large number, the summation sign can be replaced with an integral or

$$A_{ef} = \frac{\pi}{m\triangle} \int_0^l \left[ R^2 - R^2(x,t) \right] dx, \tag{3.41}$$

where $\triangle$ is the distance between the turns of the coil and $l$ is the length of the spiral coil. When $\beta = \mu_0 NI/l$ and $l = Nm\triangle$, the inductance of the SMCG is

$$L(t) = \frac{\mu_0 \pi N^2}{l} \left[ \frac{1}{l} \int_0^l (R^2 - R^2(x,t)) dx \right]. \tag{3.42}$$

If it is assumed that the liner, with radius $R_1(t)$, is concentric with the spiral coil, then by using Eq. 3.42, it can be shown that

$$L(t) = \frac{\mu_0 \pi N^2}{l} \left[ R^2 - R_1^2(t) \right]. \tag{3.43}$$

If the walls of the liner increase in thickness along its length according to

$$R_1 = R_2 - \frac{(R_2 - R_1)\, x}{l}, \tag{3.44}$$

then Eq. 3.43 becomes

$$L = \frac{\mu_0 \pi N^2 (R_2 - R_1)(2R_2 + R_1)}{3l}. \tag{3.45}$$

When designing and manufacturing MCGs, account must be taken of the fact that they obey the *law of similarity*. That is, investigations do not have to be carried out on a full-size device, but can be based on scale models, which significantly reduces the investigation time and the research costs. For example, denoting the inductance, resistance, and operating times of a model and of a real device as $L_1$, $R_{g1}$, and $t_1$ and $L_2$, $R_{g2}$, and $t_2$, respectively, and assuming that all the linear dimensions of the real device are $n$ times larger than they are of the model, then for each time $t_2 = nt_1$, the equality $L_2 = nL_1$ is valid and the flux conservation coefficients are

$$\lambda_{k1} = \exp\left[-\int_0^{t_1} \frac{R_{g1}}{L_1} dt\right], \quad \lambda_{k2} = \exp\left[-\int_0^{t_2} \frac{R_{g2}}{L_2} dt\right]. \tag{3.46}$$

If at each time $t_2 = nt_1$, $R_{g2} = R_{g1}$, then

$$\int_0^{t_1} \frac{R_{g1}}{L_1} dt = \int_0^{nt_1} \frac{R_{g2}}{nL_1} dt, \quad \lambda_{k1} = \lambda_{k2}. \tag{3.47}$$

However, this $n$-fold increase in the dimensions of the generator causes an $n$-fold decrease in the equivalent frequency, $\omega = (2/I)\, dI/dt$, which means that there is a $\sqrt{n}$ increase in the skin depth and a $\sqrt{n}$ decrease in the active resistance of the circuit. If the flux losses in the MCG are due only to the final conductivity, then

$$R_{g2} = \frac{R_{g1}}{\sqrt{n}}, \quad \lambda_{k2} = \lambda_{k1}^{\frac{1}{\sqrt{n}}}. \tag{3.48}$$

This relation works well for axially symmetric systems. Experimental investigations of highly inductive spiral MCGs have shown that if the coils are insulated and are optimally distributed along the axis, then a generator can amplify the initial energy by 1000 times or more. The specific energy, which is the ratio of the energy delivered to the load to the initial volume of the MCG, is 30–60 J/cm$^3$, and the efficiency with which high explosive energy is converted into energy delivered to the load is near 4.8%.

## 3.5 Plate MCGs

Planar MCGs are basically two plates that are explosively driven together to compress a magnetic field, and they are also referred to as strip, bellows, or plate generators [3.3,3.7]. The term *plate MCG* will be used in this book. The schematic drawing of Fig. 3.9 shows that a plate MCG (PMCG) consists of two copper plates (1), isolated from the cassette (3) that is filled with the explosive charge (4), with a single coil solenoid at one end. When the high explosives is detonated, the force drives the walls of the cassette toward the copper plates. As a result, the initial magnetic field is driven into the solenoid.

Investigations of MCGs have focused on the energy characteristics of these generators, with the goal being to deliver either the maximum amount

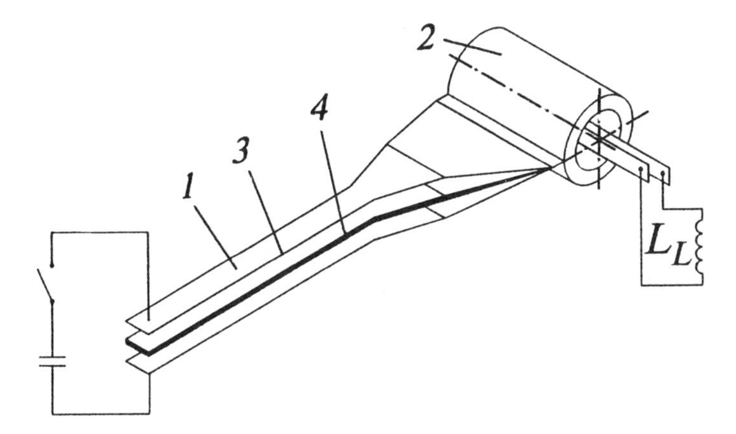


FIGURE 3.9. In plate MCGs, the explosive is used to drive to metal plates together to compress magnetic fields. The generator consists of 1 – copper plates, 2 – load, 3 – cassette filled with 4 – explosives.

of energy to an active load or the maximum amount of energy from the high explosive to the load (that is, increase the efficiency of this process). Considering these two objectives, the optimal inductance of the MCG driving an ohmic load, the resistance of which is changing as it is heated, can be determined.

The PMCG can be represented by the equivalent circuit shown in Fig. 1.10, where the generator has a decreasing inductance, $L_g(t)$, and a resistance, $R_g(t)$, and is connected to a load having an inductance $L_L$. At the moment that the magnetic field compression begins ($t = 0$), a current $I_0$ is flowing through the generator. In performing the calculations, it is assumed that the magnetic flux losses in the generator are much less than the losses in the load.

Using Faraday's law of induction, the current in the MCG can be found as

$$d\ \ln(I) = (1 - \rho_*)d\ \ln(L), \tag{3.49}$$

where $L = L_g(t) + L_L$. The current also depends on the normalized function $\rho_*$; i.e., the normalized resistance of the circuit, which, in turn, depends on the resistance $R_g(t)$ and the MCG design through $dL/dt$:

$$\rho_* = -\frac{R_g(t)}{\frac{dL}{dt}}. \tag{3.50}$$

In certain cases, the current equation can be derived explicitly, provided that the behavior of $\rho_*(t)$ is known. By changing $dL/dt$ for specific values of $R_g(t)$, different temporal curves for $\rho_*(t)$ and for the current flowing through the load can be plotted. The dependence of the current on the normalized circuit parameters in Eq. 3.50 can be used to generate these curves.

The generator is considered to be operating optimally relative to the high-explosive energy, if it converts a larger portion of its explosive energy into electromagnetic energy. The optimal conditions are given in Section 3.1 and are valid if the power created during the deformation of the circuit in any section of the generator is equal to the maximum share of the power produced during the detonation of the high explosive, that is,

$$-\frac{I^2}{2}\frac{dL}{dt} = \eta_{\text{HE}} q S_{\Phi} D. \tag{3.51}$$

If the resistance changes linearly with respect to temperature, $R_g = R_{g0}(1+\alpha_T T)$, due to Joule heating, then

$$\frac{dR_g}{dt} = \frac{R_{g0}\alpha_T}{C_\alpha} R_g I^2, \tag{3.52}$$

where $\alpha_T$ is the temperature coefficient of the resistance, $C_\alpha$ is the total heat capacitance of the load, and $R_{g0}$ is the initial value of the resistance. Using Eqs. 3.50–3.52, it can be shown that the dependence of the resistance on time is

$$R_g(t) = R_{g0}\left[1 + \frac{2\alpha_T \eta_{\text{HE}} q D}{C_\alpha} \int_0^t S(\xi)\rho_*(\xi)d\xi\right] \tag{3.53}$$

and that the optimal inductance of the generator is

$$L_g(t) = L_g(0) - \int_0^t \frac{R_{g0}}{\rho_*(\tau)} A\, d\tau, \tag{3.54}$$

where

$$A = 1 + \frac{2\alpha_T \eta_{\text{HE}} q D}{C_\alpha} \int_0^t S(\xi)\rho_*(\xi)d\xi. \tag{3.55}$$

Let us now determine the optimal width of the plates, current, power, and the behavior of the ohmic load in time. This optimization is based on the amount of energy generated by the high explosives. Assuming that the PMCG is connected to a resistive load, where the resistance increases linearly with temperature, the inductance of the PMCG is given by

$$L_g(t) = L_g(0) - \int_{-l}^{-l+Dt} \frac{\mu_0 b}{y(x)} dx. \tag{3.56}$$

In this analysis, it is also assumed that the magnetic field in the generator is uniform and that the distance between the plates, $2b = \text{const}$, is smaller than the width of the plates, $2y(x)$, where the width of the plates changes along the length of the generator, $x(t)$. The plates are accelerated toward each other at a speed equal to the detonation velocity of the high explosives. The parameter $l$ is the total length of the generator, where the origin of the coordinate system is at the end where the plates are connected to the load. The cross-sectional area of the high-explosive charge, which has a constant thickness of $2\delta_{\mathrm{HE}}$, is

$$S_{\Phi} = 4\delta_{\mathrm{HE}} y(x). \tag{3.57}$$

It follows from Eqs. 3.51 and 3.56 that the change in the current in the optimized PMCG coincides with the change in the width of the plates that are compressing the magnetic field, i.e.,

$$I = I_0 \frac{y}{y_0}, \tag{3.58}$$

where $y_0 = y(-l)$. This behavior may be explained by the fact that the PMCG was optimized on the basis that $\eta_{\mathrm{HE}} = \text{const}$ and that the kinetic energy reserve of any of the conducting elements obtained from the high-explosive charge, which has a constant thickness, remains unchanged along the length of the generator. To attain an optimal value of $\eta_{\mathrm{HE}}$, it is necessary that the force acting on the conductor from the side of the magnetic field, which decelerates the plates over the distance $2b = \text{const}$, remains unchanged over the length of the explosive charge. This means that the width of the plates must change so that the linear current density on the surface of the plates remains the same.

If it is assumed that $\rho_* = \rho_{*0} \exp(\beta_* t)$, where $\beta_* = \text{const}$, then the equation that describes the optimal width of the plates along the length of the generator is

$$y(x) = \frac{y_0 \exp\left(\beta_* \frac{x+l}{D}\right)}{\sqrt{1 + \frac{B}{2\rho_*}\left[\exp\left(2\beta_* \frac{x+l}{D}\right) - 1\right]}}, \tag{3.59}$$

where

$$B = \frac{16\alpha_T \eta_{\mathrm{HE}} q \delta_* y_0 D \rho_{*0}}{C_\alpha} \tag{3.60}$$

and the increase in the resistance of the inductor is

$$R_g(t) = R_{g0}\sqrt{1 - \frac{B}{2\beta_*}\,[\exp(2\beta_* t) - 1]}. \tag{3.61}$$

Knowing $R_g(t)$ and $I(t)$, an expression for the power delivered to the load can be derived.

If the generator is sufficiently long, the change in the width of the plates, current, resistance, and power strongly depend on the parameter $\beta_*$. When $\beta_* > 0$ and as $t \to \infty$, the width of the plates tends toward the limiting value

$$y(x) = y_0\sqrt{\frac{2\beta_*}{B}}, \tag{3.62}$$

and the current tends toward the limiting value

$$I = I_0\sqrt{\frac{2\beta_*}{B}}. \tag{3.63}$$

The resistance and power increase exponentially according to the expressions

$$R_g \to R_{g0}\sqrt{\frac{B}{2\beta_*}}\exp(\beta_* t), \tag{3.64}$$

$$P \to R_{g0}I_0^2\sqrt{\frac{2\beta_*}{B}}\exp(\beta_* t). \tag{3.65}$$

When $\beta_* < 0$, the resistance may be no greater than $R_{g0}\sqrt{1 - B/2\beta_*}$ and the width of the plates, current, and power decrease to zero. When the active load is constant, that is, $\alpha_T = 0$, then

$$y(x) = y_0 \exp\left(\beta_* \frac{x + l}{D}\right). \tag{3.66}$$

If it is assumed that $\rho_* = \rho_{*0}$, the operation of the PMCG will differ qualitatively. In this case, the width of the plates, and thus the current, will decrease according to the expression

$$y(x) = \frac{y_0}{\sqrt{1 + B\frac{x+l}{D}}}, \tag{3.67}$$

and the ohmic resistance of the load will increasc in time:

$$R_g = R_{g0}\sqrt{1+Bt}. \tag{3.68}$$

The decrease in the width of the plate is quite natural, because of the increasing resistance. The condition $\rho_* = \text{const}$ remains valid only if $-dL/dt = 4\pi bD/y(x)$ increases. This is why the requirement that $\rho_* = \text{const}$ leads to a narrowing of the plates of the generator. Assuming this geometry for the plates, the energy delivered to the load during the compression of the magnetic field is

$$W_H = \int_0^{\frac{l}{D}} R_g I^2 dt = \frac{C_\alpha}{\alpha_T}\left(\sqrt{1+B\frac{l}{D}}-1\right). \tag{3.69}$$

The energy produced by the explosive charge is

$$\begin{aligned} W_{HE} &= \int_{-l}^{0} 4q\delta_{\text{HE}}y(x)dx \\ &= \frac{C_\alpha}{2\alpha_T\eta_{\text{HE}}\rho_{*0}}\left(\sqrt{1+B\frac{l}{D}}-1\right). \end{aligned} \tag{3.70}$$

It follows from the relationship

$$\frac{W_H}{W_{\text{HE}}} = 2\eta_{\text{HE}}\rho_{*0} \tag{3.71}$$

that, when $\rho_{*0} > 1/2$, the amount of energy delivered to the resistive load may be greater than the amount of energy contained in the high explosive. This apparent contradiction can be explained as follows.

The amount of power that is used to increase the magnetic field energy and that is delivered to the active load is $-(I^2/2)\, dL/dt$. The energy equation for the system is

$$-\frac{I^2}{2}\frac{dL}{dt} = R_g I^2 - \frac{d}{dt}\left(\frac{LI^2}{2}\right). \tag{3.72}$$

It follows, that, if $R_g > -(1/2)\, dL/dt$, not only is high explosive energy, but so too is part of the initial magnetic field energy in the generator. This is why Eq. 3.71 does not make sense. Integrating Eq. 3.51 and assuming that the cross-sectional area, $S_\Phi(x)$, is known, the inductance of the MCG is

$$L_g(t) = L_g(0) - \int_0^t \frac{2\eta_{\mathrm{HE}} q S_\Phi(\xi) R(\xi)}{P(\xi)} d\xi. \tag{3.73}$$

The MCG is capable of delivering power, $P = I^2/R_g$, which has the required dependence on time, to the resistance $R(t)$.

Let us now consider methods for calculating the parameters of a profiled PMCG. Profiling refers to changing not only the dimensions, but also the shape of the plates by adding notches as shown in Fig. 3.10. If the magnetic Reynolds number, $N_p$, and the inductance trimming coefficient, $\gamma_L$, of the PMCG are known, then its basic parameters can be determined. The current gain is

$$\gamma_I = \frac{y_H}{y_0} = y_L - \frac{y_L - 1}{N_p}, \tag{3.74}$$

the energy conservation coefficient is

$$\lambda_k = 1 - \frac{1}{N_p} + \frac{1}{\gamma_L N_p}, \tag{3.75}$$

and the magnetic energy coefficient is

$$\varepsilon = \gamma_L \lambda_k^2 = \gamma_L \left(1 - \frac{1}{N_p} - \frac{1}{\gamma_L N_p}\right). \tag{3.76}$$

When the above parameters are associated with a specified plate width $2y_H$ in the last sections of the generator, the distance between the plates is

$$b = \frac{2N_p y_H R_g}{\mu_0 D}. \tag{3.77}$$

In deriving Eq. 3.77, it was assumed that the inductance of a generator having arbitrary lengths and widths of $2y_H$ is

$$L_H = \mu_0 \frac{b l_H}{2y_H}. \tag{3.78}$$

If $l_H$ and $N_p$ are known, then change in the width of the plates over their length can be determined from the following expression:

$$y(x) = y_H \exp\left[\left(1 - \frac{1}{N_p}\right)\frac{x}{l_H}\right], \tag{3.79}$$

Therefore, if the substitutions $x = l_0$ and $y = y_0$ are made and the following equation is employed,

$$y_0 = y_H \frac{1}{\gamma_L - \frac{\gamma_L - 1}{N_p}}, \tag{3.80}$$

the length of the generator can be found as

$$l_0 = \frac{l_H}{1 - \frac{1}{N_p}} \ln \left( \gamma_L - \frac{\gamma_L - 1}{N_p} \right). \tag{3.81}$$

If the operating time of the PMCG satisfies the previously defined conditions, then $y_H$ is not an independent parameter and may be determined from the solution of Eqs. 3.77, 3.78, and 3.81 and the equation

$$t_g = \frac{l_0}{D}, \tag{3.82}$$

where $t_g$ is the operating time of the generator.

If $R = 0$ and if it is assumed that $N_p \to \infty$, the working equations for a generator that have been optimized for the conversion of the maximum high explosive energy connected to an inductive load can be derived. For such a generator, it can be assumed that $\lambda_k = 1$, $\gamma_I = \gamma_L$, and $\varepsilon = \gamma_L$. An expression describing how the width of the plates should vary along the length of the generator can be derived as

$$y(x) = y_H \exp \left( \frac{x}{l_H} \right). \tag{3.83}$$

Using Eq. 3.81, an expression for the length of this generator can also be derived as

$$l_0 = l_H \ln \gamma_L. \tag{3.84}$$

When $N_p > 1$, it has been shown that profiled PMCGs can generate significantly higher output energies than PMCGs that have not been profiled. The most efficient use of the energy from the high explosive can be achieved when the generator is profiled.

When the linear dimensions of the generator are changed, the energy scale of the experiments can be widely varied. For these generators, the current gain can reach values of 22–24 and the energy gain values of 8–9. In order to increase the efficiency of the generator, one way, for example, is to increase the length of the generator, $l_0$, or the distance between the

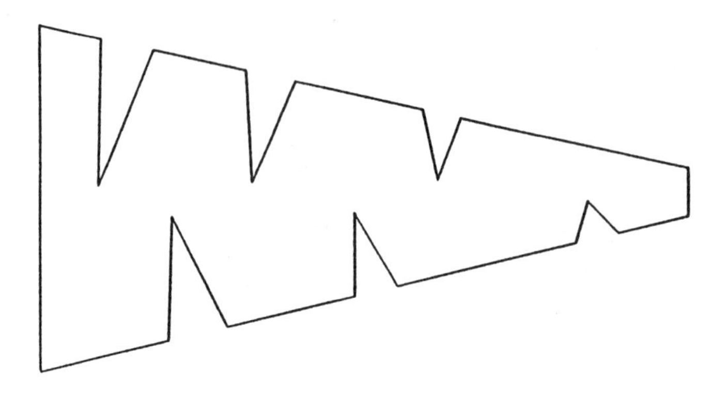

FIGURE 3.10. Diagram of profiled plate for PMCG. The inductance of the PMCG can be increased by changing the profile of the plates, without changing the current, because the highest flux losses occur in the notched regions owing to the complex topology of its fields and its current distribution.

plate and the cassette. However, an increase in the length $l_0$ leads to an increase in the amount of high explosive charge used and an increase in the distance $b$ leads to an increase in flux losses.

The inductance of the MCG can be increased by changing the profile of the plates, as shown in Fig. 3.10, without increasing the current, because the highest flux losses occur in the notched regions owing to the complex topology of its fields and its current distributions.

In summary, when designing PMCGs, it is best to increase the length of the gap between the cassette and the plates to its maximum allowed value and to decrease the length of the generator, which reduces the amount of high explosive used and increases isolation.

## 3.6 Loop MCGs

To increase the efficiency of an MCG, it is important that, first, the increase in magnetic field strength in the compression region of the generator during its operation is restricted to an upper limit, i.e., $H(t) \leq H_*$, which depends on the electrophysical characteristics of the materials used in the conductors. The material must be selected to prevent the electrical explosion of the skin layer of the conductor (i.e., electrical breakdown) at high fields, for example, $H_* \sim 1.2$ MOe in the case of copper. Second, it is also necessary that electrical breakdown between sliding contact areas, which are characteristic of MCGs, be eliminated.

The first objective can be achieved by developing specific profiles for the generator circuit, and the second by increasing the angle at which the conductors come in contact with each other during the operation of the generator. These objectives are easily achievable by using a single-turn (loop) MCG, where the loop has a parabolic shape.

The loop MCG (LMCG) [3.3,3.8] is related to the class of so called "fast" MCGs, in which relatively high powers (several terawatts) are generated by the simultaneous compression of relative high magnetic fluxes by the entire surface of a conductor. Rather high specific energy densities ($2\times10^8$ J/m$^3$) are generated in the volume of the MCG and rather high surface current densities (80 MA/m$^2$) are created in the electrical circuit of these relative easy to construct and low-cost generator designs. The LMCGs parameters can be reliably predicted. The parameters of these generators may be varied over a wide range either by changing the geometrical layout of a single generator or by using multiple generators. The output currents range up to tens of megaamperes and the output energy from hundreds of kilojoules to tens of megajoules, when the generator is connected to loads with inductances ranging from a few nanohenrys to 100 nH. The LMCGs wide range of output parameters excludes the necessity of complicated designs and determines its advantage in a wide range of applications.

A schematic drawing of an LMCG is presented in Fig. 3.11. The electrical circuit consists of the external loop (1) with an explosive-filled cylindrical shell (2) placed inside the loop. The high explosive charge (5) is initiated by a chain of electric detonators placed along the axis $O_1$, which are initiated simultaneously along the length of the generator. The cylindrical shell expands as a result of the action of the explosive and closes the contacts (6) at the moment the initial current in the LMCG reaches its peak value. Further expansion of the shell compresses the captured magnetic flux with its entire surface. The magnetic flux is pushed into the load (3) through channel (4).

The inductance of the LMCG depends on the distance between the axis of the shell and the axis of the external loop. The inductance of the generator increases and its operating time decreases, as the distance between the axis of the shell and the axis of the external loop, $\delta_{oc}$, decreases. The inductance of an LMCG with a uniform loop profile is

$$L_g = \frac{\mu_0 S_b K_\alpha}{l_b}, \tag{3.85}$$

where $S_b$ is the cross-sectional area of the gap between the loop and the internal shell, $l_b$ is the loop width, and $K_\alpha$ is a tabulated form factor (0.5–1.0), which depends on the ratio of the loop width to the length of the gap between the coil and the shell. The working equations for the LMCG are

- magnetic flux balance equation:

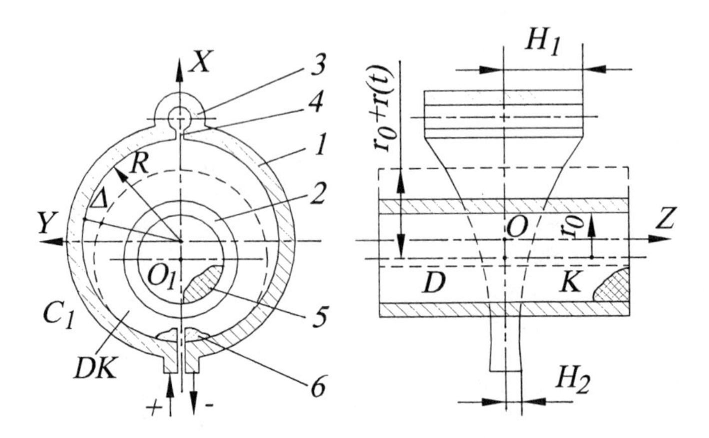


FIGURE 3.11. Diagram of a single-turn LMCG. In order to increase the efficiency of MCGs, the magnetic field strength in the compression region must be restricted to an upper limit and electrical breakdown between the sliding contact areas eliminated. The first can be achieved by developing specific generator profiles and the second by increasing the angle at which the conductors come together. Both of these objectives can be achieved by using a single-turn LMCG. The LMCG consists of 1 – external conductor, 2 – explosive-filled inner conductor, 3 – load, 4 – channel connecting compression region to the load, 5 – explosive charge, and 6 – electrical contacts.

$$(L_g + L_h)\frac{dI}{dt} + I\left(\frac{dL_g}{dt} + \frac{dL_h}{dt}\right) + V = 0, \tag{3.86}$$

which is derived from Faraday's induction law,

- the skin-layer equation,

$$\frac{d\delta}{dt} = \frac{I^2}{2.1 \cdot 10^{17}\delta l_b}, \tag{3.87}$$

- and an equation that accounts for the load parameters,

$$V = L_2\frac{dI}{dt} + R_2 I, \tag{3.88}$$

where $L_h$ is the inductance of the magnetic flux skin-layer, $V$ is the voltage on the load, and $L_2$ and $R_2$ are the inductance and resistance of the load, respectively. The inductance of the skin-layer magnetic flux in the LMCG is

$$L_h = \frac{\mu_0 P_T \delta}{l_b}, \tag{3.89}$$

where $P_T$ is the perimeter of the circuit of the LMCG. Equations 3.86–3.88 can be used when the magnetic field amplitude is greater than 430 kOe, that is, when the initial asymptotic conditions assumed in the derivation of $\delta$ are true.

The current gain of the LMCG is 7–10, its magnetic flux conservation coefficient is 0.7–0.8, and the specific energy increase in its volume is 2.0–2.5 $\times 10^8$ J/m$^3$. The magnetic field in the volume being compressed by the liner increases up to 1 MOe. The current densities and field amplitudes attained in LMCGs are close to the optimal values that can be generated by the conversion of high explosive energy into electromagnetic energy. At these field values, the magnetic counterpressure, $\mu_0 H^2/2$, decelerates the expanding liner to $\sim 0.5$ of its initial velocity, and about 70% of the kinetic energy of the liner is converted into electromagnetic energy. A further increase in current density is limited by the internal losses that arise from vaporization of the surface of the conductors and disruption of the sliding contact between the liner and the loop due to the deceleration of the liner by the magnetic pressure. To achieve efficient deceleration of the expanding liner over the length of the generator, the loop of the generator must have an azimuthally changing width.

In previous sections, it was shown that one of the factors that causes the loss of energy is breakdown in the compression region of the generator. The origin of these breakdowns is related to the preionization of the gas (air) in the generator by the shock waves that compress the magnetic field.

Since compression takes place simultaneously along the entire length of the generator, multiply reflected shock waves may form. These reflected shock waves may lead to intensive heating of the gas in the generator. Near the sliding contact between the inner and the outer conductors during the compression process, the gas is intensely compressed, and, therefore, heated to high temperatures. In the sliding contact region, the degree of compression and heating and the effects of the gas on the output parameters of the generator depend on the angle at which the liner approaches the outer conductor. Therefore, by properly selecting the placement of the internal conductor and the external loop, i.e., by properly selecting the eccentricity angle, electric field strengths may be created that practically exclude the possibility of electric breakdown in the region of the sliding contact of the current-bearing conductors when critical magnetic field strengths $H = H_*$ are achieved.

The dependence of flux conservation on the angle between the conductors as they approach each other for one model of LMCG is shown in Fig. 3.12. The angle of approach was changed by selecting the eccentricity of the liner relative to the loop. It can be concluded that the initial inductance

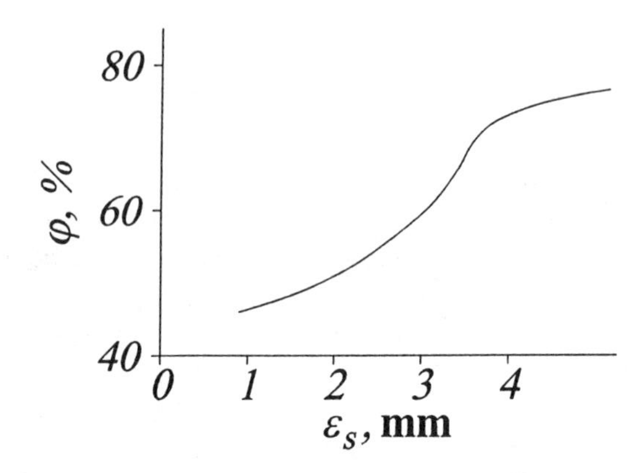


FIGURE 3.12. The dependence of magnetic flux conservation on the eccentricity, $\varepsilon_S$, of the shell relative to the conductor for one type of LMCG. The angle of approach is changed by selecting the eccentricity of the liner relative to the loop, which implies that the initial inductance of the LMCG depends on the distance between the axis of the liner and of the loop.

of a profiled MCG, as opposed to one that has not been profiled, depends on the distance between the axes of the liner and of the loop. The inductance increases and the operating time of the generator decreases as the eccentricity, $\delta_{oc}$, increases. The speed at which the inductance changes is maximum when $\delta_{oc} = 0$. However, the value of $\delta_{oc}$ is based, first of all, on the reduction of magnetic losses in the compression zone of the generator and elimination of electrical breakdown.

An important feature of LMCGs is the ease with which they can be combined into multistage generator sets, where the inductance of the connected elements is minimal, which allows one to vary the values of $I, U$, and $t$ over a wide range. The compression of the magnetic field in the various loops that make up the generator set may be performed simultaneously or serially so as to provide a tailored current increase. A generator set consisting of three identical LMCGs connected in series has generated currents of 46 MA, voltages of 120 kV, and energy pulses of 30 MJ into a 30 nH load.

## 3.7 Disk MCGs

The characteristic magnetic flux compression time of most MCGs is several tens of microseconds, which means that the generated magnetic field strengths do not exceed $\sim$ 1 MOe. In the case of the coaxial MCG, this condition restricts their output current. To generate higher currents, the radius of the coaxial MCG must be increased. (See Section 3.3.)

One version of MCG that generates higher currents is the disk MCG (DMCG) [3.9–3.13]. The earliest form of this generator consisted of a metal toroidal shell with a rectangular cross section. A disk-shaped high explosive charge is placed on one of the end surfaces of the disk. The concept of compressing a magnetic field created by DC current with a moving disk was first proposed by V.K. Chernyshev and A.I. Korolev in the 1961–1962 time frame. The basic idea behind the DMCG is to compress the field in a short cavity, where the field in compressed radially from the axis of the generator toward its perimeter. Chernyshev and Korolev experimentally compared five different configurations of the toroidal cavity. Three of the configurations consisted of the usual coaxial conductors, with the internal conductor expanding, while the other two configurations were disks, where the high explosive charge was positioned at the base of the circuit to be deformed and was initiated along the length of the axis. Drawings of two of the first types of DMCG are shown in Figs. 3.13a and 3.13b. These are coaxial designs in which the internal tube expands when the high explosive is detonated. A comparative analysis of the coaxial and disk versions of DMCG having the same design parameters revealed that the magnetic flux conservation coefficient is significantly less for the coaxial design than the disk or plate design in which a disk-shaped explosive charge is used. In Chernyshev and Korolev's experiments using the design shown in Fig. 3.13a, the energy gain, $\varepsilon$, was equal to 35 and the current gain, $\gamma_I$, was equal to 43, when the magnetic flux conservation factor was $\lambda_k = 0.82$.

Further improvements were made in the disk generator to increase the current amplitude to 100 MA and to increase the output power of the generator. The DMCG that generates the greatest energies and powers are those where the conductors are accelerated toward each other. A drawing of this generator is presented in Fig. 3.13b. To increase the output power, the width of the cavity to be compressed must be decreased, by increasing the radius of the inner conductor radially from the axis of the generator out toward its perimeter. When the magnetic flux transitions from the compression cavity into the toroidal load through the gap, the operating time of the DMCG decreases to $r_H/D$, where $r_H$ is the outer radius of the compression cavity and $D$ is the detonation velocity.

A DMCG design that generates with the greatest possible efficiency is shown in Fig. 3.13c. In this design, the release of energy by the high explosive, which depends on the time taken for the detonation front to propagate from the center of the disk to its periphery, is coordinated with the increase of electromagnetic energy in the DMCGs electrical circuit. As a result, the efficiency at which high explosive energy is converted into magnetic energy is the highest, i.e., 40%, of all the possible DMCG designs. However, it must be noted that, as the efficiency increases, the output power decreases. Coordinating the optimization of the efficiency of the generator with its other parameters is difficult to do.

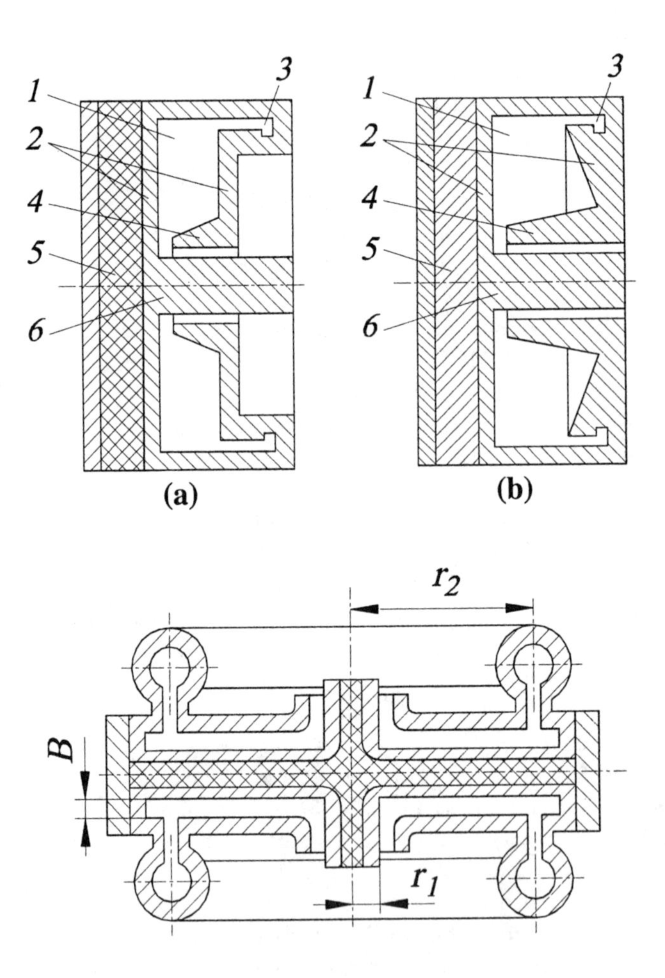


FIGURE 3.13. Disk-shaped MCGs produce higher currents than a corresponding spiral MCG. These generators consist of 1 – flux-compression cavity, 2 – compression circuit, 3 – load, 4,6 – terminals, and 5 – explosives.

The basic parameter that characterizes the operation of the DMCG shown in Fig. 3.13c can be determined by using the magnetic flux balance equation $\Phi_1(t) + \Phi_2(t) = \Phi_2(t_0)$, which, for $t_0 \leq t \leq t_1$, is

$$\lambda_k = \left[1 + \frac{L_1(t)}{L_2(t)}\right]^{-1}, \tag{3.90}$$

where the indices 1 and 2 refer to the shell and the dielectric cavity enclosed by the shell. The functions $L_1(t)$ and $L_2(t)$ are defined by the expressions

$$L_1(t) = \oint_{K(t)} \Psi(s,t) K(s) ds \tag{3.91}$$

and

$$L_2(t) = \frac{\mu_0}{2\pi} \int_{r_1}^{r_2} 2Z\left[r(t)\right] \frac{dr(t)}{r(t)} + L_H, \tag{3.92}$$

where $K(s) = B_S(0,t)/I(t)$, $K(t)$ is a circuit parameter, $L_H$ is the load inductance, $Z\left[r(t)\right]$ is the disk profile, and

$$\Psi(s,t) = \frac{\int_0^\infty B_S(x,t)dx}{B_S(0,t)}. \tag{3.93}$$

Assuming the linear diffusion approximation, the function $\Psi(x,t)$ can be found by using the equation

$$\frac{\partial B_S}{\partial t} = K(s)\frac{\partial^2 B_S}{\partial x^2}, \tag{3.94}$$

for the selected initial and boundary conditions. For example, when a power law, sinusoidal, or exponential field is excited on the boundary of a hemisphere when $x > 0$, that is

$$B_S(0,t) = B_0 \left(\frac{t}{t_0}\right)^{\frac{m}{2}},$$

$$B_S(0,t) = B_0 \sin(\omega t), \tag{3.95}$$

$$B_S(0,t) = B_0 e^{\frac{t}{\tau}}$$

and when the initial conditions are $B_S(x,0) = 0$ in the regime $0 < x < \infty$, then it can be shown that

$$\Psi(s,t,m) = \frac{2\sqrt{K(t)}}{m+1},$$

$$\Psi(s,t,m) = \sqrt{\frac{2K(s)}{\omega}}, \tag{3.96}$$

$$\Psi(s,t,m) = \sqrt{K(s)\tau},$$

respectively. $K(s) = \sigma_0^{-1}\mu_0^{-1}$ is the magnetic diffusivity, where diffusion is assumed to take place in the $x$ direction, which is perpendicular to the conductor surface. It follows from the form of Eq. 3.96 and numerical analysis of this equation that the function $\Psi(x,t)$ does not exhibit a sensitive dependence on the shape of the boundary, which is why any of the functions in Eq. 3.96 may be used as a rather good approximation. Since $r_2 - r_1 > z_0$ in the case of the DMCG, where $r_1$ is the initial contact radius and $r_2$ is the contact radius when the compression process has been completed, it can be assumed that the azimuthal magnetic field $B_\varphi$, which is expressed in cylindrical coordinates $(r,\varphi,z)$ aligned with the axis of symmetry of the DMCG, is compressed in the toroidal cavity by a cylindrical conductive piston moving at the detonation velocity $D = \text{const}$ between two fixed conductive walls at $z = -z_0/2$ and $z = +z_0/2$. The piston begins its motion at time $t_f = r_1 + (r_f - r_1)/D$ and radius $r = r_f < r_1$. The equation of motion for the piston is

$$r(t) = r_1 + D(t - t_0). \tag{3.97}$$

The magnetic flux losses related to nonuniformities in the contact surface may be neglected. Thus, Eqs. 3.91 and 3.92 can be written as

$$L_1(t) = \frac{\mu_0}{\pi}\Psi(t)\left\{2 + \frac{z_0}{2}\left(\frac{1}{r_1} - \frac{1}{r_2}\right) - 2\frac{\arctan\sqrt{\tau_*}}{\sqrt{\tau_*}} + \ln\left[\frac{r_2}{r_1(1+\tau_*)}\right]\right\} \tag{3.98}$$

and

$$L_2(t) = \frac{\mu_0}{2\pi}\ln\left[\frac{r_2}{r_1(1+\tau_*)}\right], \tag{3.99}$$

where $\tau_* = D(t - t_0)/r_1$. Substituting Eqs. 3.98 and 3.99 into Eq. 3.90, analytical expressions that describe the operation of the DMCG are derived as

$$\lambda_k(t) = \{1 + 2\Psi(t)G\}^{-1} \tag{3.100}$$

and

$$z(t) = \frac{\ln \frac{r_2}{r_1}}{\ln \frac{r_2}{r_1(1+\tau_*)}}, \tag{3.101}$$

where:

$$G = \frac{\frac{2}{z_0} + \left(\frac{1}{r_1} + \frac{1}{r_2}\right) - 2\arctan \frac{\sqrt{\tau_*}}{z_0\sqrt{\tau_*}}}{\ln \frac{r_2}{r_1(1+\tau_*)}} - z_0. \tag{3.102}$$

In one experiment, a current pulse of 150 MA with a rise time of 4 $\mu s$ was created by a DMCG having a diameter of 400 mm and an initial current of 5.7 MA. The current increased at a rate of $3.8 \times 10^{13}$ A/s and the generator operating time was approximately 20 $\mu s$. The magnetic energy delivered to the load was approximately 3.5 MJ. When the magnetic field in the load reached its peak value, the magnetic flux conservation coefficient was still high, i.e., $\lambda_k \sim 0.6$.

Investigations of the DMCG have shown that they have a high energy capacity. For those designs that have been tested, the specific magnetic energy density, which is the ratio of the stored magnetic energy to the source volume of the generator, reached a value of $\sim$ 400 J/cm$^3$. A characteristic feature of the DMCG is the high speed with which the electromagnetic energy is coupled into the load, which is on the order of units of microseconds. This permits the delivery of powers on the order of $10^{12}$ W to the load.

## 3.8 Semiconductor MCGs

The traditional MCG converts energy from high explosives into electromagnetic energy by means of conductors moving through a magnetic field. The metal parts of the electric circuit of the generator (i.e., the liner and spiral coil in the case of the spiral MCG) form these conductors. One difficulty associated with MCGs is the considerable loss of magnetic flux due to the compressibility of the liner materials, which may even occur in substances having infinite electrical conductivity. To reduce these losses, new techniques have been developed for compressing magnetic flux. It is well known that certain materials, including semiconductors and insulators, when exposed to high pressures transition from a nonconducting state into a conducting state, where the increase in conductivity can be as high as 6 to 8 orders of magnitude. Some substances that exhibit this property are silicon, germanium, gray tin, silicon oxide, cesium iodide, and germanium iodide. Based on this property, it may be possible to create highly conductive layers that propagate through these semiconducting or dielectric materials and compress magnetic flux, that is, nonmetallic liners.

It has also been observed that a shock wave moving through these materials causes semiconductor-to-metal or insulator-to-metal transitions within the shock wave itself [3.14,3.15]. This conductive region moves with the shock wave. This means that the current-bearing conductors (i.e., the liner) used to compress the magnetic fields in standard MCGs can be minimized or even eliminated, thus simplifying the problem of contact losses. This also allows magnetic compression to be achieved with almost any shape and positioning of the generator plates. In addition, the use of moving conductive waves to compress magnetic fields offers the possibilities of new methods for manufacturing low-weight, compact generators with superstrong magnetic fields. These types of MCGs are called *shock wave magnetocumulative generators* (SWMCG) or *semiconductor magnetocumulative generators.* The advantages of the SWMCG over the conventional MCGs are

- more efficient transfer of chemical energy from the explosive to the shock wave,
- practically complete compression of the magnetic field generated in the working volume,
- increased initial magnetic field in the generator because the energy is concentrated.

Solving the magnetohydrodynamics (MHD) equations, it has been shown that magnetic flux accumulation by a metalized shock wave has the added advantages of being

- practically independent of the time it takes to create the initial magnetic field in a conventional MCG,
- free of the instabilities associated with imploding liners in conventional MCGs,
- very fast, which is determined by the shock velocity.

### 3.8.1 Theory of Operation

In the schematic drawing of a SWMCG presented in Fig. 3.14, either permanent magnets or electromagnets can be used to create the initial magnetic

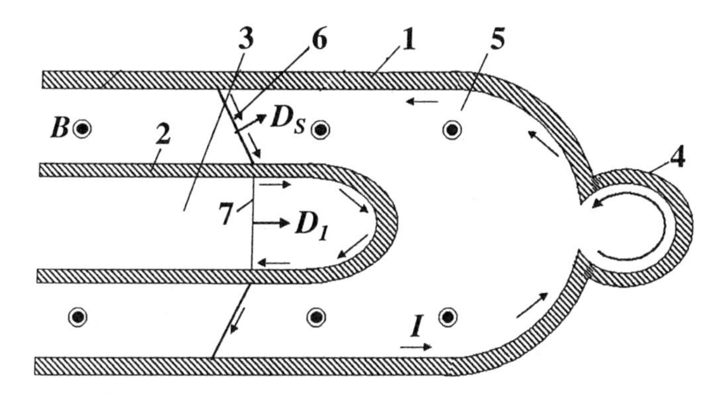


FIGURE 3.14. In shock wave MCGs, magnetic compression takes place in a monocrystalline or powdered medium. This reduces MHD losses, such as Rayleigh–Taylor instabilities. This generator consists of 1 – copper plate, 2 – copper cassette, 3 – explosives, 4 – load loop, 5 – crystal silicon powder, 6 – shock wave moving through the semiconductor, and 7 – shock wave in moving through the explosives. Either permanent or electromagnets are used to generate the seed magnetic field. The current flows through the closed loop consisting of the metal conductors and the conducting shock front.

field. The darker line (6) represents the shock wave front in the semiconductor material, which moves with velocity $D_s$, and the lighter line (7) represents the front of the shock wave moving through the explosives with velocity $D_l$. In these devices, a conductive layer is formed in the shock wave, which moves transverse to the external magnetic field, thereby exciting an EMF in this layer. Since this conductive layer and the other circuit elements form a closed circuit, the magnetic field is captured and compressed. The distribution of the magnetic field, the direction of motion of the shock wave, and the direction of current flow are all shown in Fig. 3.14. In experiments conducted in the Former Soviet Union, a 90-fold amplification of an initial magnetic field of 40 kG to 3.5 MG was achieved [3.14].

The basic features of this method for compressing magnetic fields by shock waves are that a wave of changing conductivity propagates through the medium with a velocity, $D$, equal to that of the shock wave and that work is done against the magnetic pressure by the compressed material which is moving at a velocity $U$, where $U < D$. If the shock wave is sufficiently strong and if it compresses the substance to a density greater than some critical value of $\rho_c$, then the field frozen in the highly conductive region increases proportionally to the degree of compression. However, this will not affect the field value before the shock front. The value of the ratio

$$\beta = \frac{U}{D} = 1 - \frac{\rho_0}{\rho} \tag{3.103}$$

depends on the compressibility of the working substance, where $\rho_0$ is the initial density of this substance and $\rho$ is the density of the substance in the shock wave. If the substance is not very compressible, then $\rho_0 \sim \rho$ and $U/D << 1$. On the other hand, if the substance is compressible, then $U \sim D$. The difference in the values of $U$ and $D$, even if the conductivity behind the shock wave front is ideal, leads to the capture of magnetic flux by the substance and its removal from the shock wave front, with a velocity of $D - U$ relative to the moving shock front. Owing to this effect, the magnetic flux in the compressed region decreases.

When $\beta = U/D =$ const, the magnetic flux gain in the working volume is

$$\Psi = \frac{\Phi}{\Phi_0} = \left(\frac{S_0}{S}\right)^{\frac{U}{D}-1} \tag{3.104}$$

and that of the magnetic field is

$$\gamma = \frac{B}{B_0} = \left(\frac{S_0}{S}\right)^{\frac{U}{D}}, \tag{3.105}$$

where $B_0$ is the initial magnetic field strength, $\Phi_0$ is the initial magnetic flux, $B$ is the final magnetic field strength, $\Phi$ is the final magnetic flux, $S_0$ is the initial cross-sectional area of the working volume, and $S$ is the final cross-sectional area. If $U/D << 1$, then within the limits of the approximation that the cross-sectional area approaches zero, all the flux is removed from the working volume, but the magnetic field increases. The magnetic energy in the working volume is

$$\varepsilon_H = \gamma\Psi = \left(\frac{S_0}{S}\right)^{2\frac{U}{D}-1}. \tag{3.106}$$

From this expression, it can be seen that the magnetic energy strongly depends on the parameter $\beta = U/D$. If $\beta > 1/2$, the magnetic energy in the working volume increases; if $\beta < 1/2$, it decreases; and if $\beta = 1/2$, it remains the same. When $\beta \leq 1/2$, the magnetic energy in the working volume does not increase, and the deceleration of the shock wave by the magnetic field does not lead to the extraction of large amounts of energy from the shock wave. This mode of completely "collapsing" the shock wave does not contradict the law of the conservation of energy. The magnetic energy remains finite in value, and the flux is completely captured by the conductor, so that the magnetic energy density increases without limit, though the initial energy in the system is finite. In other words, as the cross-sectional area of the generator decreases, the captured flux increases, i.e., $\Phi/S \to \infty$, as $S \to 0$. This effect has an analog in theoretical hydrodynamics known as the "phenomenon of infinite accumulation." Hence, magnetic energy accumulation in the finite compressed region of the shock wave, where the transition from a nonconducting to a conducting state occurs when $\rho/\rho_0 > 0.5$, is an interesting example of magnetic energy accumulation that possesses, theoretically at least, the potential for a strong increase in the magnetic energy density.

The effects of finite conductivity on the operating parameters of the SWMCG can be estimated. The capture of the magnetic field by the working medium in the shock, and its removal at a velocity of $D - U$ relative to the shock front, leads to a smooth distribution of the field across the shock front. Since the magnetic diffusion losses are proportional to the field gradient, that portion of the magnetic flux removed is insignificant, and it is reduced to the slow smoothing of the field distribution that is frozen into the working medium. Like the classical MCG, the restrictions due to finite conductivity are not the principal limitations, and only affect the final stages of the compression process.

Let us now consider magnetic field compression in the two-dimensional case shown in Fig. 3.15. After the compression process has been completed, the value of the magnetic field for this configuration may be obtained from

$$\left(\frac{2(1-\beta)+m}{2\beta-1}\right)\left(\gamma_*^{\frac{2\beta-1}{\beta}}-1\right)+\gamma_*^{\frac{2(\beta-1)}{\beta}}-1=\frac{em}{2\beta}, \tag{3.107}$$

where

$$m=\frac{2M}{x_0}, \quad \epsilon=\frac{\frac{Mu_0^2}{2}}{\frac{B_0^2 x_0}{2\mu_0}}, \tag{3.108}$$

$x_0$ is the initial radius of the magnetic field, and $B_0$ is the initial magnetic field strength. When the magnetic field diffusion becomes significant, the compressed distance, $x_d$, is estimated to be

$$\frac{x_d}{x_0}=\frac{D+U}{\mathrm{Re}_m(D-U)^2}, \tag{3.109}$$

where $\mathrm{Re}_m = \mu_0 \tau D x_0$ is the magnetic Reynold's number.

It can be seen from Eq. 3.109 that, when the magnetic Reynold's number is high, the nonideal conductivity of the working substance is negligibly

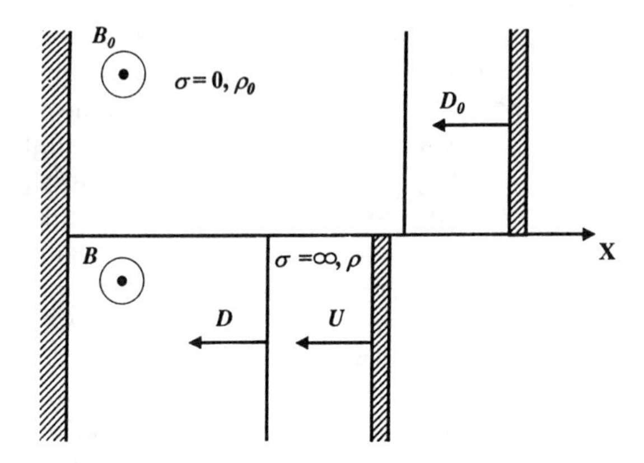

FIGURE 3.15. In the top figure, the shock wave has not entered the region containing the magnetic field and it moves with velocity $D_0$. In the bottom figure, when the shock wave enters the magnetic field, it moves with a velocity $D$ and work is done against the magnetic field by the compressed material that moves at velocity $U$. If the shock wave is sufficiently strong and if the substance is compressed to a density greater than some critical value, then the field is frozen into the highly conductive region increases proportionally.

small during almost the entire compression time, and only during the final stages of compression is there a finite increase in the field gain:

$$\gamma_f = \left(\frac{S_0}{S_d}\right)^{\frac{U}{D}}, \tag{3.110}$$

where $S_d$ is the cross-sectional area of the compressed region into which magnetic field has diffused.

The energy losses during flux compression in the SWMCG can be divided into three regimes:

- dynamic losses,
- magnetic energy losses,
- thermal losses.

The thermal processes cause melting of the working substance. The most important feature of the SWMCG is the high electrical conductivity of the melted material in the thin shock wave front. For example, experimental data show that the resistance of silicon at high shock pressures is reduced to 0.1 mΩ. Theoretical investigations have demonstrated the importance of the dependence of resistance on temperature. During the initial stages of operation of the SWMCG in which the radius of the material has decreased from $R = 1.0$ to $R = 0.5$ relative units, magnetic compression does not take place due to the low conductivity of the material. However, when $R = 0.015$, compression does occur and the ratio $B/B_0$ increases to 1950, where $B_0$ and $B$ are the initial and final magnetic field strengths, respectively. At that moment, the magnetic pressure in the material begins to counteract the shock wave, thus decreasing its energy. That is, as the field is compressed, the current increases, thus creating a force opposite to the shock wave, which causes its velocity to decrease.

Theoretical investigations were performed assuming an initial density of 4.5 g/cm$^3$ at atmospheric pressure, a temperature of 298 K, and an initial magnetic field radius of 5 mm. When the shock wave was initiated, the magnetic field increased from 0.4 T to 5 T, as the pressure increased from 0.3 Mbar to 1 Mbar. At such high pressures (1 Mbar), the density ratio $\rho/\rho_0$ increased from 1.7 to 2.2. At the same time the temperature of the working substance increased to its melting point, and, as a result, a highly conductive state was created. The melting process was suppressed when the density ratio was relaxed down to 1.67.

### *3.8.2 SWMCG Working Substances*

The working substances used in SWMCGs are typically dielectric or semiconductor monocrystals or powders. Conductive layers can be created in these materials due to the effects of shock wave pressures on these materials. However, it has been found that powdered metals coated with an oxide layer can also be used.

The transition of dielectrics and semiconductors into a metallic state has been observed when these materials are subjected to either very high static pressures or shock pressures. For example, changes of six orders of magnitude have been observed in the conductivity of both silicon and germanium. These shock induced transitions take place only in the very thin shock front.

If the conductivity in the region behind the shock is sufficiently high, the magnetic flux becomes "frozen" into this region, and moves with the particle velocity at each point. Since the flux passes through the conducting material in the shock wave front, eddy currents are induced in the surface of the shock wave, which prevents the flux from diffusing through it.

In addition to silicon and germanium, another substance that has been studied is cesium iodide (CsI) doped with approximately 1% thallium. The interest in CsI arises from the fact that it is an ideal substance for studying both insulator-to-metal transitions and melting under high pressures. Another reason for this interest is the fact that CsI is isoelectronic to the rare-gas solid xenon (Xe), which is the most studied material in insulator-to-metal transition experiments. Both CsI and Xe consist of atoms having filled $5p$ orbitals separated by a large energy gap from an empty conduction band. In both cases, it is believed that they take on metallic properties as a result of band closure. It has been observed that under high pressures, the band gaps of CsI and Xe begin to shift, until, in the case of CsI, at sufficiently high pressures, the $5d$ band of the cesium and the $5p$ band of the iodine overlap to cause metallization. The metallization pressure of CsI has been predicted to be $\sim 1$ Mbar, which is lower than that of Xe, which is $\sim 1.3$ Mbar. This is physically reasonable, since the energy gap of CsI at zero pressure is 6.4 eV, while that of Xe is 9.3 eV. In addition, if shock waves are used to cause the transition to a metallic state, high shock temperatures, on the order of thousands of degrees Kelvin, are generated, which causes a significant number of electrons to be excited across the rapidly closing gap. The shock wave converts the ionic crystal of CsI into a semiconductor with donor and acceptor levels formed by defects in the lattice. Thermal activation of electrons from these levels leads to a predominate n-type conductivity, $\sigma$:

$$\sigma(P) \sim e^{-\frac{W}{2kT}}, \tag{3.111}$$

where $T$ is the temperature and $W$ is the energy gap corresponding to the external pressure $P$. As the pressure increases from 10.3 GPa to 23.0 GPa, the energy gap of CsI decreases from 7 eV to 2 eV. This causes a significant increase in the electrical conductivity of CsI.

At higher pressures ($P > 40$ GPa) and higher temperatures, CsI is transformed into a melted state that has semiconductor properties. At even higher pressures ($P \sim 100$ GPa), CsI transitions into a highly conductive state with a conductivity of $10^5\ \Omega^{-1}\text{cm}^{-1}$.

When planar shock waves move through a sample of CsI, the material is compressed and irreversibly heated to thousands of degrees Kelvin. The hot material behind the shock front is opaque, and emits thermal radiation in the visible region of the electromagnetic spectrum. This radiation can escape in the forward direction through the unshocked and still transparent material. This provides a technique for observing the thermal radiation emitted from the electrons behind the shock wave, which are in thermodynamic equilibrium with the lattice.

The type of material used in the SWMCG depends on the acceptable pressures that cause the transition to the metallic state. While CsI appears to be acceptable, because its conductivity changes from 10 to $10^5\ \Omega^{-1}\text{cm}^{-1}$ are achievable for pressure changes from 20 to 100 GPa, it is in fact not suitable, since the material is unstable after the shock wave has passed. In addition, other physical characteristics of the substance, such as brittleness, must be considered.

Another substance that has been investigated is oxide coated aluminum (Al) powder. The Al powders demonstrate high electrical resistivity due to the aluminum oxide that coats the surface of the powder grains. The $Al_2O_3$ coating is necessary in order for the external initial magnetic field to penetrate into the space between the grains of powder. When a shock wave is applied to the powder, it breaks down the insulator coating of aluminum oxide, so that the aluminum in the various grains can make electrical contact with each other, thus trapping the magnetic field, since the magnetic field cannot propagate rapidly through the metal.

### *3.8.3 SWMCG Designs*

Both MK-1 and MK-2 types of SWMCGs have been designed, built, and tested. The three designs that will be considered in this section are coaxial, planar, and spherical. The feasibility of using shock-generated metallize waves to compress magnetic fields and generate electrical power was first demonstrated using the coaxial configuration of this generator. A diagram of this generator is presented in Fig. 3.14. The generator is made of copper tubes, welded together at one end. The inner tube is filled with high explosive, and powdered silicon is packed into the annular space between the tubes. A capacitor is discharged into an external electromagnet to create the initial magnetic field within the generator. When the current created by the capacitor reaches its peak value, the explosive charge is initiated at the end of the generator away from the load. When the shock wave caused by the explosives is coupled into the silicon powder, a conductive layer is formed, which connects the inner and outer conductors. This conductive layer moves along with the detonation wave toward the load, thus compressing the magnetic field created by the external electromagnets. Permanent magnets can be used in place of electromagnets.

A diagram of the planar generator is presented in Fig. 3.16. It consists of two metal plates separated by a cassette filled with high explosives. A single-turn solenoid serves as the load for the generator. The initial magnetic flux in the generator is created by an external electromagnet with a current rise time much longer than the operating time of the generator, or by connecting the plates directly to a capacitor. In the latter case, the circuit equation for this generator is

$$\frac{d}{dt}(LI) + \left[R_g - \left(1 - \frac{u}{v_k}\right)\frac{dL}{dt}\right] I = 0, \tag{3.112}$$

where $v_k$ is the velocity of the shock wave in the silicon, $u$ is the mass velocity behind the shock wave, and $R_g$ is the resistance of the generator circuit. This equation differs from the equations for conventional MCGs by the factor $u/v_k$, which describes the convection of the field frozen in the conductive layer formed during the phase transition of the silicon.

If it is assumed that the speed at which the inductance changes is constant, then the solution of Eq. 3.112 for the initial conditions $L(0) = L_0$ and $I(0) = I_0$ is

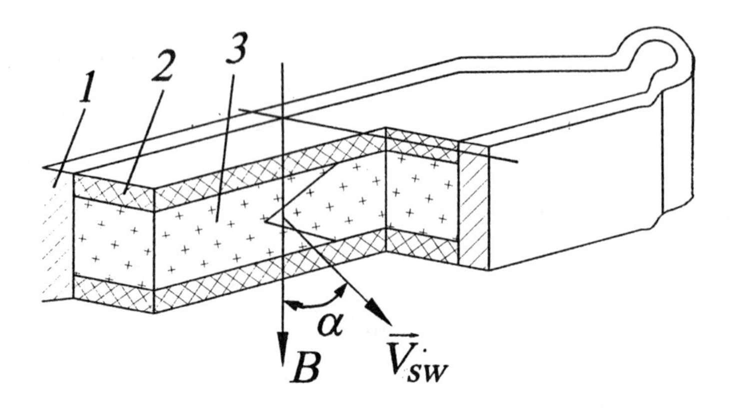


FIGURE 3.16. Plate version of the shock wave MCG: 1 – solenoid, 2 – copper plates, and 3 – explosives.

$$I = I_0 \left( \frac{L_0}{L_k} \right)^{\frac{uR_g}{v_k |\dot{L}|}} . \tag{3.113}$$

This solution conserves the power-law dependence of current amplification on the *trimming coefficient*, $L_0/L$, of the circuit, like conventional MCGs. However, the degree of the power law is determined not only by the magnetic Reynold's number of the MCG circuit, i.e., $\left|\dot{L}\right|/R$, but also by the ratio of the convection speed, $u$, of the moving conductive layer to the shock wave velocity, $v_k$. It is worth noting that the rate at which the inductance changes, $\left|\dot{L}\right|$, depends on the velocity of the boundary of the high-pressure region, $v_{k1}$, along the length of the generator. If the shock wave forms an angle with the axis, as shown in Fig. 3.16, this velocity will differ from the shock wave velocity, $v_k$. These two velocities are related to each other by the expression $v_k = v_{k1} \sin\alpha$, where $\alpha$ is the angle between the direction of motion of the shock wave and the direction of motion of the detonation wave along the axis of the generator.

The circuit equation for the above planar generator in an external magnetic field $B_0$ is

$$\frac{d}{dt}(LI) + \left[ R_g - \left( 1 - \frac{u}{v_k} \right) \frac{dL}{dt} \right] I = \frac{uaB_0}{\sin\alpha}, \tag{3.114}$$

where $a$ is the width of the generator channel. Taking into account the initial conditions given above, the solution of this equation is

$$I = \frac{B_0 h_k}{\mu_0 \left(1 - \frac{R_g v_k}{|\dot{L}| u}\right)} \left[\left(\frac{L_0}{L}\right)^{\frac{u}{v_k} - \frac{R}{\dot{L}}} - 1\right], \tag{3.115}$$

where $h_k$ is the height of the generator channel.

It follows from Eq. 3.113 that the current in the generator increases if

$$R_g < \frac{u}{v_k} \left|\dot{L}\right|. \tag{3.116}$$

To control the shape of the pulse in the planar SWMCG, the shape of the generator plates can be varied. The condition presented in Eq. 3.116 shows that the shape of the current pulse in the filled generator can be controlled within certain limits by changing the ratio $u/v_k$, which depends on how the explosive charge is distributed over the length of the generator and on the filling density.

Using Eq. 3.113 and neglecting the active resistance of the circuit, the expression for calculating the field increase over the length, $l_0$, of the generator is

$$B(t) = B_0 \left(\frac{l_0}{l_0 - v_k t}\right)^{\frac{u}{v_k}}. \tag{3.117}$$

It can seen from this equation that the energy stored in the generator will increase provided the condition $2u > v_k$ is satisfied. Likewise, it also follows from Eq. 3.115 that the energy due to the initial magnetic field in the generator will increase during the operation of the generator, provided that this last condition is satisfied.

These distinctive features of current and energy amplification are specific to those types of generators in which a substance transitions into a conductive state and in which the convective-like removal of the field from the generator takes place, even for ideal conductivities. To increase the efficiency of these types of generators, materials having high compressibility, i.e., when $u \simeq v_k$, must be used in the generator. It is for this reason that silicon powder was selected as the working substance for the SWMCGs described in this section. However, since silicon is rather expensive, experiments have been carried out with aluminum powder, which has a density of 1.6 g/cm$^3$ and a grain size of 0.03 mm.

Despite the principal limitation that the working substance must undergo a phase change, this method is still useful for developing compact sources of high pulsed currents and for generating high magnetic fields within small volumes. Current gains of $\gamma = 1.5$ have been achieved in SWMCGs.

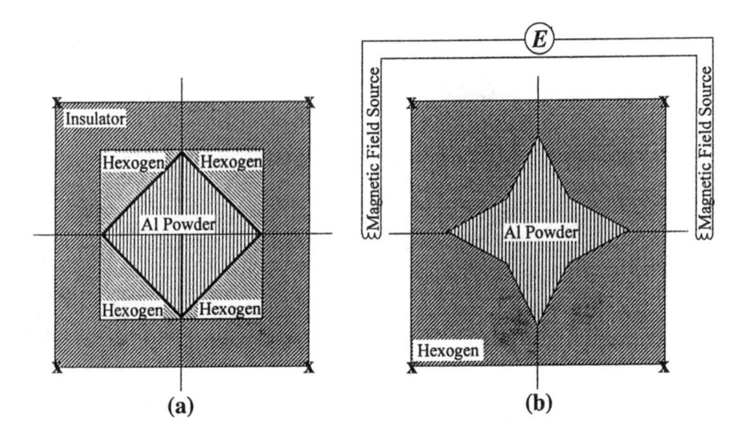


FIGURE 3.17. Different geometries of the MK-1 shock wave MCG developed by Bichenkov [3.14–3.16]. Powered aluminum is pressed into blocks and encased in hexogen. The initial magnetic field was generated by a battery of capacitors connected to a inductive coil outside the generator. The detonators were placed at the points marked with an "x".

## Examples of MK-1 Generators

E.I. Bichenkov et al. [3.15] and S.D. Gileev and A.M. Trubachev [3.16] developed rectangular-shaped SWMCGs of the MK-1 class. Schematic drawings of these generators are given in Figs. 3.17a and 3.17b. The difference between this type of generator and the MK-2 generators described in the next section is that the aluminum powder is pressed into rectangular blocks with dimensions of $50 \times 50 \times 30$ mm$^3$ and $92 \times 92 \times 30$ mm$^3$. The initial density of the powder blocks varied from 0.33 to 1.6 g/cm$^3$, depending on the nature of the experiment.

Powdered hexogen, with an initial density of approximately 1 g/cm$^3$, was placed on each surface of the aluminum block, as shown in Fig. 3.17. The mass of the explosive varied from 50 to 200 g. The initial magnetic flux in the SWMCG was generated by a battery of capacitors connected to a single inductive coil located outside the generator as shown in Fig. 3.17b. The detonators were placed at the corners of a square filled with high explosives and marked with an "x" in Fig. 3.17. The results of Bichenkov et al.'s [3.14,3.15] experiments are presented in Table 3.2. The maximum values of the final magnetic field, $B_f = 1000$ kG, and the ratio $B_f/B_0 = 46$ were achieved for the case when the mass of hexogen was 200 g. The high value of the ratio allowed Bichenkov et.al. [3.15] to demonstrate the advantage of using aluminum over silicon or germanium, which only have a maximum ratio of $B_f/B_0 \approx 10$. Bichenkov believes that if both the initial magnetic field and the dimensions of the SWMCG are increased, then the value of $B_f$ can be also increased.

| Mass of Hexogen (g) | $B_f$ (kG) | $B_f/B_0$ |
|---|---|---|
| 70 | 68 | 10 |
| 70 | 115 | 18 |
| 70 | 440 | 20 |
| 200 | 17.2 | 3.5 |
| 200 | 76 | 20 |
| 200 | 630 | 33.5 |
| 200 | 500 | 26 |
| 200 | 340 | 18 |
| 200 | 900 | 46 |
| 200 | 1000 | 26 |

TABLE 3.2. Results of Bichenkov MK-1 SWMCG experiments [3.14–3.16].

A more effective version of the MK–1 class SWMCGs is the spherical variant proposed by A.B. Prishchepenko and his colleagues [3.17]. While very little information on the parameters of this device are available, some of its parameters can be estimated from on a photograph (Fig. 3.18a) presented in the article. This device is of particular interest because of the difficulties associated with initiating uniform spherical implosions.

Owing to the poor quality of the photograph in [3.17], a conceptual drawing based on the written description is presented in Fig. 3.18. It is estimated that the diameter of the generator is somewhere between 75 and 80 mm. The generator is criss-crossed by six semicircular iron rings that are "field guides". It is assumed that the magnets are made from either Sm-Co or Nd-Fe-B-Co-Al. Based on the assumed dimensions and type of magnetic material used, the axial magnetic field is estimated to be approximately 0.4 T. It is also assumed that a single spherical monocrystal of some type of insulator (possibly silicon, CsI, or germanium) is used as the working body of the generator, the most probable diameter of the working body is 25–30 mm, and the mass is about equal to the mass of the explosives, which is approximately 320 g, if the density of the explosive is assumed to be 3.5 g/cm$^3$.

In the photograph of the spherical generator, narrow tracks having a width of 1.5–2.0 mm are visible on its surface. It is assumed that these are explosive delay lines and are used to detonate the high explosive in some prescribed manner so as to achieve a uniform implosion. The tracks at the equator of the generator have the highest density. The equator plane is probably perpendicular to the direction of the axial magnetic field, which is created by permanent magnets in the form of an ellipsoid of revolution. The shape of the working body and the magnetic field have been arranged to ensure that the shock wave propagating through the working body is perpendicular to the direction of the axial magnetic field. The resulting

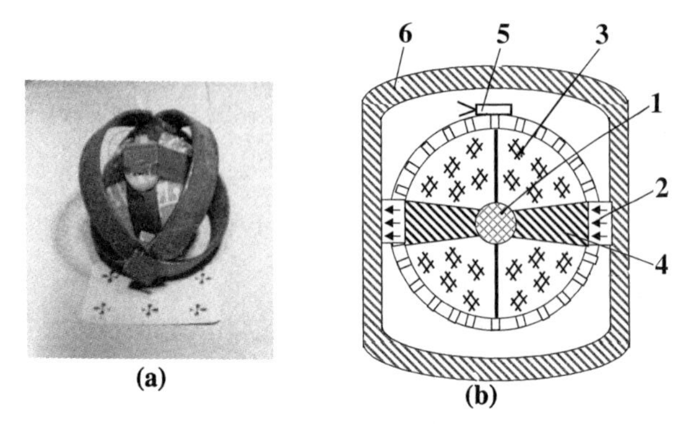


FIGURE 3.18. Photograph (a) and diagram (b) of a MK-1 type spherical version of the shock wave MCG. It consists of 1 – the working body (CsI), 2 – permanent magnets, 3 – high-explosive, 4, 5 – magentic field concentrators, and 6 – detonator. The white lines on the body of the device are explosive delay lines configured to ensure a uniform implosion.

converging cylindrical shock wave in the working body creates a metallized front that compresses the magnetic field, and thus generates electrical power. If it is assumed that 320 g of explosives is used, then it is estimated that about 8.7 J of electrical energy is generated, with a pulse length of 10–20 $\mu s$.

### Examples of MK-2 Generators

E.I. Bichenkov et al. [3.15] constructed three types of MK-2 class generators. One of these was a coaxial generator and the other two were planar generators. A schematic drawing of the coaxial generator with a capacitor as its prime energy source is similar to that in Fig. 3.14. The generator was made of two coaxial copper tubes connected at one end by a loop load with a diameter of 10 mm. The length of the outer tube was 300 mm, and its wall thickness was 2 mm. The greatest diameter of the inner tube was 18 mm, and its walls had a thickness of 15 mm. The annular space between the tubes was filled with silicon crystalline powder consisting of particles having a diameter of 0.10–0.15 mm. The pressure at which the phase transition occurs in silicon is 1200 kbar, and the conductivity of the silicon while in the metallic phase is close to the conductivity of a normal metal. In one set of experiments, an initial current of 6.7 kA was delivered to the generator. When the central tube was filled with 180 g of high explosive, a final current of 27 kA was generated. The duration of the electrical pulse was 20–30 $\mu s$. In the best case, half of the energy of the high explosive was used for flux compression. The rest of the energy is converted into heat. If it is assumed that each gram of high explosive is capable of generating 2 kJ of energy and that the density of the high explosive is 3.5 g/cm$^3$, then the total energy available for flux compression is 180 kJ.

The value of $\gamma$, which is the ratio of the final magnetic field strength to its initial value, ranges from 10 to 27 in these types of generators. The maximum magnetic field strengths achieved during the final stage of operation was 1 MGauss for an initial magnetic field strength of 38.6 kGauss.

A second example of the MK-2 type SWMCG is similar to the planar generator shown in Fig. 3.16. This generator consisted of a strip of copper having a width of 15–20 mm and a thickness of 1 mm. This strip was folded to form a cavity with a volume of $17 \times 80 \times 250$ mm$^3$. A copper capsule, having a volume of $17 \times 12 \times 250$ mm$^3$, was inserted into the cavity. The height of the capsule was 12 mm, and the distance between the capsule and the cavity walls was 32 mm. The space between the capsule and the copper foil was filled with silicon powder having a grain size of 0.10–0.15 mm. The load of the SWMCG was a single-turn solenoid coil with a diameter of 8 mm connected to the closed end of the copper cavity. The total active length of the generator was 200–270 mm. The axial magnetic field was created by electromagnets having an initial magnetic field strength of 3 kGauss. The pressure at which the phase transition occurs was 130 kbars. The initial current in the generator is 3.3 kA, and the output current was 9.1 kA with a pulse length of approximately 15 $\mu$s. If it is assumed that 180 g of high explosives was used, then the output energy is estimated to be 190 kJ. The output of the planar SWMCG can be controlled by either changing the shape of the generator or by varying the ratio $\beta$, which depends on how the explosives is distributed along the length of the generator and the density of the working substance.

## 3.9 Cascaded MCGs

In order to produce higher current gains and increase energy outputs, cascaded MCGs were developed [3.18–3.22]. A cascaded system consists of two or more MCGs connected in series with air transformers, where each succeeding MCG is the load of the previous generator. In this system, the total gain, which may reach very high values, is equal to the product of the gains of each MCG in the cascade. Any type of MCG may be used in cascaded systems. However, in the majority of cases, the spiral MCG is used as the first stage. One example of a cascaded system is a planar MCG connected in series with a spiral generator without using a transformer as a coupling device. The current gain of this system is approximately 1000 when the mass of the high explosive used is about 1 kg.

Another example of a cascaded system is the coupling of a spiral generator with a coaxial MCG, as shown in Fig. 3.19 A cylindrical transformer was used to couple the generators. In one experiment, this system delivered

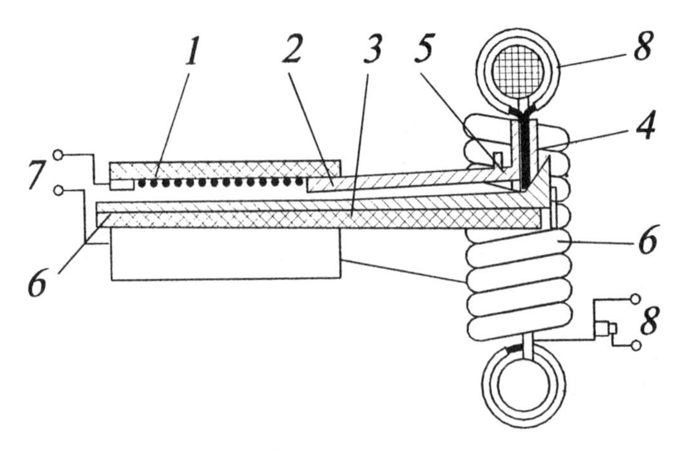


FIGURE 3.19. Cascaded MCG systems consist of two or more MCGs connected in series with air transformers, where each succeeding MCG being the load of the previous generator. As an example, a spiral–coaxial cascaded MCG is shown and it consists of 1 – spiral coil, 2 – coaxial section, 3 – liner, 4 – conducting plates, 5 – insulation, 6 – transformer, 7 – generator input, and 8 – generator output.

a current of 250 kA into a 6.5 $\mu$H load, when the initial current was 16 kA. When the same generator was connected to the passive feed circuit of a betatron, it delivered approximately 1 MJ of energy in 100 $\mu s$.

When $N$ generators are connected in series, the magnetic flux, energy (power), and inductance of the load, which is matched to the inductance of the MCG, increase $N$-fold, while the current and operating time remain the same. Figure 3.20 shows the diagram of a four-barrel (four stage) MCG battery in which the MCGs are connected in series and in parallel. In this diagram, the spiral coil is (1) and the centerline of the high explosive is (2). The generator $L_{10}$ is used to power generators $L_{11}$–$L_{14}$. The loop $L_*$ connects $L_{10}$ to $L_{14}$. $L_2$ is the load, where $L_* > L_2$. This generator operates in the following way. Initially, generator $L_{10}$ drives generator $L_{14}$, but current does not flow through the central tubes of two of the generators. After this, all four generators, $L_{11}$–$L_{14}$, act simultaneously. When four generators are connected according to this series–parallel scheme, the inductance of the load, which is matched to the inductance of the MCG system, does not change, and the energy delivered to the load is increased by a factor of 4 and the magnetic flux by a factor of 2. The magnetic flux coupled from these batteries to the load coil exceeds that of the input magnetic field, i.e., the batteries operate like magnetic flux amplifiers. These cascaded systems have generated current pulses with an amplitude of 500 MA and power densities in the realm of $10^7$–$10^{14}$ W/m$^3$.

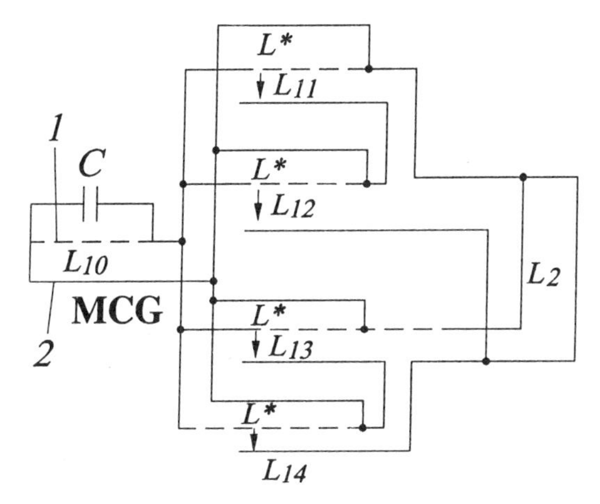


FIGURE 3.20. Diagram of four-barrel MCG battery. The MCGs are connected in series and in parallel.

## 3.10 Short-Pulse MCGs

Magnetocumulative generators produce electrical current pulses that rise from zero to a maximum in tens to hundreds of microseconds. However, many loads require that the risetime to be of the order of microseconds or less. These risetimes are usually achieved by applying combinations of opening and closing switches at the output of the generator. Switches developed for use with MCGs are hard to manufacture and are very large in size and weight. In this section, MCG designs will be described that are independent of the energy storage and coupling times relative to the load. These MCGs are called *short-pulse generators* [3.6,3.23], which are capable of coupling megaamperes of current into loads in hundreds of nanoseconds to microseconds.

The design of one such short-pulse generator is shown in Fig. 3.21. The initial magnetic flux is created by a single-turn coil, which surrounds the entire generator, using either a capacitor or another MCG. At the moment the initial magnetic field attains its peak value, the input to the coil is closed by detonation of a high explosive charge. The initial magnetic flux is captured in the region enclosed by the short-circuited single-turn coil and the cylindrical liner. The generator is designed such that the outer surface of the liner starts to move at the moment the coil closes. The explosive is initiated simultaneously along the whole axis of the generator and the liner expands radially.

When the liner begins to move, the magnetic compression process begins. However, a voltage will appear at the output of the generator only when the liner reaches the contact rings. This is a key feature in the operation of short-pulse MCGs. After the liner meets the contact rings, the flux trapped between the liner and stator coil is driven in a radial direction. The distance

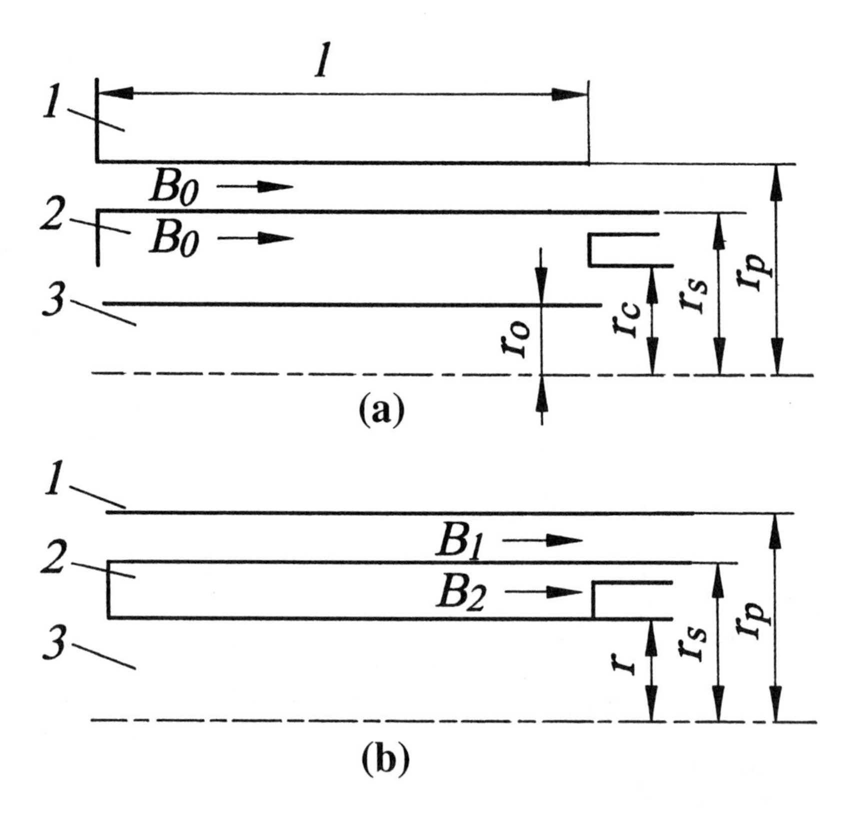


FIGURE 3.21. Diagrams describing the operation of short-pulse MCGs. The seed magnetic flux is created by a single-turn coil that surrounds the entire generator. When the explosives are detonated, the liner is accelerated and compresses the magnetic field. However, voltages only appear when the liner reaches the contact rings.The distance between the contact rings and the inner surface of the stator determines the length of the current pulse delivered to the load. Figures a and b show the position of the components of the generator prior to and just after the liner makes contact with the contact rings, respectively.

between the contact rings and the inner surface of the stator determines the length of the current pulse delivered to the load. Figures 3.21a and 3.21b show the position of the components of the generator just prior to and just after the liner makes contact with the contract rings, respectively.

Let us now determine the basic parameters of the short-pulse generator discussed above. The initial magnetic flux, $\Phi_0$, between the armature and primary coil is

$$\Phi_0 = \pi \left(r_p^2 - r_0^2\right) B_0 \tag{3.118}$$

where $B_0$ is the initial axial magnetic inductance. If it is assumed that the magnetic field inside the single-turn coil remains constant during the entire time the liner is moving, then

$$B_0 \left(r_p^2 - r_0^2\right) = B_1 \left(r_p^2 - r_s^2\right) + B_2 \left(r_s^2 - r^2\right), \tag{3.119}$$

where $B_1$ is the magnetic induction between the primary coil and the stator and $B_2$ is the magnetic induction between the stator and the liner. To simplify these calculations, the following assumptions are made:

- The stator and liner materials have infinite conductivities; i.e., the flux losses are zero $(\lambda_k = 1)$.
- The current flowing through the stator and the liner is equal to beam currents with zero density.
- The magnetic field inside the generator is constant in the axial direction.

It follows from the second assumption that

$$\frac{\mu_0 \omega_s I}{l} = B_2 - B_1. \tag{3.120}$$

Solving Eqs. 3.119 and 1.120 for $B_1$ and $B_2$, gives

$$B_1 = B_0 \left[\frac{r_p^2 - r_0^2}{r_p^2 - r^2}\right] - \frac{\mu_0 \omega_s I_s}{l} \left[\frac{r_s^2 - r^2}{r_p^2 - r^2}\right], \tag{3.121}$$

$$B_2 = B_0 \left[\frac{r_p^2 - r_0^2}{r_p^2 - r^2}\right] + \frac{\mu_0 \omega_s I_s}{l} \left[\frac{r_p^2 - r_s^2}{r_p^2 - r^2}\right]. \tag{3.122}$$

The magnetic flux inside the liner is

$$\Phi_s = \pi \omega_s \left(r_s^2 - r^2\right) B_2. \tag{3.123}$$

Introducing the following notation,

$$f = \frac{\mu_0 \omega_s^2 \pi}{l} \left(r_p^2 - r_s^2\right), \tag{3.124}$$

$$Y(r) = \frac{\left(r_s^2 - r^2\right)}{\left(r_p^2 - r^2\right)}, \tag{3.125}$$

Eq. 3.123 becomes

$$\Phi_s = Y(r)\left(\omega_s \Phi_0 + f I_s\right). \tag{3.126}$$

Using this equation, the output voltage of the generator can be found as

$$U = \frac{d\Phi}{dt} = 0, \tag{3.127}$$

for $r < r_c$, and

$$U = U_D + L_D \frac{dI}{dt} + I \frac{dL_D}{dt}, \tag{3.128}$$

for $r_c < r < r_s$, where

$$U_D = \omega_s \Phi_0 \frac{dY}{dt}, \tag{3.129}$$

$$L_D = Y f, \tag{3.130}$$

$$\frac{dL_D}{dt} = f \frac{dY}{dt}, \tag{3.131}$$

$$\frac{dY}{dt} = -2rU \frac{\left(r_p^2 - r_s^2\right)}{\left(r_p^2 - r^2\right)^2}. \tag{3.132}$$

From Eqs. 129–132, it can be seen that the output voltage is present even when it is in the passive mode.

Now let us consider a short-pulse generator connected to a load having an inductance $L_L$. Using the first assumption defined above, it follows that:

$$\Phi_s + L_L I = \Phi_0 \frac{r_s^2 - r_c^2}{r_p^2 - r_c^2}. \tag{3.133}$$

Substituting Eq. 3.126 into Eq. 3.133, it is found that

$$I = \omega_s \Phi_0 \frac{Y_c - Y}{L_L - Yf}, \tag{3.134}$$

where $Y_c = Y(r_c)$. Using the definition $U = -LdI/dt$ and taking the derivative of Eq. 3.134, an expression for the voltage on the load can be derived as

$$U = -\omega_s \Phi_0 \frac{1 + \frac{Y_c f}{L_l}}{1 + \frac{Yf}{L_L}} \frac{dY}{dt}. \tag{3.135}$$

To prevent premature closure of the stator winding by the liner, the radius of one end of the stator is greater than at the other. When designing an MCG with a specific set of parameters, the third assumption above must also be taken into account. The magnetic field is a function of only the axial distance, that is, $B_1 = B_1(z)$ and $B_2 = B_2(z)$. If the radius of one end of the stator is denoted as $r_s$ and at the other end as $r_s + \delta_s$, it can be shown that the expressions that describe the current and voltage remain the same. However, the expressions for $Y$ and $dY/dt$ are different. Since the primary coil has the same conical shape as the stator coil, where the cone opening is $\Delta$, the following equations can be derived:

$$Y(r) = 1 - \frac{\Delta}{\delta_s} \left\{ \frac{(1+a)(1+b)}{ab} - \frac{\Delta}{2r} \ln \left[ \frac{a(1+b)}{b(1+a)} \right] \right\} \tag{3.136}$$

for $r_c \leq r \leq r_s$, and

$$\begin{aligned} Y(r) &= (1-\xi) - \frac{\Delta}{\delta_s} \left\{ \frac{(1+a)(1+b)}{(\xi+a)(\xi+b)} \right. \\ &\quad \left. - \frac{\Delta}{2r} \ln \left[ \frac{(\xi+a)(1+b)}{(\xi+b)(1+a)} \right] \right\} \end{aligned} \tag{3.137}$$

for $r_s \leq r \leq r_s + \delta_s$, where

$$\xi = \frac{r - r_s}{\delta_s}, \quad a = \frac{r_p + r}{\delta_s}, \quad b = \frac{r_p - r}{\delta_s}, \tag{3.138}$$

and

$$\frac{dY}{dt} = -\frac{U}{\delta_s}\left\{\frac{\Delta}{\delta_s}\left[\frac{1-\frac{\Delta}{2r}}{b(1+b)} - \frac{1+\frac{\Delta}{2r}}{a(1+a)}\right] + \frac{\Delta^2}{2r^2}\ln\left[\frac{a(1+b)}{b(1+a)}\right]\right\} \tag{3.139}$$

for $r_c \leq r \leq r_s$, and

$$\frac{dY}{dt} = -\frac{U}{\delta_s}\left\{1 + \frac{\Delta}{\delta_s}\left[\left(1+\frac{\Delta}{2r}\right)\left(\frac{1}{1+a} - \frac{1-\frac{\Delta}{2r}}{\xi + a}\right) - \frac{1-\frac{\Delta}{2r}}{1+b}\right] + \frac{\Delta^2}{2r^2}\ln\left[\frac{(\xi+a)(1+b)}{(\xi+b)(1+a)}\right]\right\} \tag{3.140}$$

for $r_s \leq r \leq r_s + \delta_s$. In deriving these expressions, it was assumed that $\delta_s << r_s$. Calculations show that the operating time of this generator when loaded may be 0.5 $\mu$s.

Another type of short-pulse generator is shown in Fig. 3.22. To separate the energy accumulation stage from the energy coupling into the load stage during the operation of the MCG, a piston is added. This MCG consists of the liner (2) that contains the high explosive (3) and an external conductor (4) in the form of a spiral. The cable (5), which is isolated by the dielectric (6), is connected to the liner between the turns of the spiral coil. Electrical contact between the spiral and the liner is provided by the piston (7) and the switch (8) with terminal (9).

At the moment the initial magnetic field reaches its peak value, the explosive charge is initiated. The liner begins to move and compress the magnetic field stored in the generator. This leads to an increase in the current flowing through the "spiral–switch–piston–liner circuit." When the detonation wave propagating through the explosive reaches the piston, it is acted on by the explosion products and it begins to move, thus disconnecting the "spiral–switch–piston–liner" circuit. The current pulse is coupled into the load through terminals (5) and (9).

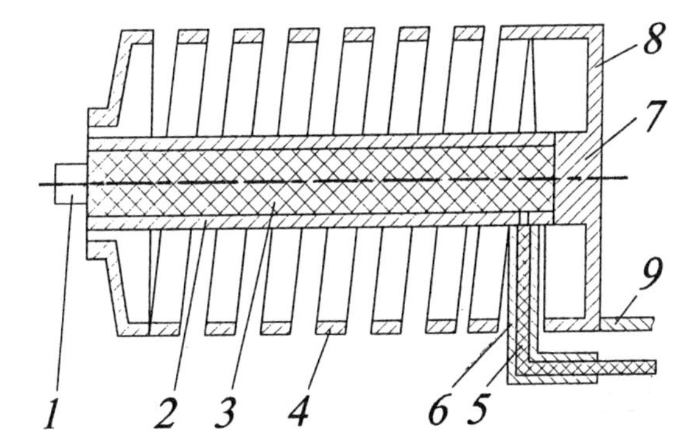


FIGURE 3.22. Short-pulse piston-type spiral MCG: 1 – detonator, 2 – liner, 3 – explosives, 4 – spiral external conductor, 5 – cable, 6 – insulation, 7 – piston, 8 – switch, and 9 – terminal. The energy-accumulation stage is separated from the coupling stage by the piston. The time it takes for the piston to separate these stages determines the length of the current pulse.

The length of the current pulse in the load may be calculated from

$$\tau_p = \frac{R_H I}{E_{br} v_p}, \tag{3.141}$$

where $R_H$ is the load resistance, $I$ is the current generated by the MCG, $E_{br}$ is the breakdown field strength of the liner–piston gap, and $v_p$ is the average velocity of the piston. For the values $R_H = 1\ \Omega$, $E_{br} = 10^7$ V/m, $I = 50$ kA, and $v_p = 5$ km/s, the time it takes for the generator to release its energy into the load is $\tau_p = 1\ \mu$s.

The time taken for the MCG to deliver its energy to the load may be reduced by either increasing the electric field breakdown threshold or by decreasing the current generated. However, it is not desirable to decrease the output current. Therefore, in order to decrease $\tau_p$, it is necessary for $E_{br}$ to be increased. The advantage of this method is that it has an insignificant effect on the mass and dimensions of the switch.

All types of MCGs use ballast inductance to separate the energy storage and the energy delivery stages during the operation of the MCG. During the "lengthy" operation of the MCG, the inductance of the generator stores energy. During the final stage of operation, the ballast inductance is connected to the load. As a result, the pulse in the load is shortened by a factor of 10 or more, and its power is increased. When an inductive load is attached to the MCG, the voltage drop in the load increases. To understand these processes, consider the PMCG. Unlike the typical PMCG, the plates of the short-pulse PMCG are connected to the ballast inductance shown in Fig. 3.23. The current-bearing conductors are positioned parallel

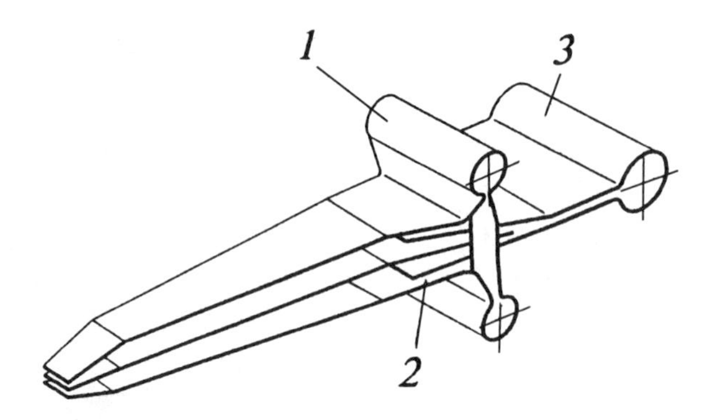


FIGURE 3.23. Short-pulse plate MCG: 1 – ballast, 2 – plates, and 3 – solenoid. The ballast is used to separate the energy-storage stage from the lengthy delivery stage during the operation of the MCG. While the MCG is operating, energy is stored in the ballast, after which it is connected to the load. As a result, the pulse delivered to the load is shortened by a factor of 10 or more.

to the standard plates at the end of the generator. The load is connected to the current-bearing conductors. A polyethylene film is used to insulate the plates and the current-bearing conductors.

This PMCG operates in the following way. When the current flowing through the generator reaches its peak value, the explosive is initiated. The cassette walls begin to move, thus closing the generator plates, compressing the magnetic field, and pushing the field into the ballast solenoid. The amount of current flowing through the plates and the ballast solenoid increases. When the point of contact between the plates and the cassette reaches the added current conductors, the "plate–ballast solenoid circuit" is switched to the additional "conductor–load solenoid circuit." In addition, magnetic flux compression will take place between the additional current conductors and the cassette. The risetime of the electromagnetic energy pulse coupled into the load is determined by the time of compression.

Let's now determine that part of the electromagnetic energy and current that is transferred from the ballast inductance into the inductive load. The energy stored in the ballast inductance is

$$W_s = \frac{1}{2} L_\delta I_1^2, \tag{3.142}$$

where $I_1$ is the current flowing through the MCG at the moment the circuit is switched. To find the residual magnetic energy, $W_\delta'$, in the ballast inductance and the energy transferred into the magnetic inductance of the load, $W_H'$, the following expressions are used:

$$W'_\delta = \frac{L_\delta^2}{(L_\delta + L_H)^2} W_\delta, \tag{3.143}$$

$$W'_H = \frac{L_\delta L_H}{(L_\delta + L_H)^2} W_\delta, \tag{3.144}$$

where $L_H = L'_H + L'_G$, $L'_H$ is the inductance of the load of the generator and $L'_G$ is the inductance of the current-bearing conductors. Using Eqs. 3.143 and 144, the ratio of energy transfer from the ballast inductance into the load is

$$\frac{W'_H}{W'_\delta} = \frac{L_\delta L_H}{(L_\delta + L_H)^2}. \tag{3.145}$$

This ratio has a maximum of 25%, when $L_H = L_\delta$. Taking into account Eq. 3.142, it can be shown that the peak current through the load as a result of the presence of the ballast inductance is equal to $0.5 I_1$.

Therefore, adding the ballast inductance to separate the energy-storage and energy-delivery stages leads to a decrease in energy delivered to the load by a factor of 4 and a decrease in current by a factor of 2. This particular generator design, despite its specific limitations, may be efficiently used if the solenoidal load of the generator is the primary winding of a pulsed transformer. It is a well-known fact that the voltage drop across an inductance is given by the expression

$$V = L_c \frac{dI}{dt}, \tag{3.146}$$

where $L_c$ is the inductance of the solenoidal load. If it is assumed that the current changes linearly in time, Eq. 3.146 can be rewritten as

$$V = L_c \frac{\Delta I}{\Delta t}. \tag{3.147}$$

By using the ballast inductance and the additional current conductors, the time required to release the energy into the load can be reduced from hundreds of microseconds down to microseconds. Therefore, for a generator of length $l_0 = 1$ m filled with high explosives with a detonation velocity of $D = 7500$ m/s and with current conductors of length $l_t = 0.1$ m, this time is shortened by approximately a factor of 60. Using Eq. 3.147 and the assumption $L_\delta = L_H$, the voltage on the solenoidal load is increased by a factor of 30.

The time taken for the MCG to deliver its energy to the load can be reduced to a minimum if the ballast inductance is switched to the load during the final stage of electromagnetic energy accumulation. With this purpose in mind, a massive metal lug is welded to the end of the high explosive cassette. This lug compresses the magnetic flux during the final stages of operation of the generator and switches the ballast inductance circuit into the inductive circuit of the load.

To determine the rate at which the current increases during the switching of the ballast inductance into the load inductance, the law of conservation of magnetic flux is applied to the circuit:

$$I_1 L_\delta = I_H L_\delta + I_H L_H. \tag{3.148}$$

Solving this equation for $I_H$ and taking the derivative with respect to time yields

$$\frac{dI_H}{dt} = \frac{L_\delta}{L_\delta + L_H}\frac{dI_1}{dt}. \tag{3.149}$$

To find the rate at which the current flowing through the generator changes during the accumulation of electromagnetic energy, the circuit equation for an MCG connected to an inductive load can be used:

$$L_G\frac{dI_1}{dt} + I_1\frac{dL_G}{dt} + L_\delta\frac{dI_1}{dt} = 0. \tag{3.150}$$

Rearranging, an expression for the rate of change of current can be found:

$$\frac{dI_1}{dt} = \frac{I_1}{L_\delta + L_G}\left(-\frac{dL_G}{dt}\right). \tag{3.151}$$

In this expression, the factor $(-dL_G/dt)$ is the rate at which the magnetic flux is compressed. Substituting Eq. 3.151 into Eq. 3.149, an expression for the rate of change of current in the load is obtained as

$$\frac{dI}{dt} = \frac{L_\delta}{L_\delta + L_H}\frac{I_1}{L_\delta + L_G}\left(-\frac{dL_G}{dt}\right). \tag{3.152}$$

At the moment of switching, $L_G = 0$ and Eq. 3.152 becomes

$$\frac{dI_H}{dt} = \frac{I_1}{L_\delta + L_H}\left(-\frac{dI_G}{dt}\right) \tag{3.153}$$

Substituting the values of the MCG parameters given above into this equation and assuming that $L_\delta = L_H = 10$ nH, the pulse length when the ballast inductance is switched into the load during the final stages of operation of the generator is calculated to be 100 times smaller than the pulse length of a generator with additional current conductors. Thus, according to Eq. 3.147, there is a 100-fold voltage drop across the load.

For the MCG designs considered in this section, it has been found that the time taken to release energy into the load may be significantly shortened by employing several design concepts, which lead to significant increases in the amount of pulsed power delivered to the load and voltage decreases in the load. However, it should be mentioned that significant energy losses arise as the result of shortening the pulse in the load, which is a significant disadvantage of short-pulse generators. When estimating the expedience of using these generators for specific applications, the output parameters of the generator must be taken into account relative to the nature and requirements of the load. The ancillary equipment required to match the output parameters of the MCG to its load is discussed in the next chapter.

# References

[3.1] A.I. Pavlovskii, "Magnetic Cumulation—A Memoir for Andrei Sakharov," *Megagauss Magnetic Field Generation and Pulsed Power Applications* (eds. M. Cowan and R.B. Spielman), Nova Science Publ., New York, pp. 9–22 (1994).

[3.2] B.M. Novac, I.R. Smith, and M.C. Enache, "Classifications of Helical Flux-Compression Generators," to be published in the Proceedings of the Megagauss VIII Conference, Tallahassee, FL (1998).

[3.3] G. Knoepfel, *Pulsed High Magnetic Fields*, North-Holland Publishing Company, Amsterdam (1970).

[3.4] G. Lehner, "Maximum Magnetic Fields Obtainable by Flux Compression: Limitation Due to the Diffusion of Magnetic Field," *Proceedings of the Conference on Megagauss Magnetic Field Generation by Explosives and Related Experiments*, European Atomic Energy Committee, Brussels, pp. 55–66 (1966).

[3.5] D.B. Cummings and M.J. Morley, "Electrical Pulses from Helical and Coaxial Explosive Generators," *Proceedings of the Conference on Megagauss Magnetic Field Generation by Explosives and Related Experiments* (eds. H. Knoepfel and F. Herlach), European Atomic Energy Community, Brussels, pp. 451–470 (1966).

[3.6] R.S. Caird and C.M. Fowler, "Conceptual Design for a Short-Pulse Explosive-Driven Generator," *Megagauss Technology and Pulsed*

*Power Applications* (eds. C.M. Fowler, R.S. Caird, and D.J. Erickson), Plenum Press, New York, pp. 425–431 (1987).

[3.7] P.J. Turchi, R.S. Caird, and J.H. Goforth, "Design of a Fast–Plate Generator Driving a Plasma Flow Switch," *Megagauss Technology and Pulsed Power Applications* (eds. C.M. Fowler, R.S. Caird, and D.J. Erickson), Plenum Press, New York, pp. 559–566 (1987).

[3.8] V.A. Vasyukov, "Explosive Magnetic Generators of Loop Type Operating within a Middle Range of Fast Current Pulses (15–45 MA)," *Megagauss and Megaampere Pulse Technology and Applications* (eds. V.K. Chernyshev, V.D. Selemir, and L.N. Plyashkevich), VNIIEF, Sarov, pp. 292–296 (1997).

[3.9] A.I. Pavlovskii, R.Z. Lyudaev, and B.A. Boyko, "Disk Magnetic Cumulation Generators Maximum Characteristics," *Megagauss Fields and Pulsed Power Systems* (eds. V.M. Titov and G.A. Shvetsov), Nova Science Publ., New York, pp. 327 – 330 (1990).

[3.10] A.I. Pavlovskii, R.Z. Lyudaev, B.A. Boyko, A.S. Boriskin, A.S. Kravchenko, V.E. Gurin, and V.I. Mamyshev, *Megagauss Fields and Pulsed Power Systems* (eds. V.M.Titov and G.A. Shvetsov), Nova Science Publ., New York, pp. 331–336 (1990).

[3.11] C.M. Fowler, R.F. Hoeberling, and S.P. Marsh, "Disk Generator with Nearly Shockless Accelerated Driver Plate," *Megagauss Fields and Pulsed Power Systems* (eds. V.M. Titov and G.A. Shvetsov), Nova Science Publ., New York, pp. 337–345 (1990).

[3.12] V.K. Chernyshev, B.E. Grinevich, V.V. Vahrushev, and V.I. Mamyshev, "Scaling Image of ~ 90 MJ Explosive Magnetic Generators," *Megagauss Fields and Pulsed Power Systems* (eds. V.M. Titov and G.A. Shvetsov), Nova Science Publ., New York, pp. 347–350 (1990).

[3.13] V.A. Demidov, A.I. Kraev, V.E. Mamyshev, A.A. Petrukhin, V.P. Pogorelov, V.K. Chernyshev, V.A. Shvetsov, and V.I. Shpagin, "Three–Module Disk Explosive Magnetic Generator," *Megagauss Fields and Pulsed Power Systems* (eds. V.M. Titov and G.A. Shvetsov), Nova Science Publ., New York, pp. 351–365 (1990).

[3.14] E.I. Bichenkov, S.D. Gilev, and L.M. Trubachev, "MCGs Utilizing Transition of the Semiconductor Material to the Metal State," PMFT, No. 5, pp. 125–129 (1980).

[3.15] E.I. Bichenkov, S.D. Gilev, and A.M. Trubachev, in *Proceedings of the 3rd International Conference on Megagauss Magnetic Field Generation and Related Topics* (eds. V.M. Titov and G.A. Shvetsov), Nauka, pp. 88–93 (1984).

[3.16] S.D. Gilev and A.M. Trubachev, Pism. V. J. Tex. Phys., **8**(5), pp. 914–916.

[3.17] A.B. Prishchepenko, V.V. Kiseljov, and I.S. Kudimov, "Electromagnetic Environments and Consequences," in *Proceedings of the European Electromagnetic International Symposium of Electromagnetic Environments and Consequences*, Part 1, pp. 267–268 (1994).

[3.18] R.Z. Lyudaev, A.I. Pavlovskii, A.S. Yuryzhev, and V.A. Zolotov, "Multibarrel MC-Cascades Generating Magnetic Flux at Autonomous Helix Fluxes Collectivization," *Megagauss Magnetic Field Generation and Pulsed Power Applications* (eds. M. Cowan and R.B. Spielman), Nova Science Publ., New York, pp. 619–628 (1994).

[3.19] A.I. Pavlovskii, L.N. Plyashkevich, A.M. Shuvalov, and E.M. Dimant, "Small Helical MCGs Cascaded Systems," *Megagauss Magnetic Field Generation and Pulsed Power Applications* (eds. M. Cowan and R.B. Spielman), Nova Science Publ., New York, pp. 629–635 (1994).

[3.20] A.I. Pavlovskii, R.Z. Lyudaev, A.S. Kravchenko, V.A. Zolotov, and A.S. Yuryzhev, "Magnetic Cumulation Generator Power Increase," *Megagauss Fields and Pulsed Power Systems* (eds. V.M. Titov and G.A. Shvetsov), Nova Science Publ., New York, pp. 385–392 (1990).

[3.21] R.F. Hoeberling, C.M. Fowler, B.L. Freeman, J.C. King, J.E. Vorthman, and R.B. Wheeler, "Electrical Performance of a Three-stage Compression Generator," *Megagauss and Megaampere Pulse Technology and Applications* (eds. V.K. Chernyshev, V.D. Selemir, and L.N. Plyashkevich), VNIIEF, Sarov, pp. 319–321 (1997).

[3.22] Ye.V. Chernykh, Ye.N. Nesterov, A.V. Shurupov, Yu.A. Karpushin, and I.O. Zolotykh, "Two Cascade MCG for Generation of Rapidly Increasing Current Pulses," *Megagauss and Megaampere Pulse Technology and Applications* (eds. V.K. Chernyshev, V.D. Selemir, and L.N. Plyashkevich), VNIIEF, Sarov, pp. 325–332 (1997).

[3.23] P.M. Mironychev, "Conceptual Design for a Short-Pulse High-Field Helical Explosive Magnetic Generator," *Megagauss Magnetic Field Generation and Pulsed Power Applications* (eds. M. Cowan and R.B. Spielman), Nova Science Publ., New York, pp. 739–743 (1994).

# 4
# Pulse-Forming Networks

The energy in magnetocumulative generators is stored in low value inductances, of the order of 10 nH, but the currents involved may reach tens and even hundreds of megaamperes. Transferring the corresponding large amount of energy into different types of loads poses a major problem, because the inductance of the MCG is smaller than that of the load. When the load is series connected to the MCG, efficient operation is only possible when the load parameters are matched to those of the generator. If pulse sharpening or matching devices are not used on the output of the MCG, it may still operate satisfactorily with an inductive load, provided that the inductance of the load is of the same order of magnitude as that of the MCG, i.e., several nanohenrys, or with an active resistance provided that its resistance is of the order of milliohms. To match the MCG to loads that do not meet these criteria, several methods have been developed to couple the energy from the MCG into the load. These methods include:

- disconnecting the MCG from a circuit branch using an opening switch, followed by its connection to the load in a parallel circuit branch (Fig. 4.1). When the switch opens, a voltage pulse is delivered to the load,

- using a transformer connected through an opening switch to the MCG and a closing switch to the load (Fig.4.2). When the opening switch is activated, energy is stored in the transformer. When the closing switch is activated, energy is delivered from the transformer to the load,

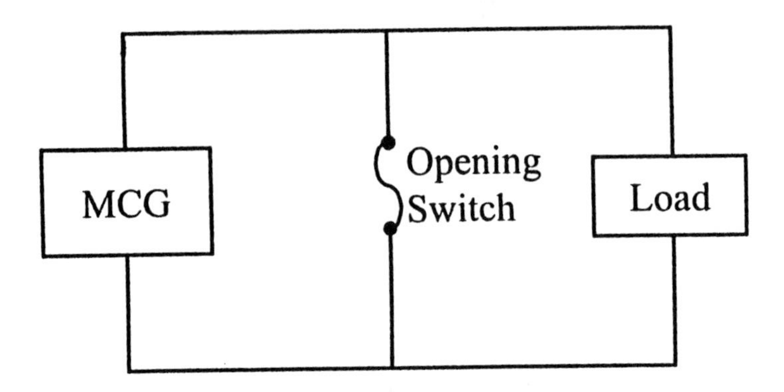


FIGURE 4.1. Prior to the opening switch activating, most of the current flows through the circuit containing the MCG and switch. When the switch opens, the current flows from the MCG into the load.

- using an opening switch to shunt the current. When the switch opens, a closing switch simultaneously connects the load to the output of the MCG (Fig. 4.3).

In order to apply these methods, the MCG and the load are usually interconnected by high-speed switches, pulse transformers, spark discharge (spark gap) switches, pulse-sharpening inductors, pulse-forming lines, and so on. These components comprise what is called the *pulse-forming network*, since it is used to provide impedance matching of the load to the MCG and to shape the electrical output pulse of the MCG to optimize the load performance [4.1]. This chapter is devoted to a study of the design of these ancillary, but very important devices.

## 4.1 High-Speed Opening Switches

The major disadvantage of inductive stores, such as MCGs and transformers, is the difficulty encountered in extracting the stored energy from them. This problem is overcome by using high-speed current breakers (opening switches), which permit efficient switching of the stored energy from the inductive store into the load. The purpose of the opening switch is to interrupt the flow of current in the circuit branch that contains the switch and to force it to transfer to a parallel circuit branch containing the load [4.2]. Various opening switches have been developed including diffuse discharge opening switches, low-pressure plasma opening switches, plasma opening switches, reflex switches, plasmadynamic opening switches, fuse opening switches, explosively driven opening switches, and solid-state

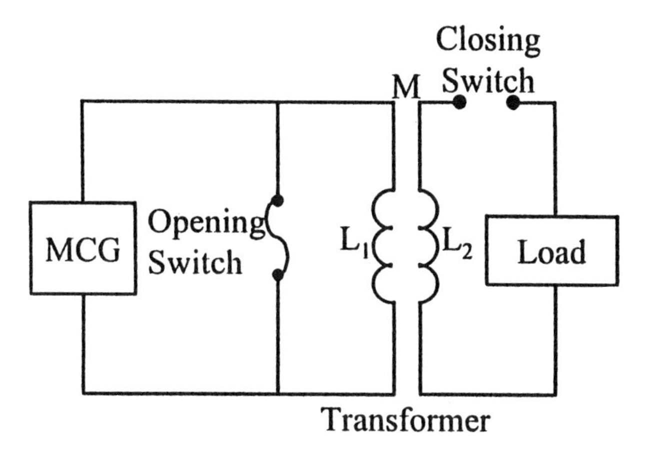


FIGURE 4.2. When the opening switch activates, the MCG delivers its energy to the transformer, where it is stored. When the closing switch activates, it delivers the energy to the load.

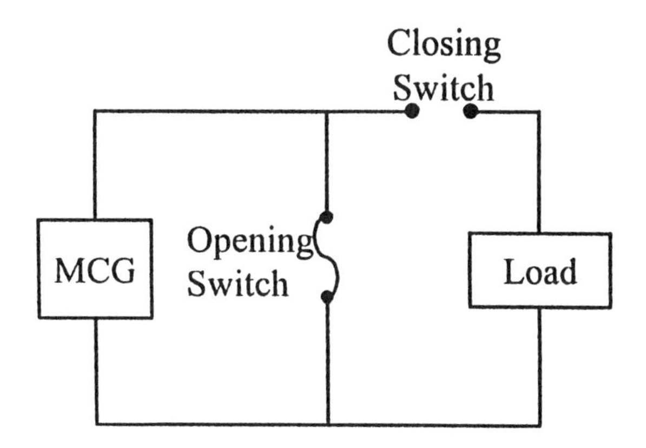


FIGURE 4.3. The branch with the opening switch initially shunts the current. When the opening switch activates, current is delivered to the branch containing the closing switch and load. When the closing switch activates, the energy from the MCG is delivered to the load.

opening switches. The family of switches can be divided into two fundamental types: *direct-interruption opening switches* and *current-zero opening switches*. The former develop the transfer voltage internally, through a mechanism in which the generalized impedance of the switch is increased. Examples include fuses, dense plasma focus devices, and superconducting switches. The latter use a voltage source external to the switch to drive the switch current to zero and thereby allow the switch to open. Examples include solid-state thyristors, vacuum switches, and liquid metal plasma valves.

The combined operation of the high-speed opening switch and the MCG allows the separation of the energy storage and the energy extraction stages, and the provision of both impedance matching and pulse shaping. The three switches that are most widely used with MCGs are explosive, electroexplosive (fuse, exploding wire, or exploding foil), and explosive plasma switches. *Explosive switches* use high explosives to break or open the electrical circuit by cutting a metal conductor with a dielectric knife blade. *Electroexplosive switches* (also called *exploding wire* or *foil switches*) use the high electrical currents generated by the MCG to heat materials that experience a very rapid rise in resistance as the result of ohmic heating. This heating leads to melting and eventually to vaporization of the conductor, which opens the circuit. *Explosive plasma switches* use the shock waves from the detonation to extinguish an electric arc to interrupt a circuit.

It should be noted that this chapter is not intended to be an exhaustive review of opening switch technologies, since excellent reviews are available elsewhere, for example, [4.2] and [4.3].

### *4.1.1 Explosive Opening Switches*

One type of explosive opening switch [4.4–4.6] is shown in Fig. 4.4. It consists of the current carrying bus (1), insulator (2), explosive charge (3), plane wave generator (4), and electrodetonator (5).

The insulating plate is usually made from Teflon, with grooves cut into this plate to create several unconnected gaps. When the explosive is detonated, a length of conductor is removed and the circuit is opened. The characteristic time required for this type of switch to decrease the current by 30% from 30 kA is 2 $\mu s$. These switches have been successfully used to switch currents of $\sim$ 100 kA at current densities of 50 kA/cm$^2$. In order to achieve a switching time of $\sim$ 2 $\mu$s, 0.1–0.3 kg of high explosives must be used. These switches may survive up to voltages of 50 kV at power levels of 5 GW.

An example of an opening switch in which massive conductors are mechanically destroyed by an explosive charge is shown in Fig. 4.5. The switch consists of a metal ring (1), one side of which is connected to the current-carrying plates of the MCG and the other side to the storage coil (3). Grooves (4) cut into the wall of the ring facilitate the breakage of the ring.

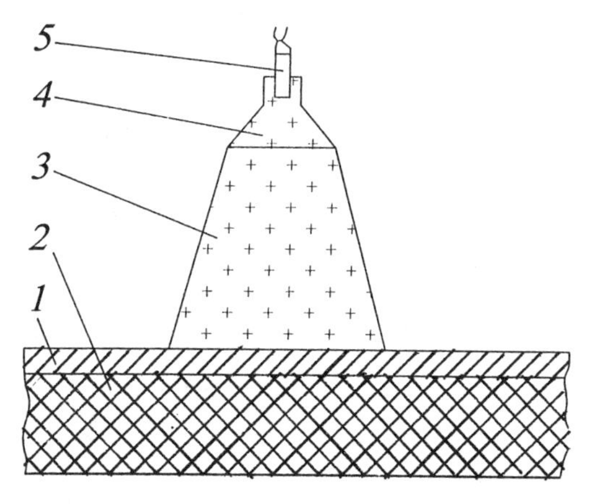


FIGURE 4.4. Schematic diagram of an explosive opening switch: 1 – current carrying bus, 2 – insulation, 3 – explosives, 4 – plane wave generator, and 5 – detonator. When the explosives is detonated, the current carrying bus is broken and interrupts the flow of current.

When the explosive charge (7) is detonated, the current source is shunted by plate (5) followed by connection to the storage coils by plate (6), and the MCG circuit is disconnected by the opening ring walls. Experimentally, currents up to $5 \times 10^5$A have been switched in 5 $\mu s$.

One disadvantage of explosive opening switches is the large amount of high explosive required, that in some cases may exceed the explosive charge in the MCG. For example, a spiral generator with an output current of $\sim$ 160 kA requires an explosive charge of 70 g in the switch. A second disadvantage is that the current switching time is of the order of microseconds, which in some cases may exceed the maximum pulse length requirement imposed by the load or the intended application.

### *4.1.2 Electroexplosive Switches*

Electroexplosive breakers (EEB) (electroexplosive opening switches) [4.7,4.8] are used when switching times are required shorter than those available when using explosive opening switches. Usually, EEBs consist of either thin wires connected in parallel or foils. To understand how they work, consider the voltage changes with time presented in Fig. 4.6. During the initial stages of operation, that is, when $t < t_1$, the resistance of the conductor increases uniformly, which implies that the voltage on the conductor changes uniformly. If thermal losses are neglected, the resistance is proportional to the amount of energy injected into the conductor. During the period $t_1 < t < t_2$, melting occurs, and when $t > t_3$, one of three possible processes can take place depending on the current density. If the injection of energy into the conductor occurs at the peak of the curve, the voltage

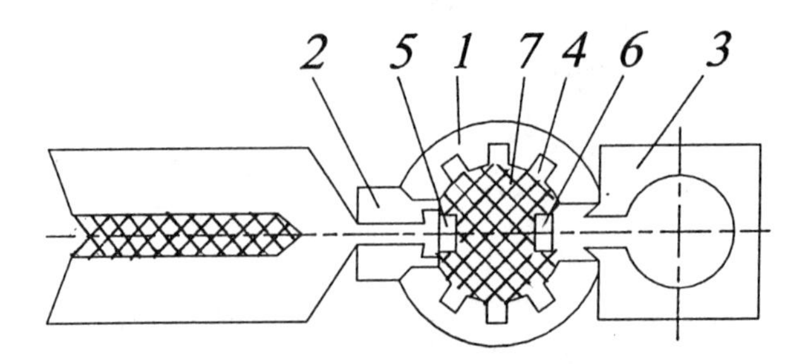


FIGURE 4.5. MCG with an electroexplosive opening switch: 1 – metal ring, 2 – current-carrying plates, 3 – storage coil, 4 – groves cut into ring, 5 – shunt, 6 – current-carrying plate, 7 – explosives. When the explosive charge is detonated, the current source is shunted by plates 5 and 6 and the circuit disconnected due to the destruction of the ring walls.

will fall to zero and the conductor remains intact. If a small amount of energy is injected beyond the peak, the conductor will be destroyed at one or more places, and the voltage will also fall. Finally, if the injected power is sufficiently high, the conductor material disintegrates and forms a vaporized wave front that moves toward the center of the conductor. During this final stage of the process, part of the electromagnetic energy is converted into kinetic energy of the vapor, depending on the current density. The voltage on the conductor increases sharply and may exceed the initial voltage by more than one order of magnitude.

If the metal used in the EEB has a boiling point that is sufficiently high, then, even before the explosion, a shunting discharge arises at the surface of the conductor and destruction of the metal does not occur. It has been shown experimentally, that, under normal conditions, tungsten, molybdenum, tantalum, and zirconium do not explode. These metals are not suitable for current switching.

To achieve a high energy transmission coefficient with an EEB, metals having as small a specific resistance as possible should be used. Some examples are silver, copper, gold, aluminum, and zinc. Since that part of the energy used for heating and vaporization lowers the efficiency of the EEB, it is best to use materials having a low thermal capacity and a low heat of vaporization. If only these parameters are taken into account, then the product of the specific resistance and the sum of the energy used to heat and vaporize the metal is an acceptable parameter for characterizing the suitability of using a particular metal in the EEB. These calculations show that silver, gold, aluminum, zinc, and copper have the smallest products of

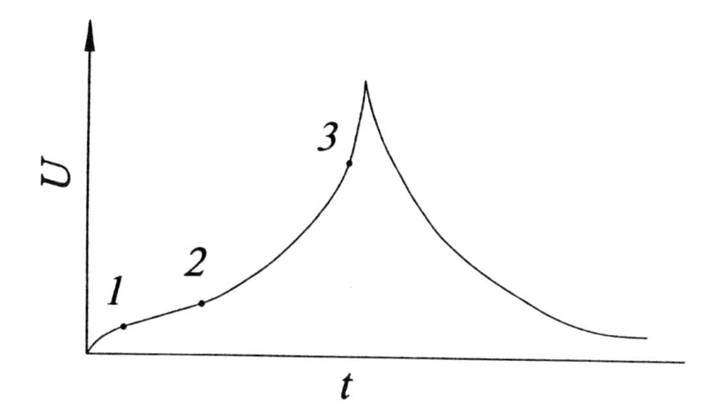


FIGURE 4.6. Voltage changes in EEB as a function of time. If the EEB consists of parallel wires, then during the initial stages ($t < t_1$), their resistance increases, which means that the votage across them also increases. During the period, $t_1 < t < t_2$, melting occurs, and when $t > t_3$, the wires disintegrate and the circuit opens.

the specific resistance and heat of sublimation. When comparing aluminum and copper wire, it has been shown that aluminum wires have less electrical strength during the explosion. As a result, the length of the aluminum wire used in the EEB must be some 1.8 times greater than the length of silver or copper.

To find an expression that describes how the resistance of the EBB changes, a model for surface vaporization, which depends on the current flowing through the conductor, is used. The speed at which the vaporization wave moves is

$$v_v = v_{v0} f\left(\theta, \alpha_v\right), \tag{4.1}$$

where

$$\begin{aligned} v_{v0} &= \frac{A T_0^{\alpha_v - 0.5}}{\rho C_V^{0.5}}, \\ f(\theta, \alpha_v) &= \theta^{\alpha_v - 0.5} e^{-\frac{1}{\theta}} (1 + \alpha_v \theta), \end{aligned} \tag{4.2}$$

$\rho$ is the specific resistance of the metal used in the EEB, $\theta = T/T_\alpha$, $\alpha_v$ is a coefficient (1.2 for copper), $C_V$ is the thermal capacity (5 J/cm$^3$·grad), and $T_0 = \Delta\theta_V / R_G$ is the ratio of the specific heat of vaporization to the gas constant. The vaporization wave velocity for copper can be approximated by the step change $v_v$, where $v_v = 0$, if $W_P < 0.27$ J/mm$^3$, and $v_v = 0.175$, if $W_P \geq 0.27$ J/mm$^3$ and where $W_P$ is the energy absorbed by the conductor.

The maximum voltage generated in the EEB is given by

$$U_b = 2.16\frac{L^2 v_v^2 I_L}{R_L d^2} \exp\left(\frac{R_L d}{2v_v L}\right), \tag{4.3}$$

where $L$ is the circuit inductance, $I_L$ is the current, $d$ is the diameter of the wire, and $R_L$ is the wire resistance. The index "$L$" refers to the instant in time at which the vaporization wave begins to move. The applicability of this surface wave model to the EEB is explained by the fact that the conductor used as the current breaker element has been heated beyond its boiling point, but does not suffer strong overheating, because the load is connected in parallel and absorbs part of the energy.

If a foil is used in the EEB instead of a wire, it may exist in two states:

- solid metal through which the vaporizing wave has not passed,
- expanding saturated vapor.

For the case of a vaporization wave moving through the foil with a constant velocity $v_v$, the size of the conducting portion changes according to the expression

$$\frac{d}{2} - \int_0^t v_v dt, \tag{4.4}$$

where $d$ is the thickness of the foil. Since the foil resistance is inversely proportional to its thickness, it can be shown that

$$R_P(t) = R_P(0)f(t)\left[1 - \int_0^t \frac{2v_v}{d}dt\right]^{-1}. \tag{4.5}$$

Neglecting the dependence of $f(t)$ on temperature and introducing the following parameters

$$\tau_P = \frac{d}{2v_v}, \quad \eta = \frac{t}{\tau_P}, \tag{4.6}$$

enables Eq. 4.5 to be rewritten as:

$$R_P(t)\frac{R_P(0)}{1-\eta}. \tag{4.7}$$

It can be shown that, for EEBs that use wires, instead of foils, the resistance changes in time according to the equation

$$R_P(t) = \frac{R_P(0)}{(1-\eta)^2}. \tag{4.8}$$

The most important parameters that characterize the capabilities of the EBB are the switching time, operating current, and maximum voltage. The first of these determines the switching quality and the losses, while the latter two determine the amount of power switched and the basic operating parameters of the switch.

It follows from Eqs. 4.7 and 4.8 that the current switching time of the EEB depends on the cross-sectional area of the exploding conductors and on the velocity at which the vaporization wave propagates through them. The cross-sectional area must be selected so that the total current flowing through it approaches the magnitude required to vaporize the conductor at the moment switching occurs. The energy required to explode the mass of the metal may exceed the energy of sublimation of the metal due to overheating of the vapor phase.

The number of parallel conductors required to switch the current $I$ flowing from the MCG can be found. The total cross-sectional area, $S$, of the wires is

$$S_P^2 = \frac{1}{\ell_c}\int_0^{t_{ex}} I^2 dt, \tag{4.9}$$

where $\ell_c$ is the *inertial integral*, which is equal to $1.95 \times 10^{17}$ $A^2{\cdot}s{\cdot}m^{-4}$ for copper and $1.09 \times 10^{17}$ $A^2{\cdot}s{\cdot}m^{-4}$ for aluminum, and $t_{ex}$ is the time at which the wires explode. If it is assumed that the inductance of the generator changes linearly, $L_0(t) = L_0 - \dot{L}_s t$, then an analytical expression can be derived to calculate the overall cross-sectional area of the switch:

$$S_p = \lambda_k I_0 \sqrt{\frac{L_0 t_{ex}}{\ell_v L_k}}, \tag{4.10}$$

where $L_k$ is the inductance of the MCG at $t = t_{ex}$.

If the EEB is connected to the secondary circuit of a pulsed transformer, where the primary circuit is the load for the MCG, the following analytical expression can be used to determine the overall cross-sectional area of the EEB:

$$S_p^2 = f_p f_W^{(2-S_t)} f_R \left[ f_W^{(S_t-1)} - f_k^{(S_t-1)} \right], \tag{4.11}$$

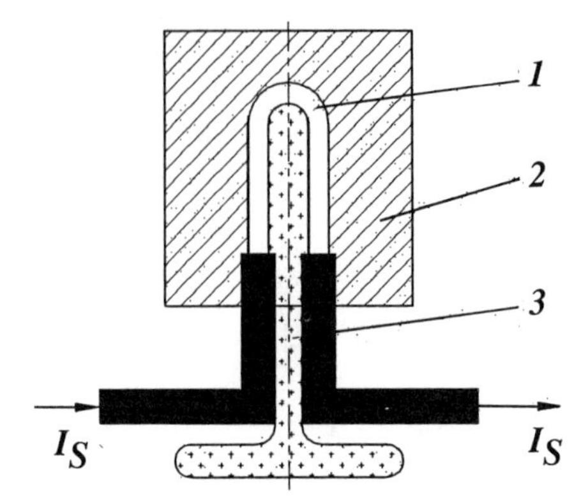


FIGURE 4.7. Another example of an EEB is the exploding foil switch. In order to reduce the opening time of the switch, the foils (1) are surrounded by powdered aluminum oxide (2), which hinders the foil vapor from spreading following the explosion of the foil and an increase in conductivity of the material. The input and output electrodes are denoted by (3).

where

$$\begin{aligned} f_W &= L_0 - \frac{M^2}{L_2}, \\ f_p &= \frac{t_{ex}}{\ell_v}\left(\frac{M}{L_2}I_0\right)^2, \\ f_R &= \frac{1}{\left(2R_1 - \dot{L}_S\right)t_{ex}}, \\ f_k &= L_k - \frac{M^2}{L_2}, \\ S_t &= \frac{2R_1}{L_1}, \end{aligned} \qquad (4.12)$$

$M$ is the mutual inductance between the windings of the transformer, and $\dot{L}$ is the rate at which the inductance of the MCG changes. If the metal vaporizes sufficiently fast, there will exist a time interval, called the *current pause,* during which the electrical conductivity of the metal is very small, i.e., the circuit is open. Following this, an electrical discharge will take place in the outer layers of the metal vapor where the pressure is low, and the high-conductivity state will begin to be restored. However, this recovery process may be slowed by surrounding the exploding wire with a powder such as $Al_2O_3$, which hinders the vapor from spreading following the explosion. This is depicted in Fig. 4.7 [4.9].

It is usually necessary to determine the critical length of the switch, which is the length that will allow the explosive products of the conductor

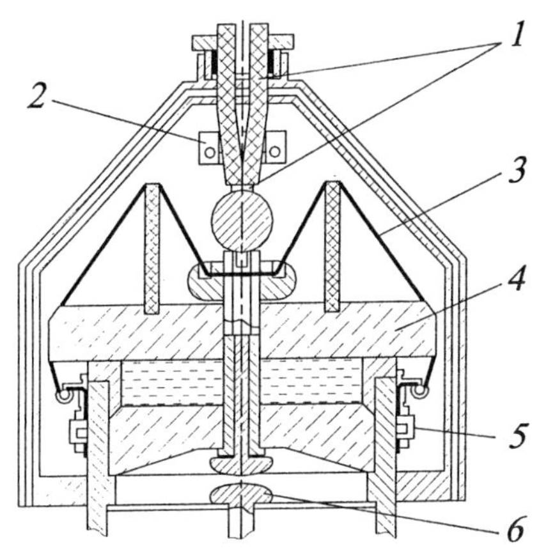


FIGURE 4.8. Example of a switch based on parallel exploding wires: 1 – high-voltage cables, 2 – Rogowski belt, 3 – copper wires in parallel, 4 – Plexiglas insulator, 5 – fastening nuts, and 6 – discharge gap.

to survive the high voltages produced within it without breakdown. On the one hand, the voltage on the switch is inversely proportional to the breakdown time, i.e., $U = LdI/dt$, and increases as the radius of the conductors decrease. On the other hand, increasing the length of the EEBs will lead to an increase in the total mass of the EEB. Heating the mass of the conductors until they explode may consume a major portion of the energy generated by the MCG, which means that only a small portion of this energy is delivered to the load. It has been observed experimentally that the electrical strength of conductors having diameters of 0.11 and 0.80 mm is 4–5 kV/cm and 8–9 kV/cm, respectively. When the conductors are longer than a critical length, a steady current flows and the rate at which energy is delivered to the load does not change. Maximum pulsed power is delivered to the load when the length of the conductors is equal to the critical length, which increases as the load resistance increases.

Figure 4.8 is a drawing of an EEB based on conductors connected in parallel, in which (1) is high-voltage cables, (2) is a Rogowski belt for measuring the current pulse, (3) are several tens of copper wire connected in parallel, (4) is a Plexiglas insulator, (5) are fastening nuts, and (6) is a discharge gap. Switches based on the electrical explosion of wires may switch energies in the megajoule range. The switched currents are rather high.

### *4.1.3 Explosive Plasma Switches*

The operation of explosive plasma switches (EPS) [4.10] is based on extinguishing an electric arc by means of a shock wave. A drawing of this type of switch is presented in Fig. 4.9. The electric arc is ignited between

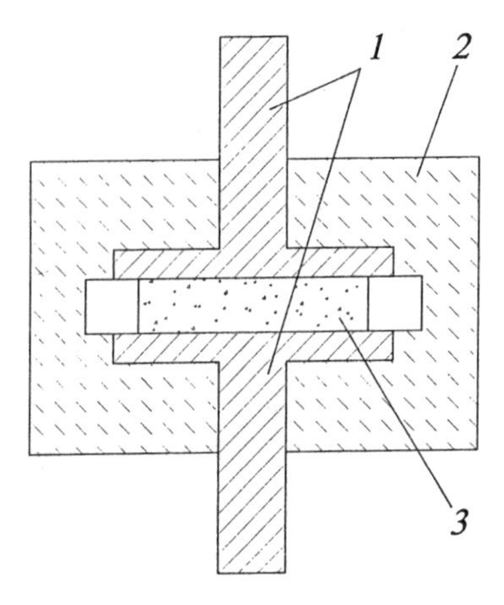


FIGURE 4.9. Explosive plasma switch: 1 – circular electrodes, 2 – insulation, and 3 – explosives. An electric arc is ignited between the circular electrodes in the ring channel along the perimeter of the explosive charge. When the explosive charge is detonated, the arc is driven toward the walls of the insulator, where it is extinguished and the circuit opened.

the circular electrodes (1) in the ring channel along the perimeter of the high-explosive charge (3). After the charge is initiated in the center, the arc is driven toward the walls of the insulator (2), where it is compressed and extinguished.. If the high-explosive charge is hexogen, with a diameter of 30 mm and a thickness of 10 mm, the EPS can switch 23 kA in 3 $\mu$s. The voltage on the electrodes is 30 kV.

The switching process is affected by the interaction of the expanding explosion products with the plasma arc, the risetime of the switched current and voltage, and the parameters of the electric circuit. The specific conductivity of the detonation products, which shunts the shock-compressed plasma arc and which decreases in time as the explosion products expand, also affect the switching process.

To find an expression that characterizes the operation of the EPS, its principles of operation, shown in Fig. 4.10, must first be analyzed. The switch serves as an energy store during the rather lengthy operating time of the MCG, followed by the rapid coupling of the energy into the load. The energy storage system consists of a metal tube and a cylindrical plasma channel, which are located in the region between the dielectric barrier and the surface of the high-explosive charge. When the current reaches its peak value in the storage portion of the switch, the plasma channel is simultaneously compressed along its entire length by the detonation of the high explosive. At the moment the plasma channel breaks, the volume of the storage system is filled with explosive by-products, which cause an increase in the inductance of the load and switching of the current into the load.

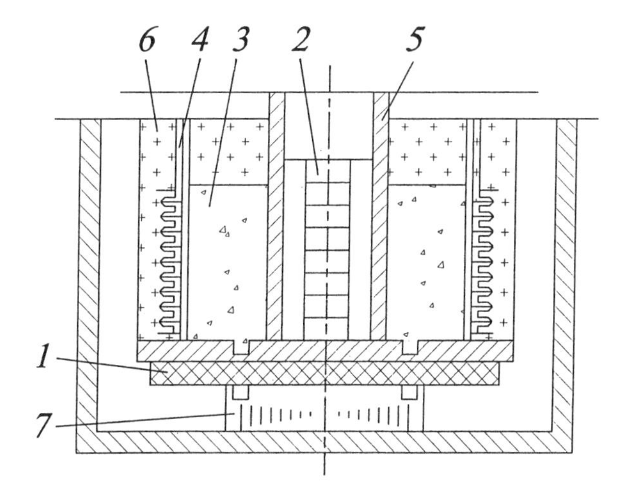


FIGURE 4.10. Switch with coaxial storage structure: 1 – dielectric, 2,5 – coaxial storage unit, 3 – explosive charge, 4 – detonator, 5 – polyethylene insulator, and 7 – load. The switch is designed to store energy during operation of the MCG, followed by the rapid coupling of this energy into the load. The energy storage systems consist of the metal tube and the cylindrical plasma channel, which are located in the region between the dielectric insulator and the surface of the explosive charge. When the current reaches its peak value, the plasma channel is simultaneously compressed along its entire length by the explosive charge. At the moment the plasma in the channel is extinquished, the volume of the storage system is filled with explosive by-products and the energy is switched into the load.

The change in the magnetic field in the storage volume of the system can be found by using the diffusion equation:

$$\mu_0 \sigma_e \frac{dH}{dt} = \Delta H, \tag{4.13}$$

where $\sigma_e$ is the electrical conductivity of the explosion products and $\mu_0$ is the magnetic permeability. The boundary condition at the radius of the inner tube, $r_0$, is

$$\frac{\partial H(r_0, t_0)}{\partial r} + \frac{H(r_0, t)}{r_0} = 0, \tag{4.14}$$

which was derived from the following equation based on the assumption that $E = 0$ at the inner tube:

$$\nabla \times H = \sigma_e E. \tag{4.15}$$

If $r_2$ is the radius of the plasma channel, the definition of $E$ is

$$E = \frac{I_* R_*}{l_k}, \tag{4.16}$$

where $R_*$ is the resistance and $l_k$ is the length of the plasma channel and $I_*$ is the current flowing through it. Rewriting $I_*$ in terms of the load current, $I_L$, and the magnetic field in the storage portion of the switch yields:

$$I_* = 2\pi r_2 H(r_2, t) - I_L. \tag{4.17}$$

From this expression, the boundary condition at the radius of the plasma channel can be derived as

$$\frac{\partial H(r_2,t)}{\partial r} + \frac{H(r_2,t)}{r_2} = \sigma_e R_* \frac{I_L - 2\pi r_2 H(r_0,t)}{l_k}. \tag{4.18}$$

or, after introducing the expressions for $I_L$ and $R_*$,

$$L_L \frac{dI_L}{dt} + R_L I_L = R_* \left[2\pi r_2 H(r_2,t) - I_L\right], \tag{4.19}$$

$$\frac{dR_*}{dt} = R_* \left(\frac{1}{S_k}\frac{dS_k}{dt} + \frac{1}{\sigma_*}\frac{d\sigma_*}{dt}\right), \tag{4.20}$$

where $S_k$ is the cross-sectional area of the channel, which is perpendicular to the flow of current, and $\sigma_*$ is the conductivity of the plasma channel. Using the following relationship between the parameters of air and plasma,

$$p = 6.65 T^{\frac{3}{2}} \left(\frac{\rho_p}{\rho}\right)^{0.12}, \tag{4.21}$$

which was derived from the equation of state

$$W = \frac{p}{(k-1)\rho}, \tag{4.22}$$

and the internal energy expression

$$W = 27.7 T^{\frac{3}{2}} \left(\frac{\rho_p}{\rho}\right)^{0.12}, \tag{4.23}$$

together with the first law of thermodynamics

$$dq = dW + pdV, \tag{4.24}$$

an equation that describes the rate at which the electrical conductivity of the plasma channel changes can be derived:

$$\frac{d\sigma_*}{dt} = 3.94T^{0.27}\ln\left(4\cdot 10^{25}n\right)\frac{dT}{dt} + 0.54\left(\frac{T}{n}\right)^{0.73}\frac{dn}{dt}, \tag{4.25}$$

where $p$ is pressure, $T$ is temperature, $\rho_p$ is the plasma density, $\rho$ is the density of air under normal conditions, $k$ is the adiabatic exponent, $dq = I_*^2/\rho S_k^2\sigma_* dt$ is the amount of heat generated per unit mass of plasma, and $V = 1/\rho$ is the specific volume of the plasma. In deriving Eq. 4.25, the formula that relates the air–plasma conductivity to temperature and density was used:

$$\sigma_* = 5.4T^{0.73}\ln(4\cdot 10^{-25}n), \tag{4.26}$$

which was constructed from tabulated data for temperatures, $T$, from 20 kT to 100 kT and for densities from $\rho_p$ to $10\rho_p$, where $n$ is the total concentration of atoms and ions.

## 4.2 Pulsed Transformers

One method for electrically matching the parameters of the MCG to those of the load is to use a step–up transformer [4.11–4.14]. The load is connected to the secondary winding of the transformer, and the MCG is connected to the primary winding.

The use of pulsed transformers to match the output of the MCG to the input of various loads was first proposed by Gaaze and Shneerson [4.15] in the FSU in 1965. The proposed design is based on the pulsed cable transformer. In this arrangement, the basic component is a high-voltage coaxial cable wound in the form of a helix, as shown in Fig. 4.11a. At each turn of the helix, the conducting shell or braiding of the cable 1 is removed from a section of length 2. These cuts are positioned above each other over the length of the spiral coil. The ends of the cuts are connected to plates or cables, which are connected to the MCG. That is, the braiding of the coaxial cable is the primary winding of the transformer. The diameter of the core conductor is $d_c \geq 20d_{sh}$, where $d_{sh}$ is the thickness of the shell of the cable.

The load is connected to both ends of the cable core, which is the secondary winding of the transformer. This lowers the requirements placed on the insulation of the cable. Variations in the inductance of the secondary winding, which depends on the inductance of the cable, due to leakage may be decreased by winding the cables in parallel, which decreases its resistance and the magnitude of the current flowing through the cable. The displacement (1) is caused by the radial stretching of the turns, the displacement (2) of the wire sections of the cable without braid is caused by the difference in the fields within the current conductor and within the transformer, and the displacement (3) of the insulation of the cable is caused by the current flowing through the cable. When using cylindrically solenoidal-shaped cable transformers, the displacement at the ends must be taken into account. Both toroidal cable transformers (shown in Fig. 4.11b) and some of its variants have been constructed. In this case, the primary winding 1 consists of several sectors of cable braid connected to the current-bearing flanges of the MCG. The transformer coefficient is approximately equal to the number of sectors [4.16].

The cable transformer possesses several advantages over other types of transformers including

- variations in the value of the inductance decreases significantly due to the fact that it is determined only by the inductance of the cable,
- the insulation problem is automatically solved,
- the cable wire is for all practical purposes not affected by the action of electromagnetic forces.

A coupling coefficient of $k_c = 0.9$ can be achieved in cable transformers. This means that up to 35% of the energy of the MCG can be transmitted to the load.

If the operating time of the transformer must be proportional to that of the MCG, then the transformer must be mounted directly on the generator. This increases the coupling factor $k_c$ and reduces the requirements placed on the transmission line to the load. In the case of high-energy spiral generators, a two-turn cable transformer (1 and 2 in Fig. 4.11c) may be used to couple the energy from the MCG to the load. This type of transformer permits the load to be connected to the gap in the braiding at any place on the cable that is desired. This is why the length of the transformer may be increased to increase the inductance of the primary winding and to decrease the overall dimensions of the transformer, as well as partially use it as the energy transmission line to the load.

The cylindrical transformers shown in Fig. 4.12 are very convenient for connecting to the plates of the PMCG. The primary winding, which is a single-turn solenoid, is the load of the generator. The secondary winding of the transformer is usually made in the form of a spiral (Fig. 4.12a) or a tape (Fig. 4.12b). The secondary winding (Fig. 4.12a) is usually a copper band having a certain thickness that is wound in the form of a spiral around

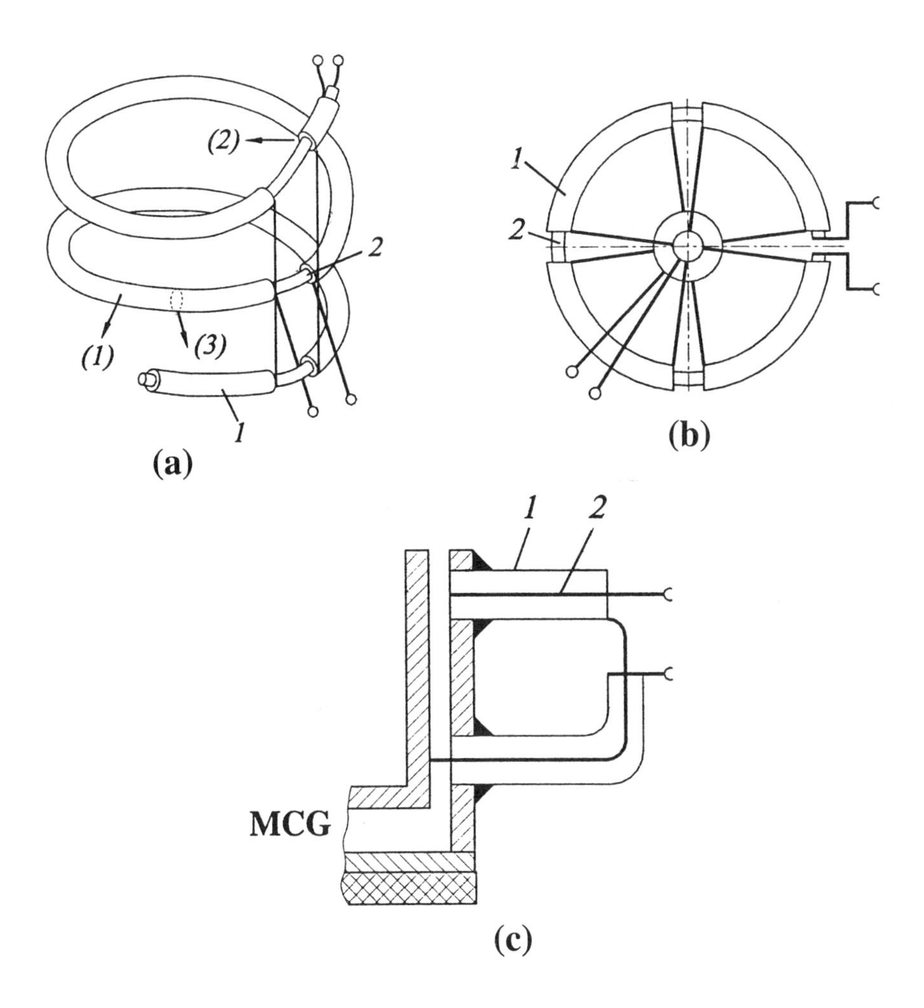


FIGURE 4.11. Diagrams of cable tranformers: 1 – conducting sheath (braid) and 2 – length of section removed. In figs. a and b, a length of conducting sheath is removed and connected by a cable. The load is connected to both ends of this cable. The primary winding is the braid, which is connected to the MCG, and the secondary winding is the center conductor of the coaxial cable. Figure c shows a two-turn cable that is mounted directly on the MCG. This increases the coupling factor, $k_c$.

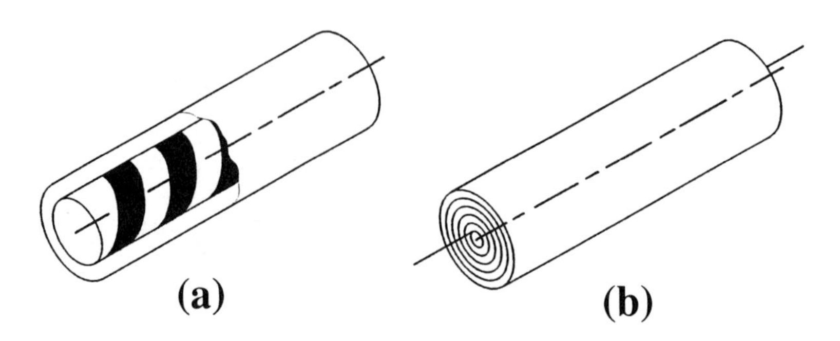


FIGURE 4.12. Diagrams of cylindrical transformers – (a) spiral wound and (b) tape wound.

a dielectric cylinder. Polyehtylene film is used as the insulator. The tape-wound transformer, shown in Fig. 4.12b, has several advantages over other types of transformers:

- the secondary winding is placed inside the primary winding,
- the output voltage is coupled into the center of the transformer,
- the change in the voltage inside the coil varies radially.

The basic problems associated with this type of transformer are breakdown between the turns and eddy currents induced in the components of the structure. However, the capacitance of roll transformers may be significant, owing to the large area and close positioning of the turns of the conducting foil. This leads to an independent capacitive equalizing of the voltage on the secondary winding of the transformer, and therefore to a lowering of the susceptibility of the transformer to damage due to unstable breakdown voltages.

The principal disadvantage of the foil and tape windings is their susceptibility to breakdown at the edges of the foil. Breakdown at the edges arises as the result of a sharp increase in the electric field that forms between the coils that are at high voltage and that bends sharply around the edges in the direction of the turns that are at low voltage. These high field strengths and their associated potentials may be significantly decreased by selecting boundaries that form coaxial equipotentials, which are almost parallel to the uniform field along the diameter of the winding.

A "leveling" technology, successfully applied by Martin and Smith to expendable transformers, consisted of filling the volume of the transformer with a resistive liquid, such as a weak solution of copper sulfate made with distilled water. This technology provides resistive leveling of the voltage,

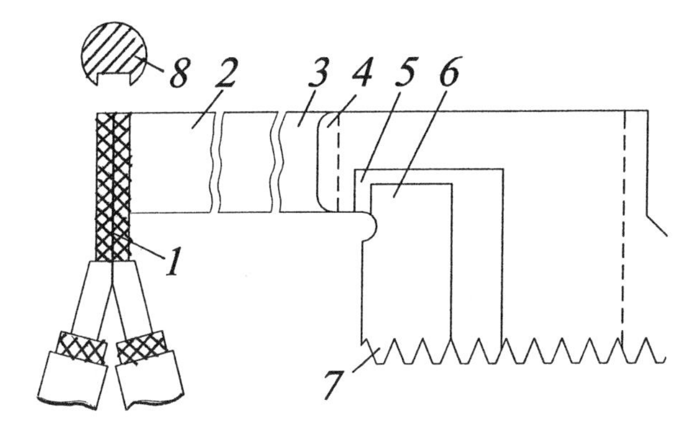


FIGURE 4.13. Diagram of winding for cable transformers: 1 – central wire of cable welded to origin of winding to couple current, 2,3 – unreeled foil roll, 4,5,6 – insulation, 7 – profiled edges of the lobes to which the cable braiding is welded, and 8 – is a top view of the welded cable. In order to prevent damage to the transformer when the output of the transformer is welded to the inner part of the secondary winding, several nodes may be welded to the the terminal as shown in this figure.

which may be applied to expendable transformers, but is not good for reusable transformers, due to changes in the properties of the liquid.

If the output of the transformer is welded to the inner part of its secondary winding, the transformer may be damaged. The problem is that the current density in this case is highest at the output end of the terminal (where it is almost zero at the input end of the terminal) and may exceed the critical current density, which would cause the foil at this point to explode and the transformer to be destroyed. This disadvantage may be eliminated by welding several lobes made of the same foil, but with decreasing area, to the end of the terminal, as shown in Fig. 4.13.

To calculate the inductance of the primary and of the secondary windings of a cylindrical transformer, the expressions used are

$$L_k = \frac{\mu_0 \pi d_k^2 k_L}{4\Delta_k} \tag{4.27}$$

and

$$L_T = \frac{\mu_0 \omega^2 \pi \overline{d}_T^2}{4\Delta_k}, \tag{4.28}$$

where $k_L$ is the trimming coefficient for the widths $\Delta_k$ and $\Delta_T$ and the diameters $d_k$ and $\overline{d}_T$. Neglecting surface effects, the average diameter $\overline{d}_T$

of the secondary winding of the transformer can be calculated using the equation

$$\overline{d}_T = \left[\frac{3d_i^2 + 2d_i d_O + d_O^2}{6}\right]^{\frac{1}{2}}, \tag{4.29}$$

where $d_i$ is the inside diameter and $d_0$ is the outside diameter of the secondary winding. The coupling coefficient of the transformer windings is calculated from

$$k_c = \frac{\overline{d}_T}{d_k}\left(\frac{\Delta_T}{\Delta_k}\right)^{\frac{1}{2}}, \tag{4.30}$$

or by using

$$k_c = \sqrt{\frac{L_T - L_T^*}{L_T}}, \tag{4.31}$$

where $L_T^*$ is the inductance of the secondary winding, when the primary winding is short-circuited.

The advantage of the tape-wound cylindrical transformer resides in the small gap between the coils, which increases the coupling factor. When the transformers operate at high currents, the linear current density must be conserved and the height of the tape must be selected proportionally to the total current. Finally, the design of the secondary winding must fit the design of the load (a cylindrical coil) of the PMCG. Only the secondary winding is inserted into the load turn of the generator.

To determine that part of the energy produced by the MCG that is transmitted into the load by a pulse transformer, consider the system depicted in Fig. 4.14. Here, the MCG is connected through a transformer to an inductive load, where $L_g(t)$ is the inductance of the MCG, $L_B$ is the inductance of the component that transitions the energy from the MCG into the transformer, $L_k$ is the inductance of the load of the MCG, which in this case is the primary winding of the cylindrical transformer, $L_L$ is the inductance of the load, and $M = k_c\sqrt{L_k L_T}$ is the mutual inductance. Solving the equations associated with the circuit shown in Fig. 4.14 for the initial conditions, $I_1(0) = I_{10}$, $L_g(0) = L_0$, $I_2(0) = I_{20}$, and $L_g(t_g) = 0$, and assuming that the generator begins operating at $t = 0$, an expression that describes how the MCG operates when it is connected to a load having an equivalent inductance without using a transformer can be derived:

$$\left[(L_g + L_B + L_1) - \frac{k_c^2 L_1 L_2}{L_2 + L_L}\right]\frac{dI_1}{dt} + \frac{dI_g}{dt}I_1 = 0, \tag{4.32}$$

where the equivalent inductance, $L_E$, is

$$L_E = L_B + L_1 - \frac{k_c^2 L_1 L_2}{L_2 + L_L} = L_2 + L_1 \left( \frac{1 + \alpha_L - k_c^2}{1 + \alpha_L} \right) \tag{4.33}$$

and where $\alpha_L = L_L / L_2$. It can be seen that the operation of an MCG connected to an inductive load through a transformer depends on $L_B$, $k_c^2$, and $L_L/L_2$. Therefore, connecting the primary winding of the transformer to the generator increases the efficiency of the MCG. To minimize the change in the diameter of the primary winding due to its expansion, it must be manufactured in the form of a massive solenoid. As to $k_c$ and the ratio of the inductances, the strongest effects are due to the coupling factor, which is a second-order effect. The equivalent inductance may change, depending on the ratio $L_L/L_2$, from $L_B + L_1 \left(1 - k_c^2\right)$ to $L_B + L_1$. If the coupling factor is close to 1 (i.e., an ideal transformer), then $L_E$ may decrease significantly due to the increase in $L_2$, provided that $L_B$ is also very small, and, therefore, increase the trimming factor of the generator and improve its operating parameters. The *transformer coefficient*, $\eta_{tr}$, may be derived without taking into account $L_B$, by using the ratio of the currents in the primary and secondary windings:

$$\begin{aligned} \eta_{tr} &= \frac{L_L}{L_E} \left( \frac{I_2}{I_1} \right)^2 \\ &= \frac{L_L (1 + \alpha_L)}{(1 + \alpha_L - k_c^2)} \left( \frac{k_c \sqrt{L_1 L_2}}{L_2 + L_L} \right) \\ &= \frac{\alpha_L k_c^2}{(1 + \alpha_L - k_c^2)(1 + \alpha_L)}. \end{aligned} \tag{4.34}$$

The magnitude of $\eta_{tr}$ attains its peak value when $k_c^2 = 1 - \alpha_L^2$. Taking into account $L_B$ leads to a more complex form of the expression for $\eta_{tr}$:

$$\eta_{tr} = \frac{\alpha_L k_c^2}{\left( \frac{L_B}{L_1} + 1 \right) (1 + \alpha_L)^2 - k_c^2 (1 + \alpha_L)}. \tag{4.35}$$

To facilitate use of this equation, Table 4.1 lists calculated values of $\eta_{tr}$ for different values of $k_C$, $\alpha_L$, and $\beta_L$, where $\beta_L = L_B / L_1$. It is clear from the table, that large amounts of energy, $\sim 90\%$, may be coupled into the load when the coupling factor is $\sim 0.999$ and $L_B = 0$.

| $a_L$ | 0.7 | 0.8 | 0.9 | 0.95 | 0.99 |
|---|---|---|---|---|---|
| | | | $\beta_L = 0$ | | |
| $k_c$ | | | | | |
| 0.02 | 0.02 | 0.03 | 0.08 | 0.15 | 0.89 |
| 0.06 | 0.04 | 0.07 | 0.16 | 0.29 | 0.91 |
| 0.10 | 0.07 | 0.13 | 0.25 | 0.42 | 0.89 |
| 0.50 | 0.16 | 0.25 | 0.39 | 0.52 | 0.66 |
| 1.00 | 0.16 | 0.24 | 0.34 | 0.41 | 0.50 |
| 2.00 | 0.13 | 0.18 | 0.25 | 0.29 | 0.33 |
| 5.00 | 0.07 | 0.10 | 0.13 | 0.15 | 0.17 |
| 10.0 | 0.04 | 0.06 | 0.07 | 0.08 | 0.09 |
| | | | $\beta_L = 0.5$ | | |
| 0.02 | 0.01 | 0.01 | 0.02 | 0.03 | 0.04 |
| 0.06 | 0.02 | 0.03 | 0.05 | 0.06 | 0.09 |
| 0.10 | 0.04 | 0.06 | 0,09 | 0.11 | 0.14 |
| 0.50 | 0.13 | 0.19 | 0.23 | 0.26 | 0.06 |
| 1.00 | 0.10 | 0.15 | 0.18 | 0.21 | 0.25 |
| 2.00 | 0.08 | 0.11 | 0.15 | 0.17 | 0.19 |
| 5.00 | 0.04 | 0.06 | 0.08 | 0.09 | 0.11 |
| 10.0 | 0.02 | 0.04 | 0.05 | 0.06 | 0.07 |
| | | | $\beta_L = 1.0$ | | |
| 0.02 | 0.01 | 0.01 | 0.01 | 0.02 | 0.95 |
| 0.06 | 0.01 | 0.02 | 0.03 | 0.04 | 0.04 |
| 0.10 | 0.03 | 0.04 | 0.05 | 0.06 | 0.08 |
| 0.50 | 0.09 | 0.12 | 0.14 | 0.16 | 0.17 |
| 1.00 | 0.07 | 0.10 | 0.13 | 0.15 | 0.17 |
| 2.00 | 0.06 | 0.08 | 0.10 | 0.12 | 0.13 |
| 5.00 | 0.04 | 0.05 | 0.07 | 0.08 | 0.08 |
| 10.0 | 0.03 | 0.04 | 0.04 | 0.04 | 0.04 |

TABLE 4.1. Energy coupling coefficients for MCGs connected through transformers to inductive loads.

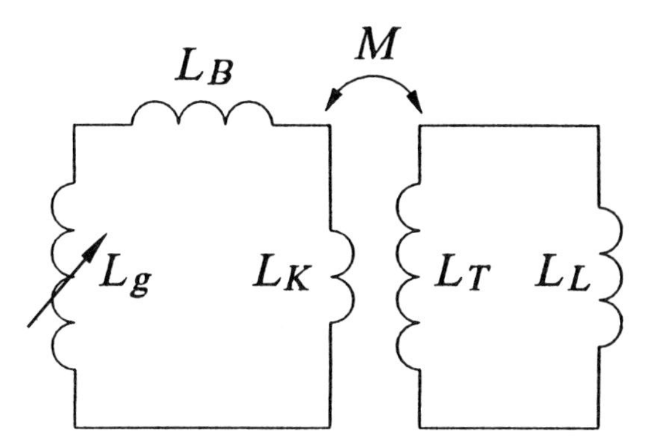


FIGURE 4.14. Equivalent circuit diagram of an MCG connected through a transformer to an active load.

If the energy is coupled into an active load through a transformer, the expression used to calculate $\eta_{tr}$ is

$$\begin{aligned}\eta_{tr} &= k_c^2 \frac{2 - \frac{k_c^2}{1-\frac{L_B}{L_1}} - \exp\left(-\frac{2T}{\alpha_R}\right)}{4\left(1+\frac{L_B}{L_1}\right) - 2k_c^2\left(1+0.5\exp\left(-\frac{2T}{\alpha_R}\right)\right)} \quad (4.36)\\ &\approx \frac{k_c^2}{2\left(1+\frac{L_B}{L_1}\right)} \approx \frac{k_c^2}{2\left(1+\beta_L\right)},\end{aligned}$$

where $T$ is the operating time of the generator and $\alpha = L_g/R_L$. It follows from Eq. 4.36 that, for this type of coupling, it is impossible to couple more than half of the energy from the MCG into the load, even when $k_c = 1$.

Numerical solutions of the equations that describe the operation of an MCG connected through a transformer to a load show that the profile of the current pulse of the MCG remains the same, even when the parameters of the load change. As $R_L$ and $L_L$ increase, there is a minimal decrease in the value of the current.

When selecting the proper thickness of the copper foil used in the cylindrical spiral transformer, its operating mode must be taken into account. The dimensions of the section of current carrying copper foil must be selected in such a way that the current flowing through it does not exceed the total current that would melt the foil.

At present, pulsed transformers that are used to electrically match MCGs to active loads of $R_L = 20\ \Omega$ at voltages of $\sim 1$ MV can couple up to 30 MJ into the load.

## 4.3 Spark Gap Switches

*Closing switches* [4.3,4.17,4.18] in very-high-power systems are designed to hold off the voltage during its build-up and to close and allow the current to flow into the load at a specified voltage level or time. The voltage stand-off of high-power closing switches are based on the voltage breakdown characteristics of some type of medium (solid, liquid, gas) or electron surface emission characteristics of separated electrodes at low pressures. Some examples of the former type of switch are gas discharge switches, liquid dielectric switches, and solid dielectric switches, and some examples of the latter type are vacuum spark gap switches and vacuum surface discharge switches.

To further sharpen the risetime of the output pulse of the MCG, both explosive opening switches and spark gap closing switches (gas discharge switches) are added to the MCG-load circuit. The basic requirements placed on these switches are that they must have

- low inductance,
- low Ohmic losses,
- jitter times that do not exceed several nanoseconds.

Sharpening switches are divided into two categories: controlled and uncontrolled. The uncontrolled switches are filled with high pressure-gas, oil, or water, which allows the switch to generate very short risetime pulses. The controlled switches are usually filled with $SF_6$.

The breakdown processes that occur in spark gap switches are very complex. The basic process is the formation of positive and negative "streamers," i.e., moving domains of spatial charge. The evolution of the breakdown process resembles the propagation of a combustion front. In spark gaps with a nonuniform field distribution, such as that associated with the edge of a plane or a blade, the average velocity of a streamer in oil at voltages less than 1 MV can be determined from the empirical expression

$$v_s = \frac{d_e}{t} \approx kU^n \text{ cm}/\mu s, \tag{4.37}$$

where $U$ is the voltage across the switch in megavolts, $t$ is time in $\mu$s, $n$ and $k$ are material constants, and $d_e$ is the distance between the electrodes in centimeters. For a positive polarity, $k_+ = 90$ and $n_+ = 1.75$, and for a negative polarity, $k_- = 31$ and $n_- = 1.28$. In the voltage range of 1–5 MV, Eq. 4.37 for both polarities becomes

$$v_s = 80U^{1.6}d_e^{-0.25}. \tag{4.38}$$

| Electrode Shape | $\frac{d_{ef}}{d_e}$ | $\mathbf{F}_p$ |
|---|---|---|
| Cylindrical | 0.115 | 1.3 |
| Spherical | 0.057 | 1.8 |

TABLE 4.2. Parameters of a spark gap filled with air or nitrogen.

When the switches are filled with water, the streamer velocities are determined from the empirical relations

$$v_s t^{\frac{1}{2}} = 8.8U^{0.6} \text{ (positive polarity)}, \tag{4.39}$$

$$v_s t^{\frac{1}{3}} = 16U^{1.1} \text{ (negative polarity)}. \tag{4.40}$$

The four gases most widely used in uncontrolled gas discharge switches are air, nitrogen, Freon, and sulfur hexafluoride. The uniform electric field strength required to cause breakdown in air or nitrogen is

$$E_{br} = \left[24.6p + 6.7\left(\frac{p}{d_{ef}}\right)^{1.2}\right] F_p^{-1}, \tag{4.41}$$

where $p$ is the gas pressure in atmospheres, $d_{ef}$ is the effective distance between the electrodes, and $F_p$ is the field gain; i.e., the ratio of the maximum field on the electrodes to the average field in the gap between the electrodes. Table 4.2 gives the values of $d_{ef}$ and $F_p$ for a discharge gap, when the distance between the electrodes is equal to the diameter of the electrodes, $d_e$.

In practice, breakdown in $SF_6$ and Freon occurs at field strengths that are 2.5–5.0 times the values given by Eq. 4.28. However, it is better to use $SF_6$, since carbon is not formed during the discharge, as it is when Freon is used as the working gas.

In gas discharge switches with a strong nonuniform field distribution, such as between a point and a plane or between the edge of a blade and a plane, the breakdown field distribution depends on both the pulse length and the pulse polarity. The breakdown field is given approximately by

$$E_{br} = k_{\pm} p^n (d_e t)^{\frac{1}{6}} \text{ kV/cm}, \tag{4.42}$$

where $E_{br} = U/d_e$ is the average value of the breakdown field, $d_e$ is the distance between the electrodes, $t$ is time in microseconds, and $p$ is pressure in atmospheres. The values of $k_{\pm}$ and $n$ are presented in Table 4.3.

| Gas | $k_+$ | $k_-$ | $n$ |
|---|---|---|---|
| air | 22 | 24 | 0.4 |
| $SF_6$ | 44 | 72 | 0.4 |

TABLE 4.3. Values of constants associated with the electrical breakdown of air and sulfure hexaflouride in gas discharge switches with strong nonuniform field distributions.

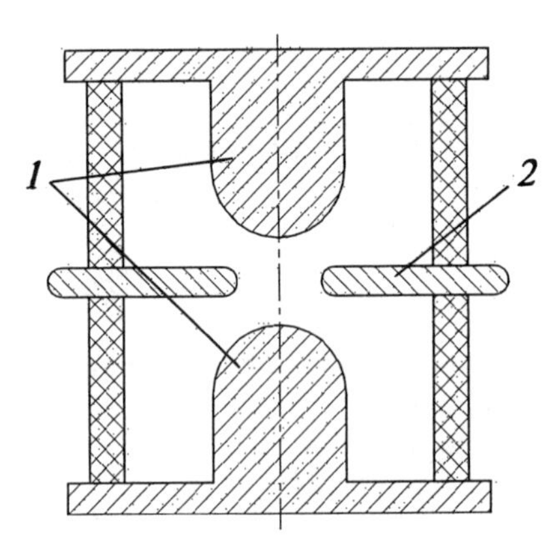


FIGURE 4.15. Schematic diagram of the side view of a field-distortion discharge switch. The control electrode (2) is placed between the spherical electrodes (1) and is shaped in the form of a disk with a central opening. When a driving pulse is applied to the electrodes, the field distribution is distorted and breakdown occurs in one-half of the switch. As a result, the full voltage is applied to the second half and self-breakdown occurs.

Eq. 4.42 is valid over the pressure range of 1–10 atm and for electrode separation distances no greater than 10 cm. At greater pressures, additional nonlinearities must be accounted for.

The most popular of the controlled gas discharge switches are the trigatron and the field-distortion switches that have a controlling central electrode. Figure 4.15 shows a schematic drawing of the side view of a field distortion discharge switch. The controlling electrode (2) is placed in between the spherical (or cylindrical) electrodes (1) and is shaped in the form of a disk with a central opening (ring). When a driving pulse is applied to this electrode, the field distribution is distorted and breakdown occurs in one-half of the discharge switch. As a result, the full voltage is applied to the second half and self-breakdown occurs there.

A schematic of the side of the trigatron is shown in Fig. 4.16. When a driving pulse is applied to the controlled needle-like electrode (2), an arc discharge forms between it and its neighboring electrode (1). This discharge

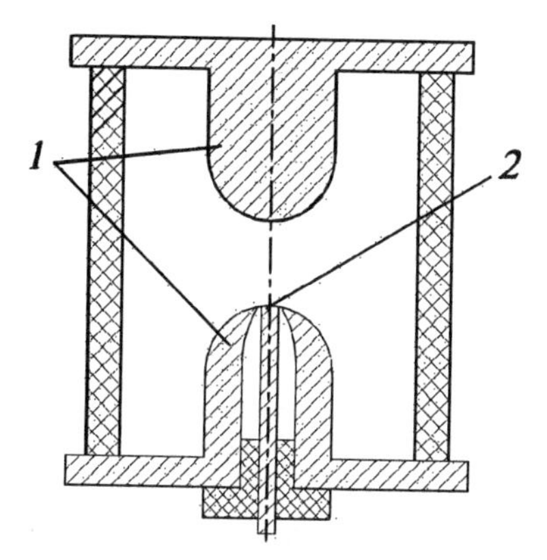


FIGURE 4.16. Schematic diagram of the longitudinal section of a trigatron with the controlling electrode embedded (2) within one of the electrodes (1). When a driving pulse is applied to the contolling needle-like electrode, an arc discharge forms between this and the neighboring electrode. This discharge distorts the field distribution near the electrode and generates electromagnetic radiation that facilitates the formation of streamers and operation of the switch.

distorts the field distribution near the electrode and, in addition to exciting the electrodes, generates photons that facilitate the formation of streamers and the closing of the discharge switch.

In addition to the controlled discharge switches discussed above, there are several other discharge switches that deserve attention. These include ultraviolet (UV), laser, soft X-ray, and electron-beam-triggered switches. In these switches, smaller amounts of energy are required to discharge the switch and the switch jitter decreases.

The operation of a discharge switch must be characterized not only by its breakdown parameters, but also by its pulse risetime, which depends on the inductance of the switch and the resistance of the heated discharge channel. The relative contribution of each of these factors is determined by the specific conditions under which the switch is operating. The time required for the current to increase $e$-fold, where $e = 2.718...$, in a switch with an inductance of $L_p$ connected to a circuit with an impedance of $Z$ (or similarly the $e$-fold voltage drop over the switch) is equal to

$$\tau_p = \frac{L_p}{Z}. \tag{4.43}$$

To calculate the channel inductance, $L_p$, the precise radius of the channel, $r_k$, must be known, although it is usually sufficient to approximate $L_p$ by

$$L_p \approx 2d_e \ln\left(\frac{r_e}{r_k}\right) \approx 14d_e \quad \text{nH}, \tag{4.44}$$

where $d_e$ is the length of the channel. Since $r_e \gg r_k$, the inductance $L_p$ has a relatively weak dependence on $r_k$ and for risetimes of several nanoseconds, the logarithm is usually assumed to have a value of 7.

A second factor that affects the pulse risetime of the switch is the dissipation of energy in the channel, which depends on the resistance of the channel. However, the resistance of the channel decreases along with its heating and expansion. The e-fold current rise time in gases, which is due to the resistance of the discharge channel, satisfies the empirical relationship

$$\tau_R = \frac{88}{Z^{\frac{1}{3}} E_{br}^{\frac{4}{3}}} \left(\frac{\rho}{\rho_0}\right)^{\frac{1}{2}} \quad \text{ns}, \tag{4.45}$$

where $\rho/\rho_0$ is the ratio of the density of the gas to the density of air under standard conditions, $Z$ is the line impedance, and $E_{br}$ is the breakdown field strength along the length of the channel. In liquids and solids, the respective risetimes can be obtained from the formula

$$\tau_R = \frac{5}{Z^{\frac{1}{3}} E_{br}^{\frac{4}{3}}} \quad \text{ns}, \tag{4.46}$$

where $E_{br}$ is measured in MV/m.

The total pulse risetime is the sum of the times described by Eqs. 4.43 and 4.45, i.e.,

$$\tau_{tot} = \tau_P + \tau_R. \tag{4.47}$$

This expression gives the e-fold risetime for the cases where the pulse increases according to an exponential law. If the rate at which the pulse rises depends on a non–exponential law, then $\tau_{tot}$ equals the ratio of the maximum voltage to the maximum time derivative of the voltage at the output:

$$\tau_{tot} \approx \frac{U_{\max}}{(\dot{U}_t)_{\max}}. \tag{4.48}$$

It follows from Eqs. 4.43, 4.45, and 4.46, that the pulse risetime in a low impedance load may significantly restrict the useful length of the pulse.

This is why a more efficient approach is proposed, based on the fact that in order to generate a pulse with a short risetime, a high-impedance pulse-forming line must be switched first to reduce the impedance to a desirable level.

Another method for decreasing the pulse risetime is based on the use of several discharge channels, instead of a single channel. If there are $n$ channels conducting current, $Z$ must be replaced by $nZ$ in Eqs. 4.43, 4.45, and 4.46. In this case, the risetime decreases by a factor of $n^{-1}$, owing to the inductance, and $n^{-1/3}$, owing to the resistance. The number of discharge channels that are closed may be predicted with high accuracy from

$$2\sigma_U \left( \frac{U}{\frac{dU}{dt}} \right) \leq 0.1\,(\tau_R + \tau_L) + 0.8T_s, \tag{4.49}$$

where $\sigma_U$ is the normalized standard deviation in the breakdown time for rapidly rising pulses. The left-hand side of Eq. 4.49 appears to be the standard deviation for the formation time of a streamer in the channel, which is accounted for by the presence of the multiplier $U/(dU/dt)$. In actuality, this is the time interval during which useful current conducting channels are formed. The time $T_s$ appears to be the time for the electromagnetic waves to propagate between the channels and is proportional to $n^{-1}$. Therefore, the right-hand side of the inequality, Eq. 4.49, corresponds to the time for the voltage in the channel to decrease from its initial value by approximately 10% and the time it takes for the information about this decrease to reach neighboring channels.

It can be seen from Eq. 4.49 that the number of current-conducting channels may be increased by one of two methods. The first is to increase $dU/dt$, and the second is to decrease $\sigma_U$. In reality, $\sigma_U$ increases along with an increase in $dU/dt$, so that both of these methods can be used to improve the operation of the switch.

Using discharge switches, experimenters have been able to switch pulses with a leading edge risetime of 30 ns from the generator to the load.

## 4.4 Pulse-Forming Lines

As shown above, to efficiently couple energy from the generator into the load, the output pulse of the MCG must be shortened. The desired effect may also be achieved by employing *pulse-forming lines*. These may take on different geometries including planar, coaxial, and radial. The two types that are normally used are single and dual pulse forming lines.

Figure 4.17 presents the schematic drawing of a single coaxial pulse-forming line. If the electrical length of the pulse-forming line is small in comparison to the time it takes for the MCG to charge the line, then the

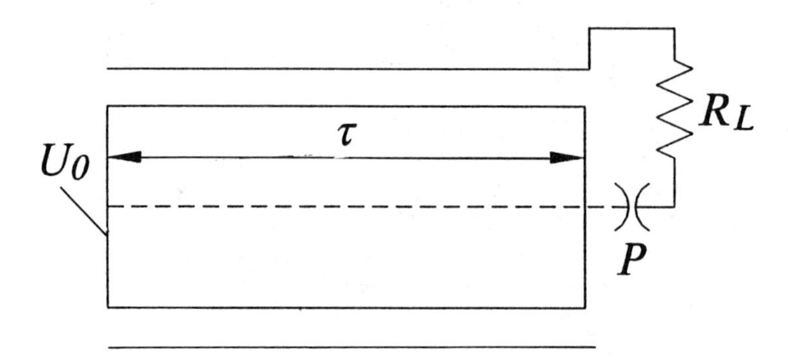


FIGURE 4.17. Diagram of a single coaxial pulse forming line. If the electrical length of this pulse forming line is short compared to the time it takes for the MCG to charge the line, then the pulse-forming line operates like a lumped capacitor shown in Fig. 4.18.

pulse-forming line operates like the lumped capacitor shown in the simple equivalent circuit in Fig. 4.18. The pulse-forming line can be charged up to the maximum voltage of $U_{\max}$ as the current in the circuit approaches zero. It can easily be shown that the value of $U_{\max}$ depends on the voltage $U_0$ at the output of the MCG:

$$U_{\max} = \frac{2U_0 C_g}{C_g + C_L}, \tag{4.50}$$

where $C_g$ is the shock capacitance of the generator and $C_L$ is the capacitance of the pulse forming line. The multiplier $2C_g/(C_g + C_L)$ is called the *voltage increase coefficient*, when resonantly charging a pulse-forming line in the absence of losses.

When charging a pulse-forming line, it can be assumed that it is an electrically long line with a wave impedance of $Z_0 = (L_1/C_1)^{1/2}$. The basic operating features of a single pulse-forming line can be easily understood by studying Fig. 4.19. An opened charged pulse-forming line with an electrical length of $\tau = (LC)^{1/2}$ may be described by the superposition of the forward and backward waves shown in Fig. 4.19a. To satisfy the boundary conditions of an opened circuit, the polarity of the voltage for each wave must be the same and the polarity of the current must change when the wave is reflected from the open ends of the pulse-forming line. This is why the total current in the transmission line is equal to zero and the voltage is a constant equal to $U_{\max}$. When the switch P in Fig. 4.19b is closed, the load, with impedance $Z_L$, is connected to the output of the pulse-forming

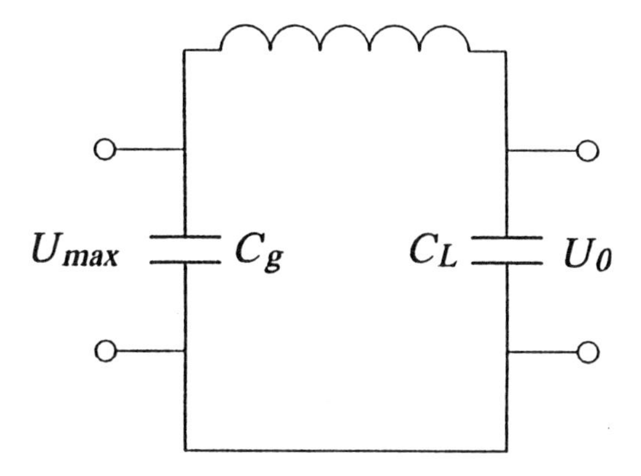


FIGURE 4.18. Equivalent circuit diagram of a single coaxial pulse-forming line.

line and the energy is transmitted from the transmission line in the form of an electromagnetic wave moving in the positive direction. If $Z_L = Z_0$ (a matched load), then the wave is not reflected at the output and all the energy stored in the pulse-forming line is coupled in time $2\tau$. Since the current flowing through the load resistance equals $I_{\max}/2$, the voltage on the load is equal to $U_{\max}/2$, i.e., half the value of the charging voltage.

Unlike the single pulse-forming line, the dual line forms a voltage pulse in a matched load with an amplitude equal to the charging voltage. The schematic of a cylindrical dual pulse-forming line is presented in Fig. 4.20. It consists of three coaxial cylindrical conductors with the central conductor being charged by the MCG. The central conductor is connected to an externally grounded conductor by means of a charging inductance coil. Under ideal conditions, the inductance coil operates like a short circuit during the charging cycle and an open circuit during the formation of the short output pulse. The operating principles of the dual pulse-forming line can be understood by examining Fig. 4.21. During the slow charging by the MCG, the inductive coil shorts the central and external conductors, so that the voltage difference between these conductors is equal to zero. When the switch connects the central and intermediate conductors, the polarity of the voltage wave propagating to the right reverses. After a time $\tau$, the total voltage at the opened output is equal to $2U_{\max}$. This voltage is conserved for a period of $2\tau$. It should be noted that the inductive coil remains in the diagram because it represents the high-impedance load for the short pulse. If a matched load with impedance $Z_L$ is connected to the output of the dual pulse-forming line after time $\tau$, then the voltage on the output decreases to $U_{\max}$; however, all the energy stored in the pulse-forming line is coupled during time $2\tau$.

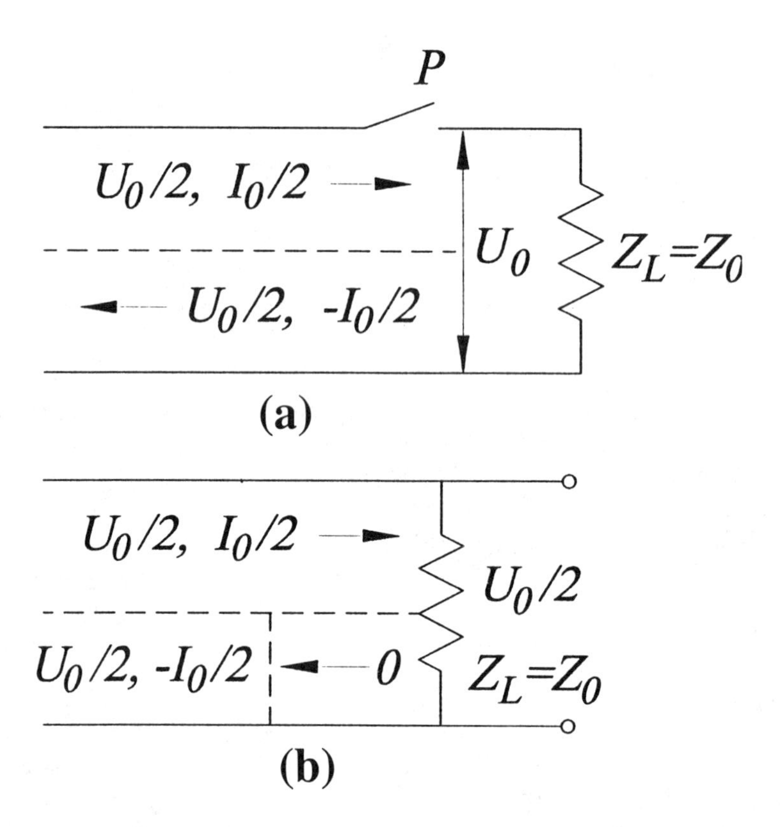


FIGURE 4.19. Single coaxial pulse-forming lines. The opened charged pulse-forming line in Fig. a can be described by the superposition of the forward and backward waves shown. To satisfy the boundary conditons, the polarity of the voltage for each wave must be the same and the polarity of the current must change, when the wave is reflected from the open ends. In this case, the total current in the transmission line is zero and the voltage is a constant equal to its peak value. When the swtich, P, is closed in Fig. b, the load is connected to the output of the pulse-forming line and energy is transmitted from the transmission line to the load.

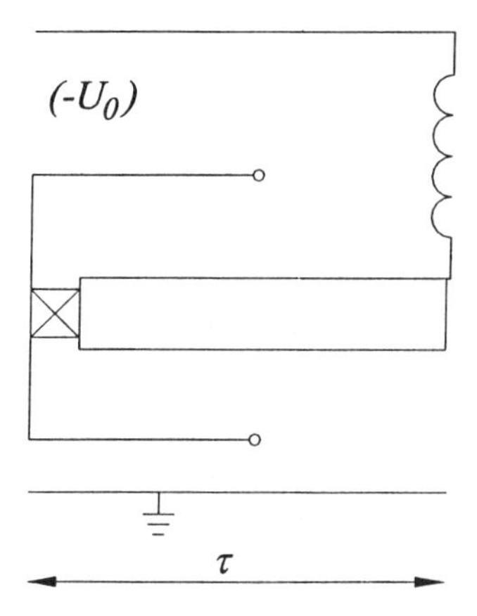


FIGURE 4.20. Schematic diagram of dual pulse-forming line. It consists of a 1 – output conductor, 2 – grounded conductor, 3 – central conductor, 4 – MCG, and 5 – charging inductance coil. Unlike the single pulse-forming line, the dual line forms a voltage pulse in a matched load with an amplitude equal to the charging voltage. This pulse-forming line consists of three coaxial cylindrial conductors, with the central conductor being charged by the MCG.

The electrical parameters of the pulse-forming line depend on its geometry and the properties of the dielectric used. For example, in the case of the single coaxial pulse-forming line, the line impedance is

$$Z = \frac{60}{\sqrt{\varepsilon}} \ln \frac{r_1}{r_2} \tag{4.51}$$

and the pulse length $2\tau$ is

$$\tau_p = \frac{2l\sqrt{\varepsilon}}{c}, \tag{4.52}$$

where $\varepsilon$ is the dielectric permittivity, $c$ is the speed of light, $l$ is the length of the coaxial pulse-forming line, $r_1$ is the radius of the internal conductor, and $r_2$ that of the external conductor. The radii of the internal and external conductors depend on the desired impedance of the pulse-forming line, the charging voltage, and restrictions related to electrical breakdown. The two most widely used dielectrics are water ($\varepsilon = 80$) and transformer oil ($\varepsilon = 2.4$). The electric field strength required to cause breakdown in these two liquids may be approximated by

$$E_{br} = \frac{k}{t_{ef}^{\frac{1}{3}} A^{\frac{1}{10}}}, \tag{4.53}$$

| Dielectric | $\mathbf{k}_+$ | $\mathbf{k}_-$ |
|---|---|---|
| oil | 0.5 | 0.5 |
| water | 0.3 | 0.6 |

TABLE 4.4. Electric breakdown constants for water and transformer oil.

where $A$ is the surface area in $cm^2$ of the electrode over which the electric field does not change by more than 10% of its maximum value and $t_{ef}$ is the time in microseconds it takes for the electric field strength to change from $0.63E_{br}$ to $E_{br}$. The values of the constant $k$ are given in Table 4.4. Note that in the case of water, the value of $k$ depends on the polarity of the voltage [4.19]. For example, if it is assumed that the conductors have a surface area of $\sim$ 1000 $cm^2$ and that the operating time is in the microsecond regime, then the electric field strength required to cause breakdown is about 15 MV/m for water (+) and 25 MV/m for transformer oil. For even shorter pulses (tens of nanoseconds), the breakdown voltage calculated by using Eq. 4.53 appears to be two times too high.

By investigating the characteristics of single and dual pulse-forming lines filled with water or transformer oil, the range of maximum allowed voltages and load impedances can be determined from Fig. 4.22. The dual pulse-forming line has advantages at higher voltages, because this pulse-forming line requires lower charging voltages. On the other hand, the single pulse-forming lines are more compact at lower voltages. When powering low-impedance loads, it is best to use water as the dielectric owing to its higher dielectric permittivity. When powering high-impedance loads, it is best to use oil as the dielectric to get better impedance matching.

## 4.5 High-Voltage MCG Systems

In order to demonstrate the role that ancillary equipment play in developing MCG driven systems, a couple of specific examples will be examined. In particular, certain applications require high voltages, which means that the relatively low voltages of MCGs must be transformed to high values. In this section, two methods are examined: flux trapping and transformers.

Magnetocumulative generators (MCG) have been used to power vacuum diodes in particle accelerators and high-power microwave sources. In order to deliver high powers to the load, the MCG must undergo a rapid change in inductance, $dL/dt$, and must also generate high voltages ($10^5$–$10^6$ V). At the present time, the highest voltages that can be achieved across loads is through the use of transformers. Two transformer experiments were carried out at Los Alamos National Laboratory (LANL). In the first, a voltage of 1.2 MV with a pulse risetime of 300 ps was achieved across a vacuum diode load impedance matched through a transformer to a loop MCG (LMCG)

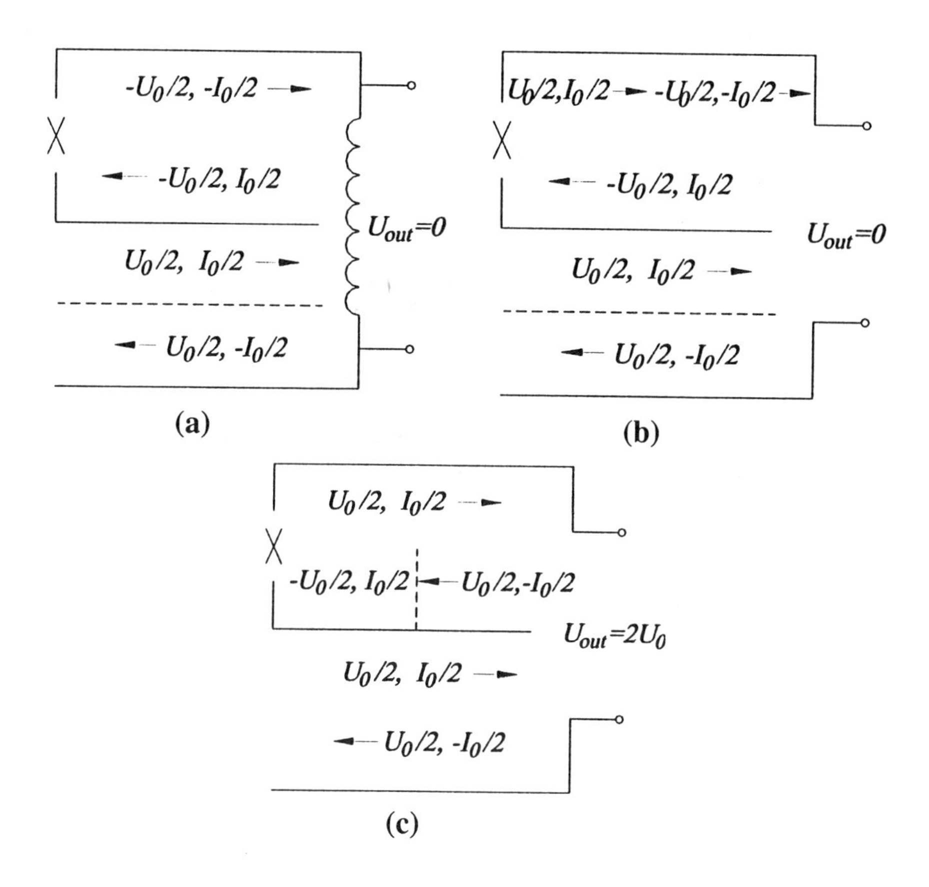


FIGURE 4.21. Principles of operation of dual pulse-forming lines. During the slow charging by the MCG in Fig. a, the inductive coil shorts the central and external conductors, so that the voltage difference between them is zero. When the switch connects the central and intermediate conductors in Fig. b, the polarity of the voltage wave propagating to the right is reversed and the amplitude of the pulse is doubled. If a matched load, as shown in the Fig. c, is connected to the output of the transmission line, the amplitude of the voltage pulse decreases to its original value. However, all the energy stored in the pulse-forming line is coupled to the load.

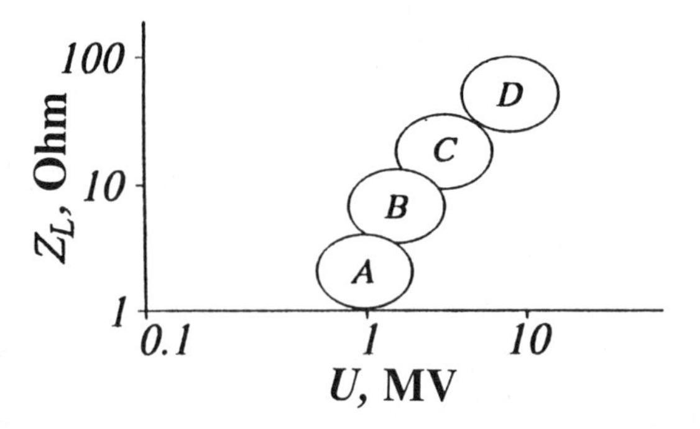


FIGURE 4.22. Domains of maximum allowed voltages and impedances in single and dual pulse-forming lines when filled with water or transformer oil: A – single pulse-forming line (water), B – dual pulse-forming line (water), C – single pulse-forming line (oil), and D – dual pulse-forming line (oil).

operating at a voltage of 35 kV [4.20]. In the second, a high speed 40 kV plate MCG (PMCG) connected through a switch and transformer generated a voltage of 1.1 MV across a 25 Ω load. When this latter system was connected to a vacuum diode, a voltage of 530 kV was generated across the diode [4.21].

In both these experiments, fast MCGs with characteristic operating times of 10 μs were used. As noted by Sheindlin and Fortov [4.22], the inductance of these generators was low ($L = 100$ nH), which means that their internal impedance was also low ($10^{-2}$ Ω). On the other hand, spiral MCGs have higher inductances (> 100 μH) and slower speeds (100 μs). The operating time of these generators are limited by the speed of the sliding electrical contact point between the liner and the coil. It is possible to increase the speed of the contact point by either increasing the detonation velocity of the explosive or by decreasing the angle at which the liner collides with the coil. Since there is an upper bound to the detonation velocities of high explosives, in practice, only the collision angle can be changed. However, the ability to reduce this angle is limited by liner and coil manufacturing accuracies and electrical breakdown. Imperfections in the construction of the liner and/or coil can lead to the contact point jumping portions of the coil ($2\pi$-clocking), which reduces the flux in the working volume. The loss of magnetic flux can be so large that electrical energy is no longer generated. Rapid expulsion of magnetic flux from the narrow gaps formed when the liner and coil collide at low angles can lead to breakdown. To avoid this, the turns of the coil must be carefully insulated by a solid dielectric with very high electrical strength. The insulation must have a minimum thickness to avoid significant flux losses in the insulator itself and to avoid reducing the cross-sectional area of the conductors in the coil. In actuality the ratio of the speed of the contact point to the speed of the liner can not be increased by more than 20–50 [4.23].

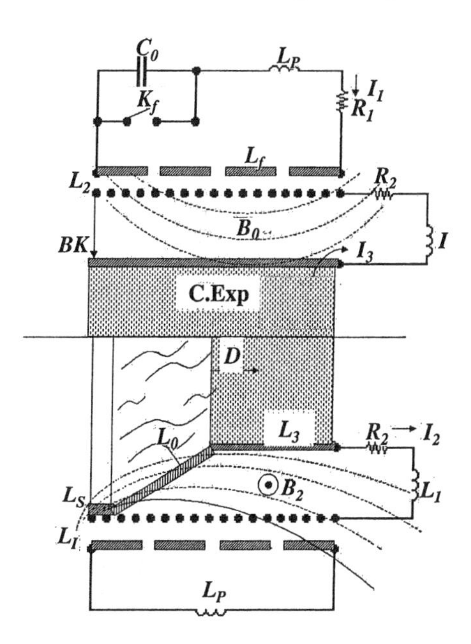


FIGURE 4.23. The top figure shows a simultaneously initiated helical MCG and the bottom figure an end-initiated helical MCG with a moving contact point. Both generators use flux trapping.

According to [4.22], helical generators cannot achieve voltages that exceed 60–70 kV. However, 1 MV generator designs have been proposed using highly evacuated working volumes and *magnetic insulation* of basic components. Similar problems must be solved when designing generators that utilize *magnetic flux trapping.*

## 4.5.1 Magnetic Flux Trapping

Multi–stage MCG systems (cascaded MCGs) that use magnetic flux trapping instead of transformers to connect the stages are used to generate and sharpen high-voltage electrical pulses. Magnetic flux trapping occurs when the magnetic energy generated in one circuit is transferred into another circuit and amplified. Both theoretical and experimental studies [4.24–4.29] show that flux trapping has advantages over systems that use conventional transformers, the most important of which is weight reduction. Cascaded systems have achieved magnetic flux amplifications up to 310 and energy increases by a factor of $10^6$ [4.27].

Two MCG models that use magnetic flux trapping were studied by Sheindlin and Fortov [4.22]. The first is a simultaneous helical generator (top half of Fig.4.23) and the second an end-initiated helical generator with a moving contact point (bottom half of Fig.4.23). An MCG with magnetic flux trapping consists of a coaxially located external feed solenoid, $L_1$, in-

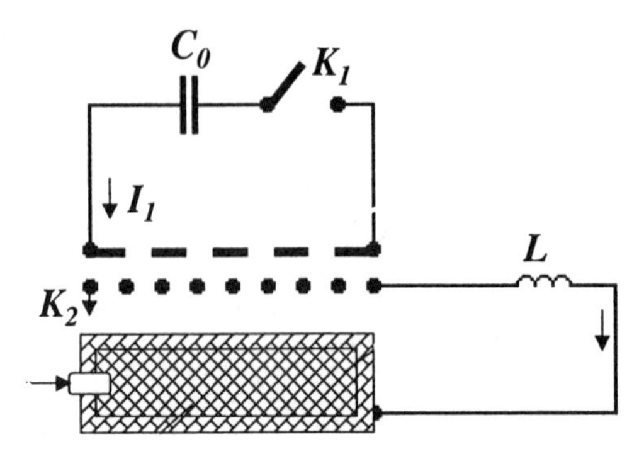


FIGURE 4.24. Diagram of end-initiated MCG with flux trapping.

ternal solenoid, $L_2$, and cylindrical liner, $L_3$. The initial magnetic flux in the external solenoid is created by discharging the capacitor, $C_0$, into it. When the current reaches its peak value, the explosives are detonated, which closes the secondary circuit, trapping the flux created by the external solenoid. In the case of the simultaneous helical generator, the liner expands radially along the entire length of the generator, while in the case of the end-initiated generator, one end expands, makes contact with the coil, and the contact point moves along the axis of the generator. In both cases, the trapped flux is forced into the load, $L_L$. The length of the pulse in the load depends on the dynamics of the expansion of the liner. Mathematical models of both types of generators have been developed and are described in [4.22]

The efficient operation of MCGs with flux trapping require a high internal helical inductances, which leads to a number of problems when the generator operates into high-impedance loads. Owing to the rapid rate at which the magnetic field increases within the working volume of the generator, strong electrical fields are generated, which could lead to breakdown and loss of energy. In estimates conducted by Sheindlin and Fortov [4.22], the output voltage of the generators he designed is no less than 50 kV. This voltage is generated between the liner and the end of the internal helix in Fig. 4.24. These high voltages require that both the electrical insulation between the turns of the coil and the insulation and between the liner and the coil be sufficient to prevent breakdown. The latter problem is difficult to solve, and for this reason, the turns of the coil in axially initiated MCGs are sometimes wound on a thick polyethylene pipe and covered with fiberglass to ensure that breakdown does not occur between the liner and the coil. This allows the generation of voltage pulses with amplitudes up to 200 kV. In the case of end-initiated generators, the insulation cannot be as thick, since this reduces the magnetic flux available for compression due to diffusion. Using an insulator with a breakdown strength of 100 kV/mm, the required thickness of the insulation is $\geq 0.5$ mm.

Generators with flux trapping place additional requirements on the insulation of the solenoids, since this insulation must prevent both volumetric breakdown and breakdown between the turns of the coil. This requires the insulation between the solenoids to be thickened, which decreases the coupling between the coils and the efficiency of the generator, since at the end of generator operation a significant portion of the flux remains trapped in the space between the solenoids.

In addition to the voltage limitations, the maximum impedance of the internal helical coil is limited by the fact that it is impossible to manufacture coils with arbitrarily small step sizes between turns. The step size cannot be less than the thickness of the wire, the minimum of which, considering the skin effect, must ensure that there is no significant resistive heating. The minimum diameter is estimated to be approximately 1 mm and the minimum step size to be 1.5 mm [4.22] at 50 kV. In order for the MCG to operate efficiently at this step size, the parts must be manufactured with a high degree of mechanical precision. Flux losses at the point of contact increase greatly if the expanding portion of the liner is not a regular cone or is not coaxial with the coil, since the contact point may jump forward and the flux trapped between contact points lost ($2\pi$-clocking). In studies done by Sheindlin and Fortov [4.22], they estimated that the center lines of the liner and coil must not differ by more than 0.1 mm, that the uniformity of the walls must not deviate by more than 0.01 mm, and that the placement of the initiators must have an accuracy of not less than 0.2 mm.

In order to investigate MCGs with flux trapping that can deliver electrical pulses to loads with the desired parameters, Sheindlin and Fortov [4.22] designed and tested three generators, with single-turn helical coils having a constant step size.

### Simultaneous Axially Initiated MCG

The axially initiated spiral MCG in Fig. 4.25 consists of a liner (1), internal solenoid (2), and external solenoid (3). The length of this generator is 200 mm. To prevent breakdown between the turns of the coil and between the liner and coil, the internal solenoid was wound with fiberglass-insulated copper strips in helical channels in a Teflon tube (4). The diameter of the tube was 104 mm, and its wall thickness was 3 mm. The channels were filled with epoxy and several layers of polyamide film (5) were placed in the channels to a thickness of 2 mm. The width of the film was 200 mm greater than the length of the solenoids, thereby ensuring no surface breakdown. The external coil was secured with several layers of glass tape (6), saturated with epoxy compound. Special ribs (7) were glued to the body of the generator to avoid surface breakdown. The liner was made of copper tube and had a length of 70 mm and a wall thickness of 2.5 mm. The tube was stretched on a mandrel and its outer surface ground. The liner was centered on the axis of symmetry of the generator by means of

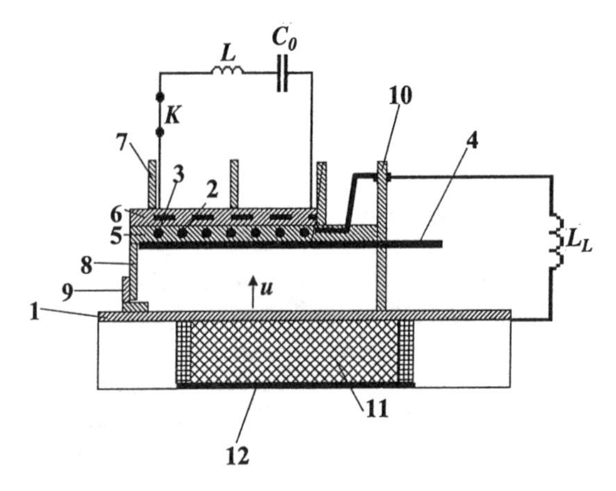


FIGURE 4.25. Axially initiated simultaneous cylindrical MCG with flux trapping: 1 – liner, 2 – internal solenoid, 3 – external solenoid, 4 – Teflon tube, 5 – polyamide film, 6 – glass tape saturated with epoxy compound, 7 – ribs glued to body of generator, 8 – retaining rings, 9 – Teflon inserts, 10 – organic glass rings, 11 – explosive charge, and 12 – copper wires.

conical retaining rings (8) with Teflon inserts (9) and rings of organic glass (10). The high explosive (11) (600 g of hexogen) was initiated by exploding copper wires (12) located along the longitudinal axis of the liner.

## Cylindrical Spiral MCGs

Cylindrical spiral MCGs with slipping contact points are the easiest to manufacture, and therefore have been designed, built, and tested with a wide range of parameters in an attempt to understand their limitations. A cylindrical MCG with flux capture (Fig. 4.26) consists of an external feed solenoid, $L_1$, internal solenoid, $L_2$, and copper liner, which contains 200–400 g of explosives. The internal solenoid is constructed with copper wire, which is uniformly laid in a single pass within slots cut into either a polyethylene or Teflon tube with a length of 60–200 mm and a diameter of 80–100 mm. The thickness of tube wall is typically 0.3 mm. Epoxy and Teflon, to which the external solenoid is attached, covers the outside of the pipe. The entire assembly is wrapped with Teflon tape saturated with epoxy. The liner is a copper tube having a diameter of 50 mm and wall thickness of 3 mm. The explosive charge is detonated at one end by the electrical detonator d. When the liner expands, it forms a cone that closes switch $K_2$ first and then shorts out each turn of the internal coil in turn, thus capturing the magnetic flux and generating current in the load.

## Conical Spiral MCGs

The operating time of the cylindrical spiral MCG depends on the length of the helical coil. In order to reduce the operating time of MCGs, it can

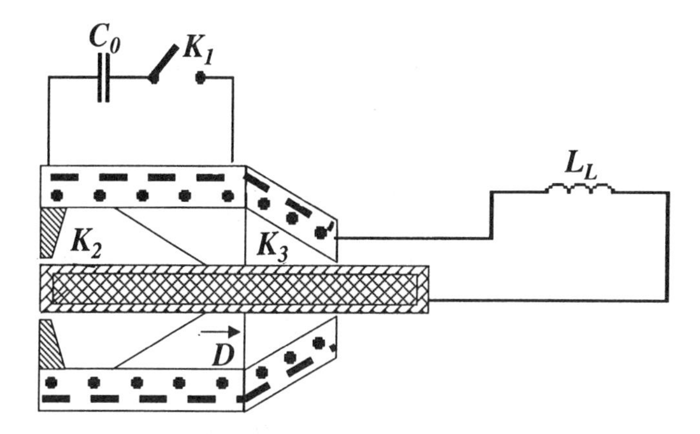


FIGURE 4.26. Two-stage cylindrical and conical MCG system. The magnetic flux is initially compressed in the cylindrical MCG generator, followed by capture in the conical and further compression to ampify the energy from the first generator.

be tapered to form a conical MCG, as shown in Fig. 4.26. In this case, the operating time of the MCG depends on the diameters at the base and tip of the conical structure. In general, the design of conical MCGs differs little from that of cylindrical MCGs. Since higher voltages are formed within conical MCGs, the copper wires are wound with 0.2-mm-thick Teflon at the base and 0.5-mm-thick Teflon at the tip of the conical structure. In order to ensure shorter operating times, a special switch is used so that the secondary circuit begins to operate just as the liner reaches the coil. The length of the solenoid is 80–100 mm, the diameter of the base end is 104 mm, and the diameter of the tip varies from 60 to 90 mm. The liner is filled with 200–300 g of explosives. This generator was tested to ensure that breakdown does not occur for average voltages up to 50 kV and peak voltages up to 150 kV.

### *4.5.2 Flux Trapping and No Transformer*

Using MCGs with flux trapping, Sheindlin and Fortov [4.22] developed a high voltage system that does not require transformers. It uses a "booster" helical MCG to drive a "fast" high-voltage MCG with flux trapping, an exploding wire switch, a gas-filled discharge switch, a peaking capacitor, and a vacuum diode as the load.

A small 100 $\mu$F, 3 kV capacitor, $C_0$, is used to feed the external coil of the booster MCG, $L_{11}$, which to 5 kJ amplifies the energy delivered to the external coil of the high-voltage axially initiated MCG. The energy output of the high-voltage MCG is 60 kJ. The explosive charge in the high-voltage MCG is initiated such that the closure of switch, $S_2$, occurs when the current is at its peak value in circuit $L_1$. As the liner in the high-voltage generator expands, the magnetic flux is trapped by the internal solenoid $L_2$, thus introducing current at the input of the electroexplosive switch $R_e$.

The high-voltage MCG operates for 6–8 $\mu$s and generates voltages of 50–200 kV, with currents up to 30 kA. The switch consists of 30–60 parallel copper wires with a diameter of 40–50 $\mu$m and length of 0.5–1.0 m, which are placed in nitrogen at 0.5 MPa. The dimensions of the wires in the switch are selected so that the maximum rate of increase in their resistance occurs at the end of the operating time of the MCG. Since the high-voltage generator was designed to withstand voltages up to 200 kV and the voltages within the MCG reach 600 kV during the detonation process, an inductance $L_l$ is connected between the MCG and the switch to decouple them and to permit energy storage. The overvoltage that occurs when the circuit is opened leads to breakdown across the spark gap P and the formation of a high voltage across the vacuum diode, thus causing the explosive emission of electrons from the cathode.

In experiments conducted by Sheindlin and Fortov [4.22], currents of 30 kA and energies up to 2 kJ were delivered to the diode. The energy required to explode the wires of the switch is 2–3 kJ, which means that the total energy generated by the MCG did not exceed 5 kJ. As a result, voltage pulses up to 600 kV with pulse lengths of 180–500 ns and rise times of 60 ns were delivered to the diode.

### *4.5.3 Flux Trapping and Transformers*

In order to reliably deliver high-voltage pulses to a load, transformers are usually used. Typically, these are air-core and tend to be bulky and massive. Recently, a new self-contained high-voltage MCG system was developed that can deliver 450 kV pulses with a pulse length of 200 ns and a risetime of 50 ns to a vacuum diode load. This system (Fig. 4.27) consists of a spiral MCG, pulse transformer (PT), electroexplosive switch (EES), and gas discharge switch (P). The transformer required to provide impedance matching between the MCG and the inductive load is a relatively compact and lightweight foil transformer. In order to obtain impedance matching between the MCG and pulse transformer, an explosive switch can be installed between them. The primary shortcoming of adding this switch is that the MCG is extremely susceptible to the accuracy of the switch. To overcome this shortcoming, a two-stage MCG with flux trapping (Fig. 4.28), similar to the one in Fig. 4.26, can be used [4.30]. The first stage is a conventional end-initiated helical MCG, and the second stage is an axially initiated helical MCG. As the detonation wave in the first generator approaches the point labelled "O," flux is trapped and compressed radially. This technique shortens the operating time of the second generator and increases the overall efficiency of the MCG.

The MCG [4.30] consists of a multiturn spiral coil (6), liner (8) with explosive charge (7), metal band (5), seed current input couplers (2), liner

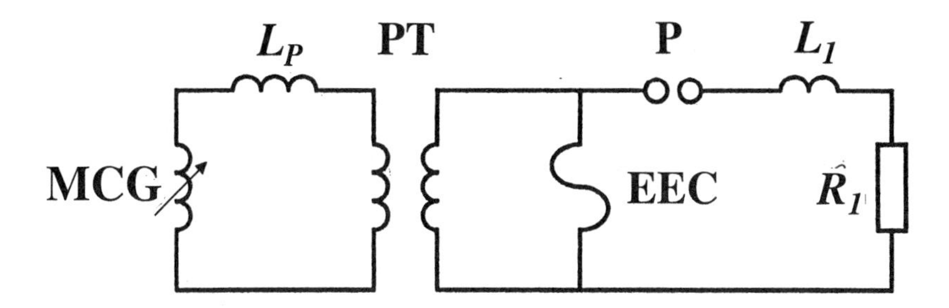


FIGURE 4.27. Equivalent circuit diagram for a high-voltage MCG and pulse-forming line. This system consists of a spiral MCG, pulse transformer (PT), electroexplosive switch (EES), and gas discharge switch (P). The transformer is required to provide impedance matching between the MCG and the inductive load.

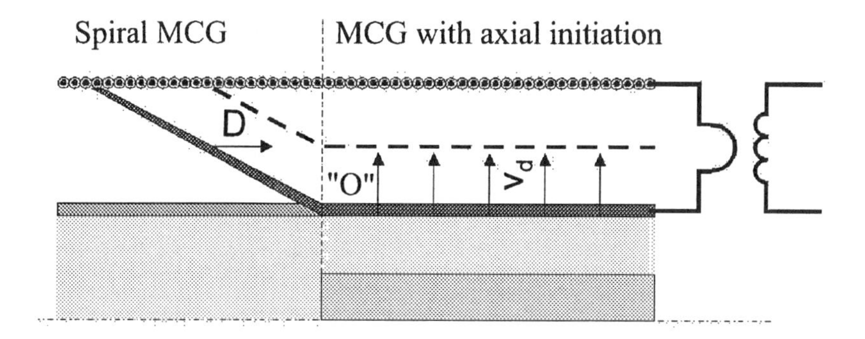


FIGURE 4.28. Two-stage MCG with flux trapping. The first stage is a conventional end-initiated helical MCG, and the second stage an axially initiated helical MCG. As the detonation wave in the first generator approaches the point labeled "O," its flux is trapped in the second genertor and compressed. This shortens the operating time of the second generator and increases the overall efficiency of the MCG.

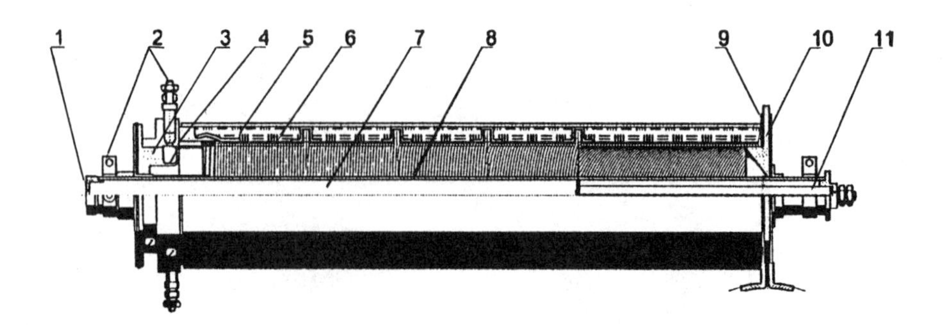


FIGURE 4.29. Schematic diagram of a two-stage spiral MCG: 1 – detonator for end-initiated MCG, 2 – seed current input couplers, 3 – liner separators, 4 – source contact, 5 – metal band, 6 – multiturn spiral coil, 7 – explosive charge, 8 – liner, 9 – output couplers to load, 10 – output structure, 11 – detonator for axially initiated MCG.

separators (3), source contact (4), and output couplers (9) to the load. The spiral coil has an inner diameter of 80 mm. The coil consists of four 80-mm-long sections with pitches of 5, 10, 20, and 40 mm, respectively, and a final section that has a length of 160 mm and a winding pitch of 120 mm. The first section is wound with two insulated wires per turn, where each wire has a diameter of 2.5 mm. In the next three sections the number of wires per turn is doubled, and in the final section the number is tripled. The detonator cord used for end-initiated MCG (1) is made from plastic explosives. The detonator for the axially initiated MCG is an exploding wire (11). The liner is made of copper that has an outer diameter of 40 mm and a wall thickness of 2 mm.

The axial detonator in the second stage is a thin-walled metal tube filled with nonflegmatized high explosive powder (TNT, hexagen). A copper wire with variable diameter is placed along the axis of the tube. By using variable diameter wire, the demands on the capacitive power supply and its switch are lessened. The tube and wire are connected to the capacitive store, which provides sufficient current to explode electrically the wire and detonate the explosive along the length of the tube with no delay. Calculations have shown that if the composition and packing density of the hexagen powder and the length and diameter of the wire are optimized, the specific energy required to activate the detonator is greater than twice the sublimation energy of the wire. The variability in the diameter of the wire, which is basically a string of bridgewires, is chosen to generate a smooth cylindrical detonation wave at the internal surface of the liner. Calculations also show that the equivalent length of the wire should be no less than 50 mm when its diameter is 0.14 mm. The capacitive stores used to activate both detonators should have a voltage of 16 kV, wave resistance of 1 $\Omega$, capacitance of 1 $\mu$F, and stored energy of 0.15 kJ.

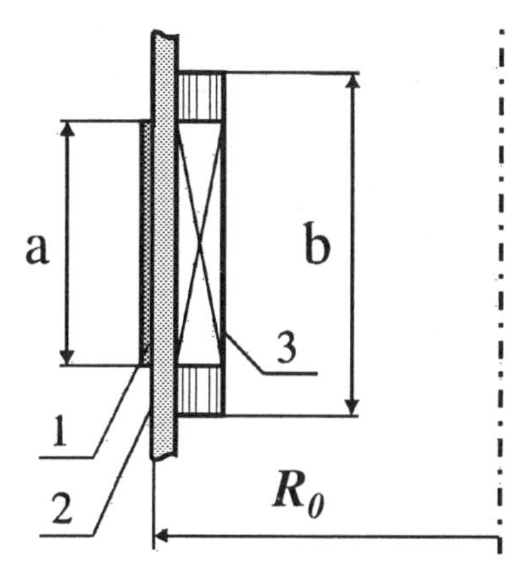


FIGURE 4.30. Foil transformer: 1 – primary winding, 2 – insulating layer, and 3 – secondary winding. The primary winding is a copper ring, and the secondary is a reel of thin foils, where the foils are separated by layers of plastic film and capacitor paper impregnated under vacuum with liquid insulator.

When designing the pulse step-up transformer, the maximum coupling factor, electrical strength, reliability, and restrictions on size and mass must be taken into account, as well as the output current of hundreds of kA. Therefore, the transformer selected has a foil-type construction (Fig. 4.30). The primary winding is a copper ring with a diameter of $a$. The secondary winding is a reel of thin foil, where the foils are separated by several layers of plastic film and capacitor paper impregnated under vacuum with a liquid insulator. The inner coil of the secondary winding is designed to sustain high voltages. The external coil sustains lower voltages and shields the high-voltage end from the grounded primary winding. This design is based on paper–film capacitor technology, enabling high discharge voltage gradients, exceeding 2 MV/cm, to be obtained.

Impregnating the insulation requires that the paper be in a vacuum prior to impregnation, thus the secondary winding is placed in a vacuum-tight insulated container. The insulator between the windings (2) reduces coupling between them and must have a certain minimum thickness to avoid damage when the casing is pumped to low pressures. Therefore, the primary winding is also placed in the same container, which means it requires vacuum-tight connectors that increases the parasitic inductance in the primary winding. Calculations show that these parasitic inductances should not have a significant impact on the efficiency of the transformer.

The diameter and width of the primary winding of the transformer are 196 mm and 50 mm, respectively. The estimated coupling factor of the transformer is approximately 0.89. The effective inductance of the primary winding is 45 nH. There are four input cables connecting it to the pulse-forming line. The transformer is housed in a vacuum-tight chamber made

| Parameter | Value |
|---|---|
| Peak Current in Load | 0.209 MA |
| Peak Voltage in Load | 436 kV |
| Peak Power in Load | 88.1 GW |
| Voltage Pulse Length | 215 ns |
| Voltage Pulse Rise Time | 50 ns |

TABLE 4.5. Output parameters of the high-voltage MCG-transformer system.

of corrosion-proof steel. The chamber is pumped down and then back filled with transformer oil and sealed.

The electroexplosive switch has a length of 400 mm and permits the use of wires with lengths up to 650 mm by using a criss-cross arrangement. The switch cartridge is designed to operate at atmospheric pressure, but can be maintained at pressures greater than 2 atm to increase the electric strength at the internal surface of the cartridge.

The discharge switch is filled with pure nitrogen at a pressure of 3–6 atm. To control the pressure, the unit casing has a gas intake valve. The interelectrode gap is 14 mm long. The field distribution across the gap is weakly nonuniform.

The output parameters of this high-voltage MCG are given in Table 4.5. It has been shown in this section, that a variety of techniques—switches, transformers, and flux trapping—are needed to electrically match the MCG to the load. As the requirements of the load change, a variety of ancillary equipment are available to optimize the operation of the MCG–load system. How the characteristics of the load affect the operation of the MCG and influence what ancillary equipment is needed are discussed in the next chapter.

# References


[4.1] J. Benford and J. Swegle, *High-Power Microwaves*, Artech House, Boston (1991).

[4.2] *Opening Switches* (eds., A. Guenther, M. Kristiansen, and T. Martin), Plenum Press, New York (1987).

[4.3] I. Vitkovitsky, *High Power Switching*, Van Nostrand Reinhold Company, New York (1987).

[4.4] V.K. Chernyshev, V.V. Vakhrushev, G.I. Volkov, and V.A. Ivanov, "The Operation of Explosive Current Peaker with Ribbed Barrier at the Variation of the Breaking Circuit Inductance in Constant Magnetic Flux," *Megagauss Technology and Pulsed Power Applications,* (eds. C.M. Fowler, R.S. Caird, and D.J. Erickson), Plenum Press, New York, pp. 525–529 (1987).

[4.5] V.K. Chernyshev, G.I. Volkov, A.N. Demin, V.A. Ivanov, M.I. Korotkov, N.N. Moskvichev, and A.F. Rybakov, "Foil Behaviour in Explosive Commutators at the Stage Prior to its Burst by Detonation Front," *Megagauss Technology and Pulsed Power Applications,* (eds. C.M. Fowler, R.S. Caird, and D.J. Erickson), Plenum Press, New York, pp. 531–533 (1987).

[4.6] V.K. Chernyshev, G.I. Volkov, V.A. Ivanov, and O.D. Mikhailov, "Experimental Investigation of Explosive Opening Switch Operation," *Megagauss Magnetic Field Generation and Pulsed Power Ap-*

*plications* (eds. M. Cowan and R.B. Spielman), Nova Science Publ., New York, pp. 731–738 (1994).

[4.7] I.R. Lindemuth, J.H. Goforth, K.E. Hackett, E.A. Lopez, W.F. McCullough, H. Oona, and R.E. Reinovsky, "Exploding Metallic Fuse Physics Experiments," *Megagauss Technology and Pulsed Power Applications* (eds. C.M. Fowler, R.S. Caird, and D.J. Erickson), Plenum Press, New York, pp. 495–504 (1987).

[4.8] J.E. Brandenburg, R.E. Terry, and N.R. Pereira, "Mass Erosion in Foil Switches," *Megagauss Technology and Pulsed Power Applications* (eds. C.M. Fowler, R.S. Caird, and D.J. Erickson), Plenum Press, New York, pp. 543–550 (1987).

[4.9] G. Schenk, "Exploding Foils for Inductive Energy Transfer," *Proceedings of the 5th Symposium on Fusion Technology*, Oxford, 1968.

[4.10] A.E. Greene and I.R. Lindemuth, "Modeling of High-Explosive Driven Plasma Compression Switches," *Megagauss Technology and Pulsed Power Applications* (eds. C.M. Fowler, R.S. Caird, and D.J. Erickson), Plenum Press, New York, pp. 505–511 (1987).

[4.11] R.E. Reinovsky, R.G. Colchaser, J.M. Welby, and E.A. Lopez, "Energy Storage Transformer Power Conditioning Systems for Megajoule Class Flux Compression Generators," *Megagauss Technology and Pulsed Power Applications* (eds. C.M. Fowler, R.S. Caird, and D.J. Erickson), Plenum Press, New York, pp. 575–582 (1987).

[4.12] B.L. Freeman and W.H. Bostick, "Transformers for Explosive Pulsed Power Coupling to Various Loads," *Megagauss Technology and Pulsed Power Applications* (eds. C.M. Fowler, R.S. Caird, and D.J. Erickson), Plenum Press, New York, pp. 583–591 (1987).

[4.13] B.L. Freeman, D.G. Rickel, A. Ramrus, and B.E. Strickland, "High-Voltage Pulsed Transformer Development," *Megagauss Fields and Pulsed Power Systems* (eds. V.M. Titov and G.A. Shvetsov), Nova Science Publ., New York, pp. 587–594 (1990).

[4.14] Yu. Bragin, A.S. Kravchenko, R.Z. Lyudaev, A.S. Yuruzhev, and V.A. Zolotov, "HMCG-320 with the Loop and Solenoidal Transformer," *Megagauss Magnetic Field Generation and Pulsed Power Applications*, Nova Science Publ., New York, pp. 493–499 (1994).

[4.15] V.B. Gaaze and G.A. Shneerson, Pribory i Tekh. Eksper. No. 6, p. 105 (1965).

[4.16] A.I. Pavlovskii, R.Z. Lyudaev, L.N. Pljashkevich, A.M. Shuvalov, A.S. Kravchenko, Yu. I. Plyushchev, D.I. Zenkov, V.F. Bukharov,

V.Ye., Gurin, and V.A. Vasyukov, "Transformer Energy Output Magnetic Cumulative Generators," in *Megagauss Physics and Technology* (ed. P.J. Turchi), Plenum Press, New York, pp. 611–626 (1979).

[4.17] *Gas Discharger Closing Switches* (eds. G. Schaefer, M. Kristiansen, and A. Guenther), Plenum Press, New York (1990).

[4.18] I. Vitkovitsky, *High Power Switching*, Van Nostrand Reinhold, New York (1987).

[4.19] G.A. Mesyats, *Generation of High-Power Nanosecond Pulses*, Sov. Radio, Moscow, p. 189 (1974).

[4.20] D.J. Ericson, R.S. Caird, C.M. Fowler, et. al., "A Megavolt Pulse Transformer Powered by Fast Plate Generator," in *Super Strong Magnetic Fields. Physics, Technology, Application* (eds. V.M. Titov and G.A. Shvetsov), Nauka, Moscow, p. 333 (1985).

[4.21] B.L. Freeman, D.J. Ericson, C.M. Fowler, et.al., "Magnetic Flux Compression Powered Electron Beam Experiments," in *Megagauss Technology and Pulsed Power Application* (eds. V.M. Titov and G.A. Shvetsov), Plenum Press, New York, p. 729 (1987).

[4.22] A.E. Sheindlin and V.E. Fortov, *Pulsed MHD-Converters of Chemical Energy into Electrical Energy*, Enrgoatomizdat, Moscow (1997).

[4.23] A.I. Pavlovskii, V.V. Druzhinin, et.al., "Magnetic-Optical Studies of Superstrong Magnetic Fields," in *Super Strong Magnetic Fields: Physics, Technology, Application* (eds. V.M. Titov and G.A. Shvetsov), Nauka, Moscow, p. 130 (1984).

[4.24] E.I. Bichenkov, S.D. Gilev, V.S. Prokopiev, et.al., "Cascade MC-Generator with Flux Trapping," in *Super Strong Magnetic Fields: Physics, Technology, Application* (eds. V.M. Titov and G.A. Shvetsov) Nauka, Moscow, p. 377 (1984).

[4.25] V.K. Chernyshev and V.A. Davydov, in *Megagauss Physics and Technology* (ed. P.J. Turchi), Plenum Press, New York, p. 651 (1980).

[4.26] A.I. Pavlovskii, R.Z. Lyudaev, V.A. Zolotov, et.al., in *Super Strong Magnetic Fields: Physics, Technology, Application* (eds. V.M. Titov and G.A. Shvetsov, Nauka, Moscow, p. 557 (1984).

[4.27] V.A. Davydov and V.K. Chernyshev, PMTF, No. 6, p. 112 (1981).

[4.28] E.C. Cnare, R.J. Kaye, and M. Cowan, in *Super Strong Magnetic Fields: Physics, Technology, Applications* (eds. V.M. Titov and G.A. Shvetsov), Nauka, Moscow, p. 50 (1984).

[4.29] V.K. Chernyshev, E.I. Zharinov, V.E. Veneev, et.al., in *Megagauss Fields and Pulsed Power Systems* (eds. V.M. Titov and G.A. Shvetsov), Nova Science Publ., New York, p. 355 (1990).

[4.30] E.V. Chernikh, V.E. Fortov, K.V. Gorbachev, E.V. Nesterov, S.A. Roschupkin, and V.A. Stroganov, "High-Voltage Pulsed MCG-Based Energy Source," to be published in the Proceedings of the Megagauss VIII Conference, Tallahassee, FL (1998).

# 5
# Electrical Loads

In this chapter, various schemes used to connect magnetocumulative generators to loads are discussed, where the loads may be complex, with inductive, capacitive, and/or resistive characteristics. It is necessary that the characteristics of the load are fully taken into account when designing a system, in order to achieve good matching between it and the generator and in order to achieve efficient operation of the generator. The nature of the loads vary widely. For example, MCGs have been use to rapidly charge inductive and capacitive energy stores, initiate detonator arrays, and drive such devices as plasma focus machines [5.1,5.2], neutron generators, high-power lasers [5.3–5.5], charged particle accelerators [5.6], high-power microwave sources [5.7], and electromagnetic launchers [5.8]. How the MCG is connected to a particular load depends on the characteristics of both the MCG and the load and usually dictates what type of ancillary equipment needs to be used [5.9,5.10].

## 5.1 Direct Connection to a Load

In this section, the direct connection of an MCG to a load with resistive, capacitive, and/or inductive characteristics is examined. The phrase *direct connection* means that components such as switches and transformers are *not* used to connect the MCG to the load. The circuit diagram for an MCG connected to a *complex load* is presented in Fig. 5.1, where the load impedance is described by the expression $Z = R + iX$, where $R$ is the

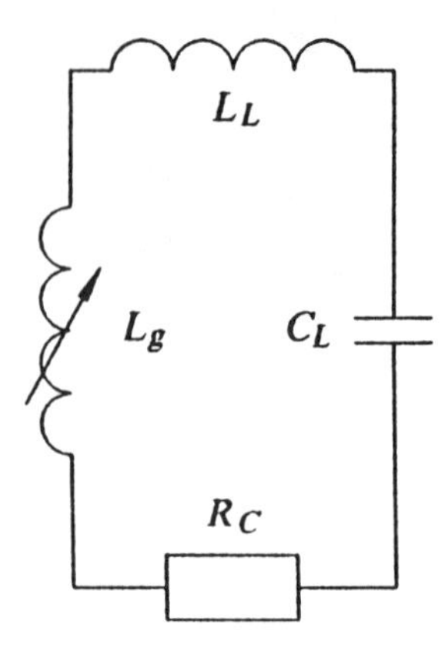


FIGURE 5.1. Equivalent circuit diagram for an MCG connected directly to a complex load. No pulse conditioning devices are used.

active (time-dependent) resistance and $iX$ is the imaginary component of the load impedance (i.e., the reactance). In this diagram, $L = L_g(t) + L_L$, where $L_g(t)$ represents the changing inductance of the MCG and $L_L$ is the inductance of the load; $R_C$ is the circuit resistance including all magnetic losses and the resistance of the load; and $C_L$ is the capacitance of the load. The decreasing inductance of the MCG significantly affects the characteritics of the series-connected $LRC$ circuit. Since the energy originates from within the circuit itself, the circuit may be described by the equation

$$L\frac{d^2I}{dt^2} + \left(2\frac{dL}{dt} + R_c\right)\frac{dI}{dt} + \left(\frac{d^2L}{dt^2} + \frac{dR_c}{dt} + \frac{1}{C_L}\right)I = 0, \tag{5.1}$$

with the voltage on the capacitor being given by

$$L\frac{d^2U}{dt^2} + \left(\frac{dL}{dt} + R_c\right)\frac{dU}{dt} + \frac{U}{C_L} = 0. \tag{5.2}$$

The functional form of the inductance $L(t)$ and the resistance $R_c(t)$ depends on the type of MCG used. In the rest of this section, Eqs. 5.1 and 5.2 will be solved for five cases, where the resistance and inductance are described by different relationships [5.11].

### 5.1.1 *Case 1:* $R_c = 0$, $L(t) = L_0 \exp(-\alpha t)$

In the first case, it is assumed that $R_c = 0$, $C_L \neq 0$, and that $L(t)$ varies exponentially according to the expression $L(t) = L_0 \exp(-\alpha t)$, where $\alpha$ is a positive constant with dimensions of $s^{-1}$. The exponential behavior of $L(t)$ is similar to that of a spiral MCG with changing step sizes between the coils. Eq. 5.1 can now be rewritten as

$$\frac{d^2I}{dt^2} - 2\alpha\frac{dI}{dt} + \alpha^2(1+\theta^2 e^{\alpha t})I = 0, \tag{5.3}$$

where $\theta = 1/\left(\alpha\sqrt{L_0 C_L}\right)$. The solution of Eq. 5.3 is

$$I = [C_1 J_0(x) + C_2 N_0(x)]\, e^{\alpha t}, \tag{5.4}$$

where $J_0(x)$ and $N_0(x)$ are zero-order Bessel functions of the first and second kind and have the argument $x = 2\theta e^{\alpha t}$. When $t = 0$, it can be shown that $x = 2\theta$, but, as $x$ increases, it is found that $x = x_0\sqrt{\gamma_L}$, where $\gamma_L = L_0/L$. Determining the value of the integration constants $C_1$ and $C_2$ from the initial conditions, Eq. 5.4 becomes

$$\gamma_I = [-N_1(x_0)J_0(x) + J_1(x_0)N_0(x)]\,\frac{\pi x^2}{2x_0} \tag{5.5}$$

where $\gamma_I = I/I_0$. Eq. 5.5 can be rewritten in terms of time, since $t = 2\alpha^{-1}\ln(x/x_0)$. In a similar manner, it can be derived that

$$\gamma_U = [-N_1(x_0)J_1(x) + J_1(x_0)N_1(x)]\,\frac{\pi x}{2}, \tag{5.6}$$

where $\gamma_U = U/\left(I_0\sqrt{L_0 C_L}\right)$.

The functions $J_1(x_0)$ and $N_1(x_0)$ indicate what processes occur during the initial phases of the system operation. If the zeros of the functions $J_m$ and $N_m$ are represented by $\mu_{mn}$ and $\eta_{mn}$, respectively, where $m$ is the order of the function and $n$ is the order of the zero of the Bessel function, then when $x < \eta_{01}$, the function $N_0(x)$ behaves monotonically and the function $\gamma_I(x)$ behaves aperiodically. When $x > \mu_{01}$, $\gamma_I(x)$ becomes oscillatory in nature. $\gamma_U$ behaves similarly for $\eta_{11}$ and $\mu_{11}$. The nature of the processes that occur during the initial stage of operation of the MCG is indicated by the value of $x_0$. If the operating time of the MCG, $\tau_P$, is sufficiently long and if the inductance decreases exponentially, then the initial aperiodic processes become oscillatory in nature.

For small values of $x$, only the first terms of the expanded Bessel function series need to be considered. The error in doing this is less than 1% when $x < 0.1$ and several percent when $x < 0.3$. For small values of $x_0$, Eqs. 5.5 and 5.6 can be rewritten as

$$\gamma_I = \gamma_L\left[J_0(x) + \frac{\pi x_0^2}{4}N_0(x)\right] \tag{5.7}$$

$$\gamma_U = \sqrt{\gamma_L}\left[J_1(x) + \frac{\pi x_0^2}{4} N_1(x)\right]. \tag{5.8}$$

If $x_f$ is small, where the subscript $f$ refers to values of the parameter at the moment the MCG has completed its operation, then $I = \Phi_0/L$, $\gamma_I \approx e^{\alpha t} = \gamma_L$, and $\gamma_U \approx x_0(\gamma_L - 1)/2$. In this case, the magnetic flux in the circuit is conserved, and the value of the current does not depend on the capacitance.

For larger values of $x$, more precise approximations can be derived for the oscillatory mode of operation, where

$$\gamma_I = \sqrt{\gamma_L}\left[-N_1(x_0)\cos\left(x - \frac{\pi}{4}\right) + J_1(x_0)\sin\left(x - \frac{\pi}{4}\right)\right], \tag{5.9}$$

$$\gamma_U = \left[-N_1(x_0)\sin\left(x - \frac{\pi}{4}\right) - J_1(x_0)\cos\left(x - \frac{\pi}{4}\right)\right]. \tag{5.10}$$

If the parameter $x_0$ is also large, then $\gamma_I \approx \gamma_L^{3/4}\cos(x - x_0)$ and $\gamma_U \approx \gamma_L^{1/4}\sin(x - x_0)$, i.e., current and voltage oscillations, with increasing amplitude and frequency, are generated. The final values, $\gamma_{If}$ and $\gamma_{Uf}$, depend on the phase of the oscillations. The functions $\gamma_I(x)$ and $\gamma_U(x)$ were calculated for various values of $x$ using Eqs. 5.5 and 5.6, over the range from $x_0 = 0.1$ to $x_f = 10$, and the resulting plots are presented in Fig. 5.2. It can be seen that initially the system behaves aperiodically, but it subsequently becomes oscillatory in nature. The plot of the magnetic flux conservation factor $\lambda_k$ is presented in the same figure. In the aperiodic mode, $\lambda_k = 1$, but in the oscillatory mode, $\lambda_k \approx \gamma_L^{-1/4}\cos(x - x_0)$, the magnetic flux oscillates in phase with the current and with an amplitude that decreases by a factor of $\gamma_k^{-1/4}$. Therefore, when the MCG is connected to a capacitive load, the magnetic flux in the generator is not conserved, even when there is no Ohmic resistance.

Denoting the energy stored in the capacitor by $W_C$ and that in the inductor by $W_L$, the following expressions can be written: $\varepsilon_C = \gamma_U^2$, $\varepsilon_L = \gamma_L^2/\gamma_I$, and $\varepsilon = \varepsilon_C + \varepsilon_L$, where $\varepsilon_L = W_L/W_0$, $\varepsilon_C = W_C/W_0$, and $\varepsilon = (W_C + W_L)/W_0$. Plots of these functions are presented in Fig. 5.3 over the range $x_0 = 0.1$ to $x_f = 10$. In the aperiodic regime, i.e., when $x_f$ is small, energy is primarily stored inductively, $\varepsilon_L >> \varepsilon_C$. When the MCG has completed its operation, energy is primarily stored capacitively, but in this case, when $2\pi\sqrt{L_f C_L} >> \tau_p$, where $\tau_p$ is the operating time of the MCG, the magnetic flux is noticeably not conserved.

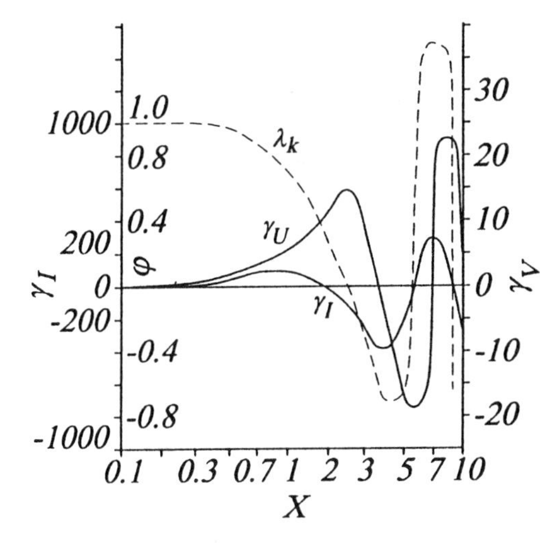


FIGURE 5.2. Plots of voltage, current, and magnetic flux conservation coefficient for a complex load in which the resistance is zero and the inductance change is exponential. Initially, the system behave aperiodically, but than becomes oscillatory. When the load MCG is connected to a capacitive load, magnetic flux is not conserved, even when there is no Ohmic resistance.

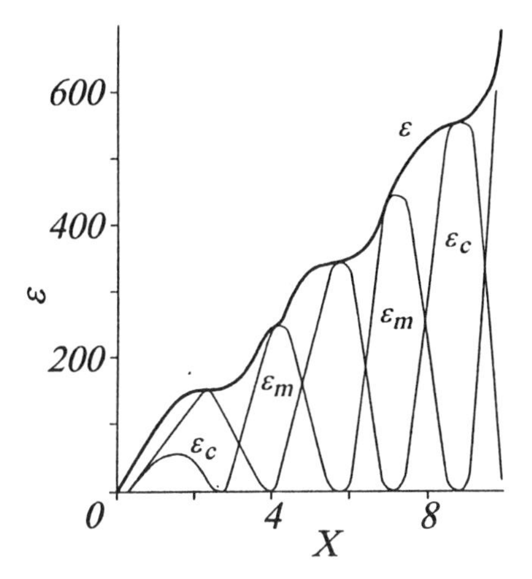


FIGURE 5.3. Plots of energy delivered to complex load by MCG. In the aperiodic regime of operation, the energy is primarily stored inductively, but when the MCG has completed operation, the energy is primarily stored capacitively. In the latter case, magnetic flux is not conserved.

### 5.1.2 *Case 2:* $R_c = 0,\ L = L_0(1 - \alpha t)$

In this case, it is assumed that $R_c = 0$, $C_L \neq 0$, and that the inductance varies linearly, with the form of the function $L$ depending on the inductive coupling characteristics of the spiral MCG, the coil step size, and so forth. Using these assumptions, Eq. 5.1 becomes

$$(1 - \alpha t)\frac{d^2 I}{dt^2} - 2\alpha\frac{dI}{dt} + \alpha^2\theta^2 I = 0. \tag{5.11}$$

The solution of this equation is

$$I = [C_1 J_1(y) + C_2 N_1(y)]\,(1 - \alpha t)^{-\frac{1}{2}}, \tag{5.12}$$

where the argument of the Bessel function is $y = 2\theta\,(1 - \alpha)^{1/2}$, which, unlike $x$ in the previous case, decreases with time. The function $I(y)$ is inversely proportional to time because $t = \left(1 - (y/y_0)^2\right)/\alpha$, but is also directly proportional to the inductance $L$. Solving for $C_1$ and $C_2$ and substituting into Eq. 5.12 yields

$$\gamma_I = [N_0(y_0)J_1(y) - J_0(y_0)N_1(y)]\,\frac{\pi y_0^2}{2y}. \tag{5.13}$$

In like manner, a similar expression can be derived for the voltage:

$$\gamma_U = [N_0(y_0)J_0(y) - J_0(y_0)N_0(y)]\,\frac{\pi y_0}{2}. \tag{5.14}$$

When $y \leq 0.3$, Eqs. 5.13 and 5.14 become

$$\gamma_I = \left[\left(\frac{\pi y_0}{2}\right)^2 N_0(y_0) + \gamma_L J_0(y_0)\right], \tag{5.15}$$

$$\gamma_U = \left[\left(\frac{\pi y_0}{2}\right) N_0(y_0) - y_0\left(0.577 + \ln\left(\frac{y}{2}\right)\right) J_0(y_0)\right]. \tag{5.16}$$

In this case, the final values $\gamma_{If}$ and $\gamma_{Uf}$ depend on the initial values of $N_0(y_0)$ and $J_0(y_0)$. If $J_0(y_0) \neq 0$, then $\gamma_I \to \infty$, as $y \to 0$. If $y_0 = \mu_{0n}$, then even if $y_f = 0$, the current is finite:

$$\gamma_{If} = [N_0(\mu_{0n})]\left(\frac{\pi\mu_{0n}}{2}\right)^2. \tag{5.17}$$

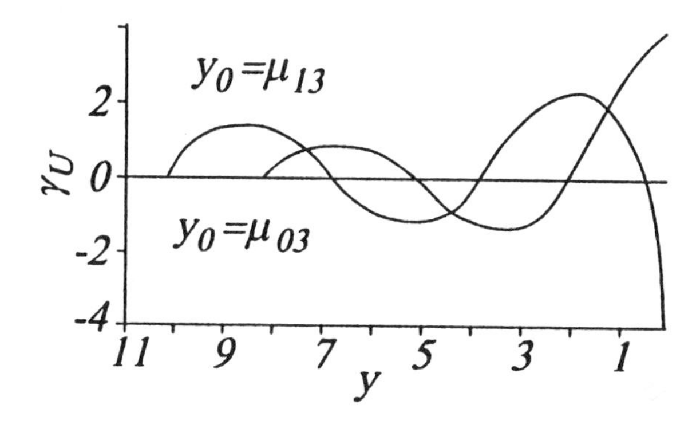


FIGURE 5.4. Plot of voltage versus the Bessel function argument when the resistance is zero and the inductance varies linearly (Case 2). Unlike inductive coupling, linear coupling does not allow aperiodic operation to transition into an oscillatory mode during the initial stages of operation.

If $y_0$ is small, then $I = \Phi_0/L$, $\gamma_I = \gamma_L$, and $\gamma_U = y_0 \ln(y_0/y)$. During the oscillatory mode at high values of $y_f$,

$$\gamma_I = \gamma_L^{\frac{3}{4}} \cos(y_0 - y), \quad \gamma_U = \gamma_L^{\frac{1}{4}} \sin(y_0 - y). \tag{5.18}$$

Plots of $\gamma_U(y)$ for $y_f = 0.1$, when $y_0 = \eta_{03}$ and $y_0 = \eta_{13}$, are presented in Fig. 5.4. For this case, there is no growth as $y \to 0$. For small $y$, the magnetic flux conservation coefficient is

$$\lambda_k \approx \left[\left(\frac{\pi y}{2}\right)^2 N_0(y_0) + J_0(y_0)\right]. \tag{5.19}$$

When y decreases, the magnetic flux conservation coefficient also decreases. When $y$ tends toward zero, $\lambda_k$ tends toward $J_0(y_0)$. If $y_0$ is small, then $\lambda_k \sim 1$. In the oscillatory mode, $\varphi = \gamma_L^{-1/4} \cos(y_0 - y)$.

In summary, unlike exponential coupling, linear coupling of the inductance does not allow the aperiodic mode of operation to transition into an oscillatory mode of operation during the initial stages of operation of the MCG. In addition, for sufficiently long operating times, $\tau_p$, the oscillating operation changes to an aperiodic process during the final stage of operation.

### 5.1.3 Case 3: $R_c \neq 0, L = L_0(1 - \alpha t)$

In this case, it is assumed that $R_c = \text{const}$, $C_L \neq 0$, and that the inductance varies linearly. Letting $R_c = \text{const}$ and $L = L_0(1 - \alpha t)$, Eq. 5.1 becomes

$$(1-\alpha t)\frac{d^2 I}{dt^2}-\alpha(v-2)\frac{dI}{dt}+\alpha^2\theta^2 I=0, \tag{5.20}$$

where $v = R_c/\alpha L_0$. Introducing the appropriate boundary conditions, the solution of this equation is

$$\gamma_I=[N_v(y_0)J_{v-1}(y)-N_v(y)J_v(y_0)]\frac{\pi y_0}{2}\left(\frac{y}{y_0}\right)^{v-1}. \tag{5.21}$$

The index of the Bessel function depends on $R_c$, which is why it is necessary that $J_v$ and $N_v$ depend on $v$ and $y$.

In like manner, similar expressions can be derived for the voltage by using Eq. 5.2, the solution of which is

$$\gamma_U=[N_v(y_0)J_v(y)-N_v(y)J_v(y_0)]\frac{\pi y_0}{2}\left(\frac{y}{y_0}\right)^{v}. \tag{5.22}$$

When $R_c = 0$, the index $v = 0$, and when $R_c = -dL/dt$, the index $v = 1$. In the interval, $0 \le v \le 1$, the criterion for the current to transition into the aperiodic mode, depending on the value of $v$, lies between $\mu_{01}$ and $\mu_{11}$. When $v > 1$, the amplitude of the current does not increase, and the transition to an aperiodic mode shifts to higher values of $y$ as $v$ increases. For the oscillating mode to occur, then $y >> 1$ and $y >> v$, so that $\gamma_I = \gamma_L^{(3-2v)/4}\cos(y_0 - y)$ and $\gamma_U = \gamma_L^{(1-2v)/4}\sin(y_0 - y)$. As can be seen, the amplitude of $I$ increases if $v < 3/2$ and the amplitude of $U$ increases if $v < 1/2$.

Assuming that $L = L_0 e^{-\alpha t}$ for the case when $R_c/L = \text{const}$, Eq. 5.1 can be rewritten as

$$\frac{d^2 I}{dt^2}+\alpha(v-2)\frac{dI}{dt}+\left(\alpha^2\theta^2 e^{\alpha t}-v+1\right)I=0 \tag{5.23}$$

Solving for the current $I$ yields

$$I=[C_1 J_v(x)+C_2 N_v(x)]\,x^{-\frac{v}{2}}, \tag{5.24}$$

where $v = R_c/\alpha L$. Introducing the initial conditions, gives the integration constants as

$$C_1=I_0\left[AN_v(x_0)-BN_{v+1}(x_0)\right], \tag{5.25}$$

$$C_2 = I_0 \left[ A J_v(x_0) + B J_{v+1}(x_0) \right], \tag{5.26}$$

where $A$ and $B$ are constants expressed in terms of complex polynomials that are functions of $v$ and $x_0$. For large values of $x_0$,

$$\gamma_I \approx \left[ A \cos(x_0 - x) + B \sin(x_0 - x) \right] \left( \frac{2}{\pi} \right)^{\frac{1}{2}} x^{\frac{1-v}{2}}. \tag{5.27}$$

This expression will be simpler for other initial conditions.

If Eqs. 5.1 and 5.2 are solved numerically, the maximum possible value of $I_0$ must be selected based on the type of MCG used, irrespective of how the MCG is coupled to the load, in order to optimize the parameters of the MCG. If, for a constant current $I_0$, $C$ is decreased, then the amplitude of the first oscillation will increase. Further increases in the amplitude may be less significant. In practice, this mode of operation determines the optimal value of the initial current.

The frequency of the oscillations in the circuit, $1/\sqrt{LC_L}$, may exceed the equivalent frequency, which is characteristic of the chosen MCG connected to inductive or resistive loads and which is determined by the basic properties of the generator. This will increase the resistance of the generator and cause additional energy losses. In order to use MCGs with capacitive loads, special designs must be created.

### 5.1.4 Case 4: $C_L = 0$

In this case, it is assumed that the load has no capacitive component. When $C_L = 0$, Eq. 5.1 takes on the same form as Eq. 1.67, the solution of which is

$$I = \frac{L_0}{L} \exp\left[ - \int_0^t \frac{R_c}{L} dt \right]. \tag{5.28}$$

During the final stage of operation of the generator, this equation becomes

$$I = \gamma_L \lambda_k. \tag{5.29}$$

The increase in the magnetic energy stored in the generator can be determined from

$$\varepsilon = \frac{L_0}{L_L \lambda_k^2}, \tag{5.30}$$

where $L_0$ is the inductance at the moment it begins to operate ($t = 0$).

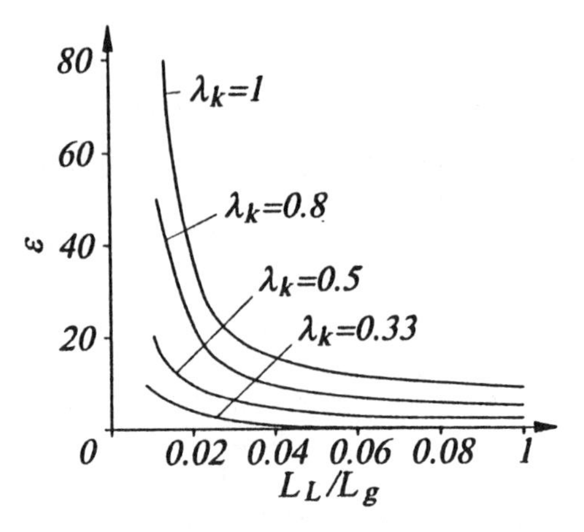


FIGURE 5.5. Plots of energy delivered to the load when the resistance and load capacitance are both zero (Case 5).

### 5.1.5 *Case 5:* $C_L = 0,\ R_c = 0$

In this case, it is assumed that the load has no capacitive component and that the circuit resistance is zero. If $C_L = 0$ and $R_c = 0$, then the MCG is connected to an inductive load, and

$$L_g \frac{d^2 I}{dt^2} + L_L \frac{dI}{dt} + \frac{dL_g}{dt} I = 0. \tag{5.31}$$

Solving this equation for the current flowing through the load, the solution is

$$I = I_0 \frac{L_0}{L_g + L_L}. \tag{5.32}$$

The ratio of the energy coupled to the load to the initial energy is

$$\varepsilon_L = \frac{W_L}{W_0} = \lambda_k^2(t) \left[1 + \frac{L_g}{L_L}\right]^2. \tag{5.33}$$

From this equation, it can be seen that the efficiency with which the MCG couples energy into a purely inductive load, depends first on the magnetic flux conservation coefficient and second on the ratio of the load inductance to the inductance of the generator. The plot of the normalized energy $\varepsilon$ versus the normalized inductance $L_L/L_g$, for different values of $\lambda_k$, is presented in Fig. 5.5. The parameters of a generator with specific energy amplification coefficients can be found from these plots.

## 5.2 Connection Through Pulsed Transformers

In this section, seven cases in which an MCG is connected through a pulsed transformer to different types of loads are discussed [5.12–5.15].

### 5.2.1 *Case 1: Complex Loads*

The equivalent circuit of an MCG with a complex load connected to the secondary coil of a transformer is shown in Fig. 5.6. In this diagram, $L_1$ and $L_2$ and $R_1$ and $R_2$ are the inductances and resistances of the primary and secondary circuits, respectively; $L_1 = L_g + L_{1T}$, where $L_g$ is the operating inductance of the MCG and $L_{1T}$ is the inductance of the primary winding (which for a cylindrical transformer is $L_{1T} = L_k$, where $L_k$ is the inductance of the primary winding of the cylindrical transformer); $L_2 = L_{2T} + L_L$, where $L_{2T}$ is the inductance of the secondary winding; $M$ is the mutual inductance; and $I_1$ and $I_2$ are the currents in the primary and secondary circuits, respectively. The circuit equations are

$$L_1 \frac{dI_1}{dt} + \left( \frac{dI_1}{dt} + R_1 \right) I_1 + M \frac{dI_2}{dt} = 0, \tag{5.34}$$

$$L_2 \frac{d^2 I_2}{dt^2} + R_2 \frac{dI_2}{dt} + \frac{I_2}{C} + M \frac{d^2 I_1}{dt^2} = 0. \tag{5.35}$$

Assuming that $R_1 = 0$, $R_2 = 0$, $U_0 = 0$, and $I_{20} = 0$, it can be shown that

$$I_1 = \frac{\Phi_0}{L_1} - \frac{M I_2}{L_1}. \tag{5.36}$$

Since the influence of the second term of this equation only becomes noticeable during the final stage of operation, the effects of $I_2$ can be neglected, even if there are high frequency oscillations in the secondary circuit. The current, $I_2$, can be assumed to be the sum of the current in the MCG and that in the secondary circuit, when the capacitance is zero and when the current oscillates during the initial stages of operation of the MCG with a frequency of approximately $1/\sqrt{L_2 C_L}$, which decreases in time. When $\tau_p << \sqrt{L_E C_L}$, where $L_E = \left( L_1 - M^2/L_2 \right) L_2^2/M^2$, then oscillations are not excited in the secondary circuit.

Assuming that $L_1 = L_0/(1+\alpha t)$, an inductance that is not characteristic of MCGs, but is suitable for coupling the inductance of a sectioned spiral coil, Eqs. 5.34 and 5.35 may be solved analytically to give

$$\gamma_{I2} = \left[ N_1(Z_0) J_1(Z) - N_1(Z) J_1(Z_0) \right] \frac{\pi Z_0 l_L (1 + \alpha)}{k_c^2 Z}, \tag{5.37}$$

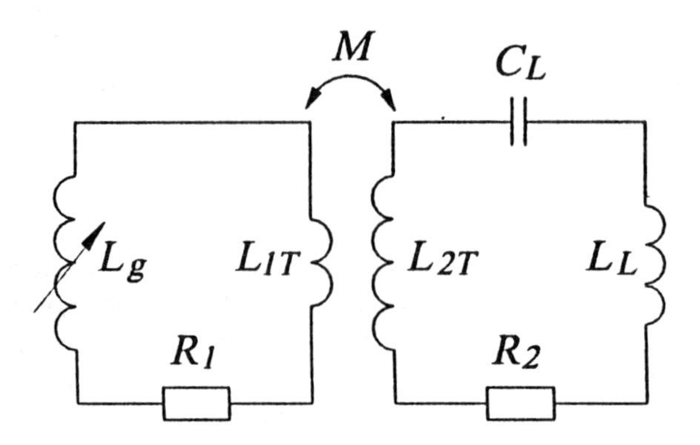


FIGURE 5.6. Equivalent circuit diagram for MCG connected through a pulsed transformer to a complex load.

where $\gamma_{I2} = -I_2L_2/MI_{10}$; $l_L = L_0/L_{1T}$; $\alpha = L_L/L_{2T}$; $k_c$ is the coupling factor of the transformer; $I_{10}$, $I_{20}$, and $L_0$ are the initial values of $I_1$, $I_2$, and $L$, respectively; $\theta = 1/\alpha\sqrt{L_2C_L}$; and

$$Z = \frac{2l_L}{\theta k_c^2}\sqrt{(1+\alpha)\left(1+\alpha-(1+\alpha t)\frac{k_c^2}{l_L}\right)}. \tag{5.38}$$

While the MCG is operating, the value of $Z$ decreases from

$$Z_0 = \frac{2l_L}{\theta k_c^2}\sqrt{(1+\alpha)\left(1+\alpha-\frac{k_c^2}{l_L}\right)} \tag{5.39}$$

to

$$Z_f = \frac{2l_L}{\theta k_c^2}\sqrt{(1+\alpha)\,(1+\alpha-k_c^2)}. \tag{5.40}$$

If $Z_f > \mu_{11}$, then $\gamma_{12}$ has an oscillating component for large $Z_f$ :

$$\gamma_{I2} = \frac{2l_L(1+\alpha)Z_0^{\frac{1}{2}}}{k_c^2 Z^{\frac{3}{2}}}\sin\left(Z_0 - Z\right). \tag{5.41}$$

If $Z_0 < \eta_{11}$, then

$$\gamma_{I2} \approx \left(1-\frac{Z_0^2}{Z^2}\right)\frac{l_L(1+\alpha)}{k_c^2}, \tag{5.42}$$

i.e., the current is the same as for the MCG operating into a secondary circuit with no capacitance.

Integrating Eq. 5.41, an expression for $\gamma_U$ can be found as, i.e., $\gamma_U = F/E$, where

$$E = \frac{-U\,(L_2 C_L)^{\frac{1}{2}}}{I_{10} L_2} \tag{5.43}$$

and

$$F = [N_1(Z_0)J_0(Z) - N_0(Z)J_1(Z_0)]\,\frac{\pi Z_0}{2} + 1. \tag{5.44}$$

For large $Z_f$,

$$\gamma_U \approx \frac{\left(\frac{Z_0}{Z}\right)^{\frac{1}{2}} \cos\,(Z_0 - Z) - 1}{E} \tag{5.45}$$

and for small $Z_0$

$$\gamma_U \approx \frac{(Z_0 - Z)^{\frac{1}{2}}}{2E}. \tag{5.46}$$

Therefore,

$$\varepsilon_c = \frac{\gamma_U^2 k_c^2}{l_L(1+\alpha)}, \tag{5.47}$$

$$\varepsilon_m = \frac{\gamma_{I2}^2 \alpha k_c^2}{l_L(1+\alpha)^2}. \tag{5.48}$$

Several plots of $I_2(t)$ and $U(t)$ calculated for a generator coupled through a transformer into a load with different capacitances are presented in Fig. 5.7. The secondary circuit of the generator is closed 80 $\mu s$ prior to the end of its run, at which time

$$L_1(t) = 9.1 \times 10^{-9} - 10.6 \times 10^4 t - 1.95t^2 + 6.2 \times 10^4 t^3 \tag{5.49}$$

and $R = 8.8 \times 10^5 - 0.83t$. The secondary winding has 16 turns, $L_{1T} = 6.8$ $\mu$H, $M = 0.4$ $\mu$H, $k_c = 0.96$, $\alpha = 0.44$, $R_2 = 0.02$ Ohm, and $I_{10} = 5.4$ MA.

The above examples show that under certain conditions the MCG may operate efficiently into capacitive loads. The operation of the MCG may be significantly different from when it operates into an inductive or resistive

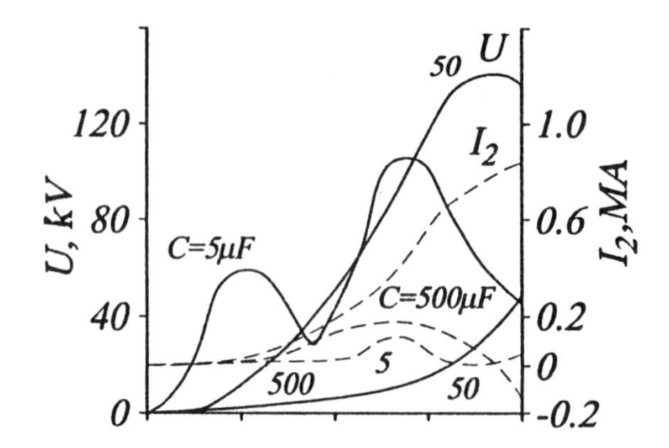


FIGURE 5.7. Plots of voltage and current for an MCG connected through a transformer of loads of different capacitances. When the MCG operates in an oscillatory mode, the energy yield increases slowly, whereas in the aperiodic mode, it is mainly stored inductively.

load. The basic difference is related to the possible existence of an oscillating component in the current. Even when there is no load resistance, the magnetic flux in the MCG circuit may not be conserved.

When the MCG operates in an oscillatory mode, its energy yield increases slowly, whereas in the aperiodic mode, the energy is accumulated mainly in the inductance. There are other possible methods by which a capacitive load may be connected to an MCG: for example, it may be connected in parallel to an inductive or resistive load or it may be introduced via a switch.

To operate an MCG into a capacitive load, the MCG must be specially designed. Such an MCG with a capacitive load generates a train of current pulses. In Chapter 8, it will be shown that an open tuned circuit with an MCG as its power source can convert the MCGs energy into radio waves. The operation of MCGs into transmission lines with distributed parameters has also been studied.

### 5.2.2 *Case 2: Resistive and Inductive Loads*

The equivalent circuit for an MCG connected through a transformer to a load with resistance $R_L$ and inductance $L_L$, closing switch $K$, and constant $R_2$ and $L_2$ is presented in Fig. 5.8. The system of equations for this equivalent circuit is

$$\frac{d(L_1 I_1)}{dt} + R_1 I_1 + M\frac{dI_2}{dt} = 0, \tag{5.50}$$

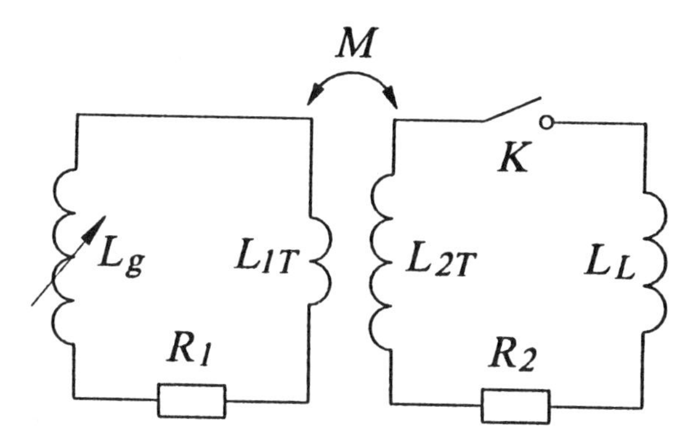


FIGURE 5.8. Equivalent circuit diagram for an MCG connected through a pulsed transformer to an active inductive load.

$$L_2 \frac{dI_2}{dt} + R_2 I_2 + M \frac{dI_1}{dt} = 0, \tag{5.51}$$

where $M = k_c \sqrt{L_{1T} L_{2T}}$ and $L_1 = L_g + L_{1T}$. This system of equations has no analytical solutions for general cases and may only be solved numerically. Since these equations describe the operation of the MCG from the moment when the discharge current from the battery has achieved its peak value in the generator, steady-state initial conditions can be used; i.e., when $t = 0$, $I_1(0) = I_{10}$ and $I_2(0) = I_{20} = M/L_2 \sqrt{1 + \frac{1}{4}(R_2 \tau_c / L_2)}$, where $\tau_c$ is the discharge time of the battery.

For the limiting cases of low and high resistances, Eqs. 5.50 and 5.51 can be simplified. In the case of high resistances, i.e., when $R_2 >> L_2/\tau_p$, where $\tau_p$ is the operating time of the MCG, the term $L_2 dI_2/dt$ may be neglected, because

$$\begin{aligned} L_2 \frac{dI_2}{dt} &\approx \frac{L_2 I_2}{\tau_p} << R_2 I_2, \\ I_2 &= \frac{M}{R_2} \frac{dI_1}{dt}. \end{aligned} \tag{5.52}$$

Substituting Eq. 5.52 into Eqs. 5.50 and 5.51, yields

$$\frac{d}{dt}\left[L_1 I_1 - \frac{M^2}{R_2}\frac{dI_1}{dt}\right] + R_1 I_1 = 0. \tag{5.53}$$

Assuming that

$$L_1 I_1 >> \frac{M^2}{R_2}\frac{dI_1}{dt} \tag{5.54}$$

and that

$$\frac{M^2}{R_2}\frac{dI_1}{dt} \approx \frac{M^2}{R_2}\frac{I_1}{\tau_p} << \frac{M^2}{L_2} I_1 << L I_1, \tag{5.55}$$

the above differential equation becomes

$$\frac{d(L_1 I_1)}{dt} + R_1 I_1 = 0. \tag{5.56}$$

Assuming that the inductance of the generator is linearly coupled, i.e., $L_g = L_0\,(1 - \alpha t)$, and that $R_1 = \text{const}$, the solution of Eq. 5.56 is

$$\gamma_{I1} = \left(\frac{L_0}{L_1}\right)^{1-\frac{R_1}{L_0 \alpha}}, \tag{5.57}$$

when $t < \tau_p$, and

$$y_{I1} = \left(\frac{L_0}{L_1}\right)^{1-\frac{R_1}{L_0 \alpha}} \exp\left[-\frac{R_1 \tau_p}{L_{1T}}\left(\frac{t}{\tau_p} - 1\right)\right], \tag{5.58}$$

when $t \geq \tau_p$, where

$$\gamma_{I1} = \frac{I_1}{I_{10}}. \tag{5.59}$$

The current flowing in the secondary circuit of the pulsed transformer may be found from $\gamma_{I2} = (M/R_2 I_{10}) dI/dt$ as

$$\gamma_{I2} = \frac{M L_0 \alpha \tau_p}{L_{1T} R_2}\left(1 - \frac{R_1}{L_0 \alpha}\right)\left(\frac{L_0}{L_{1T}}\right)^{2-\frac{R_1}{L_0 \alpha}}, \tag{5.60}$$

when $t \leq \tau_p$, and as

$$\gamma_{I2} = -\frac{M R_1}{L_{1T} R_2}\left(\frac{L_0}{L_{1T}}\right)^{1-\frac{R_c}{L_0 \alpha}} \exp\left[-\frac{R_1 \tau_p}{L_{1T}}\left(\frac{t}{\tau_p} - 1\right)\right], \tag{5.61}$$

when $t \geq \tau_p$, where $\gamma_{I2} = I_2/I_0$. When $x = 1$, Eqs. 5.60 and 5.61 become

$$\gamma_{I2m-} = \frac{M}{R_2\tau_p}\frac{L_0\alpha\tau_p}{L_{1T}}\left(1 - \frac{R_1\tau_p}{L_0\alpha\tau_p}\right)\left(\frac{L_o}{L_{1T}}\right)^{2-\frac{R_1}{L_0\alpha}} \quad (5.62)$$

and

$$\gamma_{I2m+} = -\frac{M}{L_{1T}}\frac{R_1}{R_2}\left(\frac{L_0}{L_{1T}}\right)^{1-\frac{R_1}{L_0\alpha}} . \quad (5.63)$$

The energy released into the active load is

$$\varepsilon_{R_1} = \frac{2M^2\alpha}{L_0R_2}\left(1 - \frac{R_1}{L_0\alpha}\right)^2\left(\frac{L_0}{L_{1T}}\right)^{3-\frac{R_1}{L_0\alpha}}\left(3 - \frac{2R_1}{L_0\alpha}\right)^{-1} , \quad (5.64)$$

$$\varepsilon_{R_2} = \frac{M^2}{L_0L_{1T}}\frac{R_1}{R_2}\left(\frac{L_0}{L_{1T}}\right)^{2\left(1-\frac{R_1}{L_0\alpha}\right)} . \quad (5.65)$$

Examining these two equations, it can be seen that if the load has a high resistance, $R_2$, the energy it receives is inversely proportional to $R_2$, independent of the load inductance, and directly proportional to the inductance of the secondary winding, $L_{2T}$, of the transformer.

When an MCG, with a specific set of parameters, operates into a high-resistance load, the energy transmitted to the load can be increased by increasing the inductance of the secondary winding, since

$$\varepsilon \approx \frac{M^2}{R_2} \approx \frac{L_{2T}}{R_2} \approx \frac{L_2}{R_2} . \quad (5.66)$$

In this case, the operating mode of the MCG is similar to that of an EMF source (open circuit mode). The increase in $L_{2T}$ is limited by the condition: $R_2 >> L_{2T}/\tau_p$.

### 5.2.3 *Case 3:* $R_1 = 0$ *and* $I_{20} = 0$

If it is assumed that $R_1 = 0$ and that $I_{20} = 0$, i.e., that the switch K is closed at the moment the MCG begins operating, then Eqs. 5.50 and 5.51 may be solved by the method of quadratures for both linear and exponential $L_1$ coupling. When the inductance of the generator, $L_g$, varies exponentially, the following expression can be derived:

$$\gamma'_{I2} = l_L(1+\alpha_L)\left[l_L(1+\alpha_L) - k_c^2 e^{\alpha t}\right]^{v-1} \cdot e^{\alpha t} \int_1^Z \left[\frac{z}{l_L(1+\alpha_L) - k_c^2 z}\right]^v dz. \tag{5.67}$$

The final form of this function is

$$\gamma'_{I2f} = (1+\alpha_L)(1+a_L-k_c^2)^{v-1} \cdot \int_1^l \left[\frac{z}{l_L(1+\alpha_L) - k_c^2 z}\right] dz, \tag{5.68}$$

where $\gamma'_{I2} = -I_2L_2/MI_{10}$, $z = L_0/L_1 = e^{\alpha t}$, $v = R_2/\alpha L_2$, and $a_L = L_L/L_{2T}$.

Similar expressions may be derived for both exponential and linear coupling of $L_g$. The final values of the energy coefficient for both cases are

$$\varepsilon_{mf} = \frac{\alpha k_c^2(\gamma_{I2f})^2}{l_L(1+\alpha_L)^2}, \tag{5.69}$$

$$\Psi_{kf} = \frac{\alpha k_c^2(\gamma'_{I2f})^2}{l_L(1+\alpha_L^2) + k_c(1+\alpha_L^2 - k_c^2)(\gamma'_{I2f})^2}, \tag{5.70}$$

$$\varepsilon_{Rf} = \frac{2vk_c^2}{l_L(1+\alpha_L)} \int_1^l \frac{(\gamma'_{I2})}{z^2} dz\ , \tag{5.71}$$

where $\varepsilon_{Rf} = W_R/W_0$, $W_R$ is the energy released into $R_2$ during the operating time of the MCG, $\Psi_k = L_LM^2/L_2(L_1L_2 - M^2)$, and $W_{Rf}R_L/R_2$ is the energy released into the load. The only difference in these expressions for exponential and linear coupling is in how $\gamma'_{I2}$ is defined. Therefore, the operation of this system depends on the normalized parameters: $k_c$, $l_L$, $\alpha$, and $v$. Plots of $\varepsilon_{mf}$ and $\Psi_{kf}$ versus $\alpha$ for various values of $v$ for both linear and exponential coupling are presented in Fig. 5.9. The curves are calculated for the typical case when $l_L = 10$ and $k_c = 0.9$. The two cases differ by how the characteristics of $\alpha$ and $v$ affect the peaks of the curves. The curves in Fig. 5.9(a) are for exponential coupling $[L_1 = L_0(\exp(-\alpha t))]$ and the curves in Fig. 5.9(b) for linear coupling $[L_1 = L_0(1-\alpha t)]$.

Plots of $\varepsilon_{Rf}$ versus $v$ for different values of $\alpha$, with $l_L = 10$ and $k_c = 0.95$, are presented in Fig. 5.10. The solid lines are for exponential coupling, and the broken lines for linear coupling.

When the condition $R_2 << L_2dI_2/I_2dt$ is satisfied, Eqs. 5.50 and 5.51 has the approximate solutions

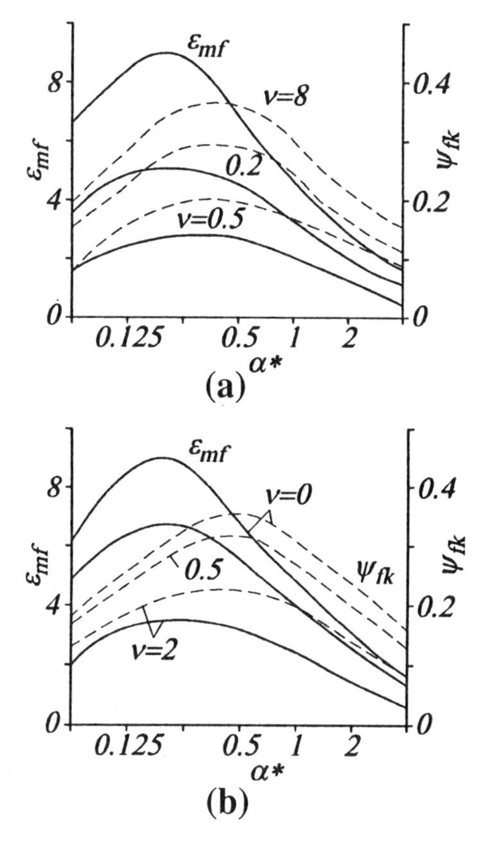


FIGURE 5.9. Plots of energy delivered to (a) exponentially and (b) linearly coupled loads through a transformer when $R_1 = 0$ and $I_{20} = 0$ (Case 3).

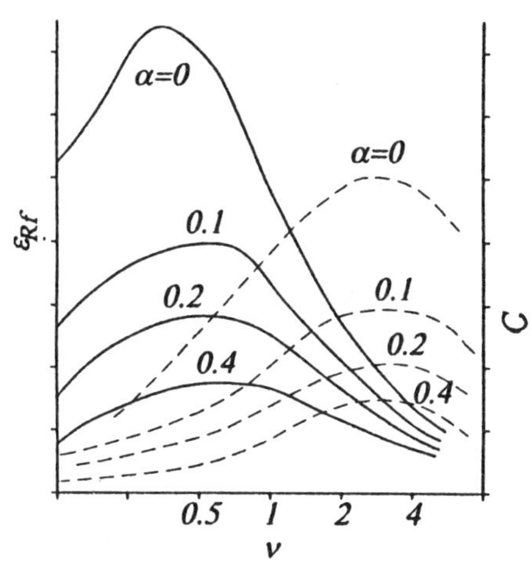


FIGURE 5.10. Plots of energy delivered to $R_2$ through a transformer when $R_1 = 0$ and $I_{20} = 0$ (Case 3) for (—) exponential and (- - -) linear coupling.

$$I_1 = \Phi_0 \frac{\lambda_k}{\lambda_E}, \quad (5.72)$$

$$I_2 = -M \frac{I_1}{L_2}, \quad (5.73)$$

where $\lambda_k = \exp\left(-\int_0^t R_E/L_E dt\right)$, $R_E = R_1 + R_2 M^2/L_2^2$, and $L_E = L_1 - M^2/L_2$.

It is worth noting that it is possible to obtain an analytical solution for Eqs. 5.50 and 5.51 provided that $L_1 = L_0(1+\alpha t)$ and $R_1 = 0$, but inductive coupling is not characteristic of real MCGs.

### *5.2.4 Case 4: Low-Resistance Loads*

When an MCG is connected to a load having low resistance, and it is assumed that $R_2 = 0$, Eqs. 5.50 and 5.51 can be rewritten as

$$\frac{dI_1}{dt} + \frac{R_1 + \frac{dL}{dt}}{L_1 - \frac{M^2}{L_2}} I_1 = 0, \quad (7.74)$$

$$\frac{dI_2}{dt} - \frac{M}{L_2} \frac{dI_1}{dt} = 0. \quad (7.75)$$

In deriving Eqs. 5.74 and 7.75, it was assumed that the terms containing $M$ had a negative sign in Eqs. 5.50 and 5.51. When $t/\tau_p < 1$, $dL_1/dt = 0$, and $L_1 = L_{1T}$, the solution of Eqs. 5.74 and 5.75 is

$$I_1 = I_{10} \frac{L_0 - \frac{M^2}{L_2}}{L_1 - \frac{M^2}{L_2}} \exp\left[-\int_0^t \frac{R_1}{L_1 - \frac{M^2}{L_2}} dt\right]. \quad (5.76)$$

Assuming linear inductive coupling and that $R_1$ is constant in the domain $t \leq \tau_p$, the solution of this equation is

$$I_1 = I_{10} \left[\frac{L_0 - \frac{M^2}{L_2}}{L_1(t) - \frac{M^2}{L_2}}\right]^{1 - \frac{R_1}{\alpha L_0}}. \quad (5.77)$$

Assuming that $L_1 = L_{1T,}$ the solution of Eq. 5.76 in the domain $t \geq \tau_p$ is

$$I_1 = I_{10}\left[\frac{L_0-\frac{M^2}{L_2}}{L_{1T}-\frac{M^2}{L_2}}\right]^{1-\frac{R_1}{\alpha L_0}} \exp\left[-\frac{R_1\tau_p}{L_{1T}-\frac{M^2}{L_2}}\left(\frac{t}{\tau_p}-1\right)\right]. \tag{5.78}$$

Taking into account the initial conditions, the current flowing through the load can be determined from Eqs. 5.74 and 5.75 as

$$I_2 = \frac{M}{L_2}I_1. \tag{5.79}$$

The energy coupled into the load is

$$W_R = R_2\int_0^t I_2^2 dt = \frac{R_2M^2}{L_2^2}\int_0^t I_1^2 dt. \tag{5.80}$$

The ratio of $W_R$ to the initial energy $W_0$ is

$$\varepsilon_R = \frac{2W_R}{L_0I_0^2} = \frac{2R_2M^2}{L_0L_2^2}\int_0^t\left(\frac{I_1}{I_{10}}\right)^2 dt. \tag{5.81}$$

The total energy delivered to the active load is

$$\begin{aligned}\varepsilon_{Rm} &= \frac{2R_2M^2}{L_0L_2^2}\left[\int_o^{\tau_p}\left(\frac{I_1}{I_{10}}\right)^2 dt + \int_{\tau_p}^{\infty}\left(\frac{I_1}{I_{10}}\right)^2 dt\right] \\ &= \frac{W_{1R}+W_{2R}}{W_0},\end{aligned} \tag{5.82}$$

where

$$\begin{aligned}\frac{W_{1R}}{W_0} &= \frac{2R_2M^2}{L_0L_2^2}\int_0^t\left(\frac{L_0-\frac{M^2}{L_2}}{L_1-\frac{M^2}{L_2}}\right)^{2-\frac{2R_1}{\alpha L_0}} dt \\ &= -\frac{2R_2M^2}{\alpha L_0^2L_2^2}\int_{L_0}^{L^{1T}}\left(\frac{L_0-\frac{M^2}{L_2}}{L_1-\frac{M^2}{L_2}}\right)^{2-\frac{2R_1}{\alpha L_0}} dL \\ &= -\frac{2R_2M^2}{\alpha L_0L_2^2}\frac{1-\frac{M^2}{L_2L_0}}{1-\frac{2R_1}{\alpha L_0}}\left\{\left(\frac{L_0-\frac{M^2}{L_2}}{L_{1T}-\frac{M^2}{L_2}}\right)^{1-\frac{2R_1}{\alpha L_0}}-1\right\};\end{aligned} \tag{5.83}$$

$$
\begin{aligned}
\frac{W_{2R}}{W_0} &= \frac{2R_2M^2}{L_0L_2^2}\left(\frac{L_0-\frac{M^2}{L_2}}{L_{1T}-\frac{M^2}{L_2}}\right)^{2-\frac{2R_1}{\alpha L_0}} \\
&\quad\times\int_{\tau_p}^{\infty}\exp\left[-\frac{2R_1\left(\frac{t}{\tau_p}-1\right)}{L_{1T}-\frac{M^2}{L_2}}\right]dt \\
&= \frac{R_2}{R_1}\frac{L_k-\frac{M^2}{L_2}}{L_2}\frac{M^2}{L_2L_0}\left(\frac{L_o-\frac{M^2}{L_2}}{L_{1T}-\frac{M^2}{L_2}}\right)^{2-\frac{2R_1}{\alpha L_0}}.
\end{aligned}
\tag{5.84}
$$

Analysis of this mode of operation of the MCG shows that the inductance of the secondary circuit decreases from $L_1$ to $L_E = L_1 - M^2/L_2$. Thus the generator does not see the load $L_k = L_{1T}$, but rather $L_{Ef}$. This means that the experimental data for specific MCGs, depending on the value of the final inductance, can be used to select $L_{Ef}$ for new optimized models of MCGs. The magnetic energy in the load is $W_L = \Psi_k W$, where $W$ is the total magnetic energy in the MCG. By examining the asymptotic expressions given in Eqs. 5.83 and 5.84, it can be seen that for loads with low resistance, $R_2$, the energy transmitted to the load is directly proportional to $R_2$ and strongly depends on the inductance of the load, $L_L$.

If $k_c = 1$ and $a_L = L_L/L_{2T} \neq 0$, then $\Psi_{kf} = 1/(1+a_L) < 1$ ; i.e., even a transformer with ideal coupling will not deliver all the energy created by the generator to an inductive load. When $a_L \to 0$, then $\Psi_{kf} \to 1$, which means that $L_{Ef} = L_{1T}(1-\Psi_{kf}) \to 0$ and may become smaller than the optimal value of the final inductance of the MCG.

If $k_c \neq 1$, then $L_{Ef} = L_{1T}\left(1 - k_c^2/(1+a_L)\right)$. Plots of $L_{Ef}/L_{1T}$ versus $\alpha$ for different values of $k_c$ are presented in Fig. 5.11. The value of $\Psi_{kf}$ is maximum when $L_{Ef}/L_{1T} = \left(1-k_c^2\right)^{1/2}$.

For low resistances $R_2$, the generator operates like a current source (short circuit mode). As the resistance of the load increases, the energy coupled into the load increases to a peak value and then begins to fall off. Therefore, for a given value of inductance in the secondary winding of the transformer, there is an optimal value of $R_2$, that permits the maximum transfer of energy.

In general, the ratio $W_R/W_0$ for a MCG with a specified set of parameters is a function of $R_2/L_T$ and $L_L/L_T$. As the ratio $L_L/L_T$ increases, the ratio $W_R/W_0$ will always increase except for those cases where the resistances are high and the effects of the inductive impedance of the load may be neglected. Plots of $W_R/W_0$ versus $R_2/L_T$ for specific values of $L_L/L_T$ have a peak value, which determines the optimal conditions for matching the MCG to the load.

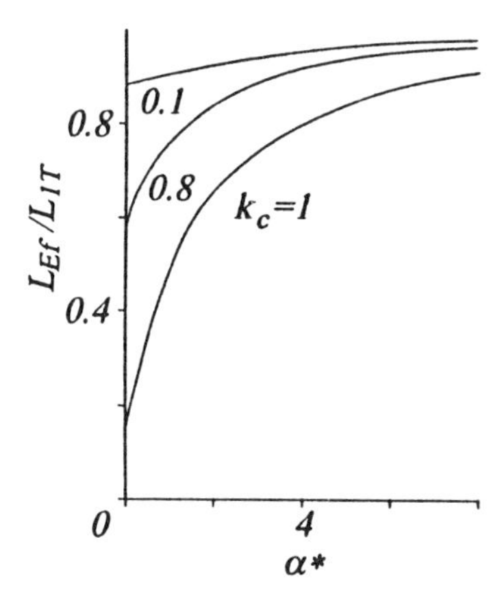


FIGURE 5.11. Plots of inductance versus $\alpha$ for different coupling coefficients, $k_c$, when the load resistance is small (Case 4).

### 5.2.5 Case 5: $R_1 = 0$, $R_2 = 0$, and $C_L = 0$

Consider the case when $R_1 = 0$, $R_2 = 0$, $C_L = 0$, and the switch $K$ is kept closed. This problem can be simplified by considering two simpler independent problems:

- Operation of the MCG into loads with specified finite equivalent inductances.
- Transfer of energy through a transformer from an external EMF source into an inductive load.

Using this approach and possessing data on the MCG such as final inductance and amplitude and shape of the current pulse, as well as having specified the dependence of the energy coupling factor of the transformer on the parameters of the transformer and MCG, the amplitude and shape of the current pulse in the load can be found. In this case, the number of independent variables that determine the energy transfer may be reduced. The simplest system will be considered, i.e., when the active resistance of the circuit, which consists of a transformer (see Fig. 5.12) connected to an external EMF source, $\xi(t)$, is negligibly small. The secondary winding of the transformer, with inductance $L_{2T}$, is connected to the load with inductance $L_L$. For this case, there are four independent variables, $L_{1T}$, $L_{2T}$, $L_L$, and $M$, which unambiguously determine the nature of the system. However, expressing the coupling factor, $\eta_{tr}$, in terms of more than two variables, while desirable, leads to two disadvantages. First, $\eta_{tr}$ tends to depend on a parameter that has dimensions (in particular length), which are inconvenient. Second, when there are four independent variables, there is no clear understanding about which of the parameters has the strongest impact on

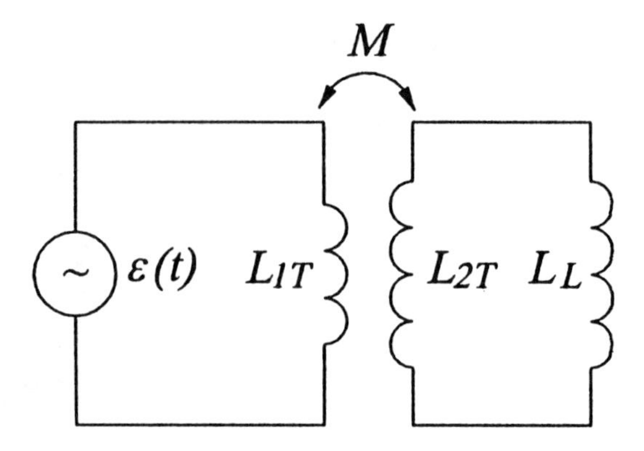


FIGURE 5.12. Equivalent circuit diagram for an MCG connected through a pulsed transformer to an inductive load.

the coupling factor and what relationships among the parameters should be used.

It is possible to get a clearer idea of what factors affect the coupling factor by reducing the number of independent variables by making some of them small in value. For example, if the active resistances of the primary and secondary circuits are neglected, then the system of equations that describe the circuit becomes extremely simple:

$$L_{1T}\frac{dI_1}{dt} + M\frac{dI_2}{dt} = \xi(t), \tag{5.85}$$

$$(L_{2T} + L_L)\frac{dI_2}{dt} + M\frac{dI_1}{dt} = 0. \tag{5.86}$$

Rearranging Eqs. 5.85 and 5.86, yields

$$\frac{dI_2}{dt} + \frac{M}{L_{2T} + L_L}\frac{dI_1}{dt} = 0, \tag{5.87}$$

$$L_{1T}\frac{dI_1}{dt} - \frac{M^2}{L_{2T} + L_L}\frac{dI_1}{dt} = \xi(t). \tag{5.88}$$

Taking into account that $k_c = M/\sqrt{L_{1T}L_{2T}}$, enables Eqs. 5.87 and 5.88 to be rewritten as

$$\frac{dI_1}{dt}\left(L_{1T} - \frac{k_c^2 L_{1T}L_{2T}}{L_{2T} + L_L}\right) = \xi(t). \tag{5.89}$$

From this last expression, the optimal inductance of the external EMF source is found to be

$$L_{1E} = L_{1T} - \frac{k_c L_{1T} L_{2T}}{L_{2T} + L_l} = \frac{L_{1T}(1 + a_l - k_c^2)}{1 + a_L}, \tag{5.90}$$

where $a_L = L_L/L_{2T}$.

Taking into account the initial conditions at $t = 0$, i.e., $I_1 = I_2 = 0$, the solution of Eq. 6.88 is

$$I_2 = -\frac{M}{L_{2T} + L_L} I_1 \tag{5.91}$$

or

$$\frac{I_2}{I_1} = -\frac{k_c\sqrt{L_{1T}L_{2T}}}{L_{2T} + L_L}. \tag{5.92}$$

From these expressions, the energy transfer coupling factor of the transformer is

$$\varepsilon_{Lf} = \frac{W_L}{W_1} = \frac{L_L}{L_{1E}}\left(\frac{I_2}{I_1}\right)^2 = \frac{a_L(1 + a_L)k_c^2}{(1 + a_L - k_c^2)(1 + a_L)^2}. \tag{5.93}$$

Since for any $a_L$ the sum $1 + a_L \neq 0$, this equation can be rewritten as

$$\varepsilon_{Lf} = \frac{a_L k_c^2}{(1 + a_l - k_c^2)(1 + a_L)}. \tag{5.94}$$

Taking the derivative of Eq. 5.93 with respect to time and setting the result equal to zero, the peak value of the coupling factor can be found provided the following condition is met:

$$1 - a_L^2 = k_c^2. \tag{5.95}$$

Substituting this condition into Eq. 5.93, it is found that $\varepsilon_{Lf_{\max}}$ is equal to

$$\varepsilon_{Lf_{\max}} = \frac{(1 - a_L^2)a_L}{2a_L^2 + 2a_L - (1 - a_L^2)a_L}. \tag{5.96}$$

Since $a_L \neq 0$, it can be shown that

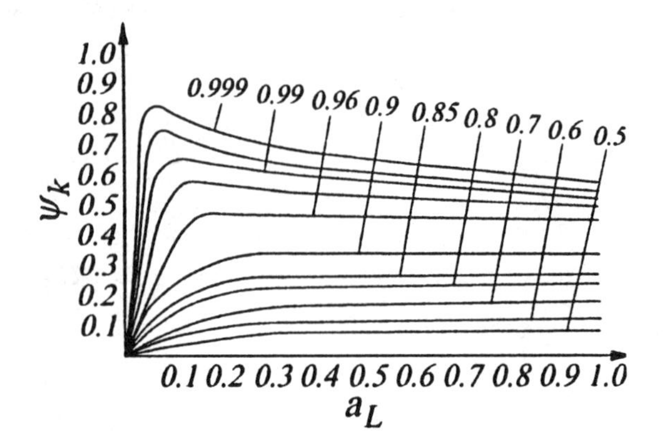


FIGURE 5.13. Plots of $\Psi_k$ versus inductance for different coupling factors (Case 5).

$$\varepsilon_{Lf_{\max}} = \frac{1 - a_L}{1 + a_L} = \frac{1 - \sqrt{1 - k_c^2}}{1 + \sqrt{1 + k_c^2}}. \tag{5.97}$$

In Fig. 5.13, in which $\Psi_k(a_L, k_c)$ is plotted versus $a_L$, it can be seen that the position of the peak value of $\varepsilon_{Lf}$ depends on the ratio $L_l/L_{2T}$. It can also be seen that high values of the energy coupling coefficient, i.e., $\varepsilon_{Lf} = 0.9$, require that the coupling factor of the transformer be extremely high, i.e., $k_c = 0.999$, which is unlikely in practice. When the value of $k_c$ is high, i.e., $k_c = 0.9$, then only 40% of the energy generated by the MCG can be coupled into the load by the transformer. If the value of $k_c$ decreases by nearly 9%, i.e., from 0.99 to 0.9, the amount of energy delivered to the load decreases by almost a factor of two, i.e.,. from 75% to 39%. The influence of the second normalized parameter of the system, i.e., $a_L$, on the energy coupling factor is much weaker. For example, when using practical values of the coupling factor ($0.9 \leq k_c \leq 0.95$), even noticeable deviations in the parameters, i.e., nearly 20–25%, from those values that yield the peak value of $\varepsilon_{Lf}$, in practice, do not lower the amount of energy coupled to the load. To verify the dependency of $\varepsilon_{Lf}$ on $\alpha$ and $k_c$, experiments were conducted, and these showed that experimental and numerical results agree to within approximately 5%.

When the inductance of the MCG is taken into account, the energy coefficient, which is the ratio of the energy delivered to the load to that energy stored in the generator, is

$$\varepsilon_f = \frac{l_L(1 + a_L) - k_c^2}{1 + a_L - k_c^2}, \tag{5.98}$$

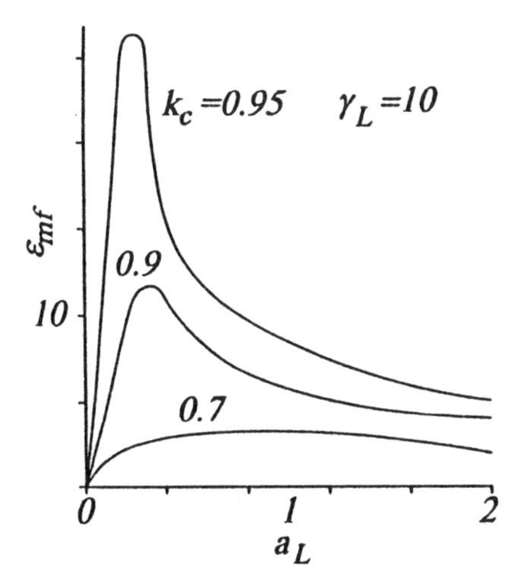


FIGURE 5.14. Plots of $\varepsilon_{mf}$ versus inductance for different coupling factors (Case 5).

$$\varepsilon_{Lf} = \frac{a_L k_c^2}{1 + a_L - k_c^2}\left(l_L - \frac{k_c^2}{1 + a_L}\right), \tag{5.99}$$

where $l_L = L_0/L_{1T}$, $\varepsilon_f = W_f/W_0$, $\varepsilon_{Lf} = W_{Lf}/W_0$, and $W_f$ is the energy in the generator during the final stages of its operation. Plots of $\varepsilon_{mf}$ versus $a_L$ for different values of $k_c$ at $\gamma_l = 10$ are presented in Fig. 5.14. At the peak value of $\varepsilon_{mf}$, it can be shown that $L_{Ef}/L_{1T} = 1 - k_c^2/(1 - k_c^2/2)$ and that $a_L = 1 - k_c^2$. When $R_1 \neq 0$, the location of the peak depends on the magnetic flux attenuation. For example, when $L_1 = L_0(1 - \alpha t)$ and $R_1 =$ const, $\varepsilon_{mf}$ has a peak value at $a_L = (\sqrt{\delta_L^2 k_c^4 + 4(1 - k_c^2)} - \delta_L k_c^2)/2$, where $\delta_L = (1 + 2R_1)/(dL_1/dt)$. This is also true when the inductance decreases exponentially, i.e., $L_1 = L_0 e^{-\alpha t}$ and when $L_1/R_1 =$ const.

The required value of $a_L$ for a fixed value of $L_{2T}$ depends on the selected value of $L_L$. When choosing a transformer for a predetermined value of $L_L$, the value of $a_L$ is determined by the values of $L_{1T}$ and $M$, i.e., by adjusting the number of turns in the secondary winding. If the value of $k_c$ can be made constant by suitable design, which is not always possible, then for $\Psi_{kf}$ to be at its peak value it is necessary that $M = k_c[L_{1T}^2 L_L^2(1 - k_c^2)]^{1/4}$, and for $\varepsilon_{Lf}$ to be at its peak value, that $R_1 = 0$, when $M = k_c\sqrt{L_{1T} L_L (1 - k_c^2)}$. The relationship between $M$ and $L_{1T}$ depends on the design of the transformer. If, for example, $L_{1T} = M/k_c\omega$, where $\omega$ is the number of secondary winding turns, then, at fixed values of $L_{1T}$ and $k_c$, for $\Psi_{kf}$ to be at its peak value, it is necessary that $\omega = (L_L L_{1T})^{1/2}(1 - k_c^2)^{1/4}$ and at fixed values of $\omega$ and $k_c$ that $L_{1T} = L_L/\omega^2(1 - k_c^2)^{1/2}$.

### 5.2.6 *Case 6: Active Load, When $R_1 = 0$*

Consider the case when the MCG is connected through a transformer to an active load with $R_1 = 0$. An *active load* is one that is considered to act like a current or voltage source or sink. The circuit equations for this case are

$$\frac{d(L_1 I_1)}{dt} + M\frac{dI_2}{dt} = 0, \tag{5.100}$$

$$L_{2T}\frac{dI_2}{dt} - R_L I_2 + M\frac{dI_1}{dt} = 0, \tag{5.101}$$

where $L_1 = L_{1T} + L_B + L_g$. The energy transmitted to the load during the time the generator is operating is given by

$$W_L = R_L \int_0^{\tau_p} I_2^2(t)dt, \tag{5.102}$$

where $\tau_p = l_g/D$, $l_g$ is the length of the generator, and $D$ is the velocity of the detonation wave. The mechanical work done by the explosion products against the electromagnetic forces is

$$W_M = -\frac{1}{2}\int_0^{\tau_p} I_1^2(t)dt. \tag{5.103}$$

The initially stored electromagnetic energy, i.e., at $t = 0$, is

$$W_{M0} = \frac{1}{2}(L_{g0} + L_{1T})I_{10}^2 + \frac{1}{2}L_{2T}I_{20}^2 + MI_{10}I_{20}, \tag{5.104}$$

where $\mathrm{L}_{g0} = L_g(0)$, $I_{10} = I(0)$, and $I_{20} = I_2(0)$. The energy stored when $t = \tau_p$ is

$$W_M = \frac{1}{2}L_{1T}I_{1m}^2 + \frac{1}{2}L_{2T}I_{2m}^2 + MI_{1m}L_{2m}, \tag{5.105}$$

where $I_{1m} = I_1(\tau_p)$ and $I_{2m} = I_2(\tau_p)$ are the peak values of $I_1$ and $I_2$, respectively. The energy balance equation, while the system it is operating, is

$$W_{M_0} + W_M = W_{M_f} + W_M. \tag{5.106}$$

While the generator is operating, part of the energy is stored in the form of magnetic energy in the windings of the pulsed transformer. After the operating cycle of the generator is complete, the energy stored in the transformer is discharged into the load. The discharge current may be found by letting $L_g = 0$ in Eqs. 5.100 and 5.101:

$$I_2 = I_{2m} e^{\frac{t}{\tau_p}}, \tag{5.107}$$

where $\tau_p = \tau\left(1 - k_c^2/(1 + L_B/L_g)\right)$ is the discharge time constant, $\tau = L_{2T}/R_L$, and $L_B$ is the inductance of the transmission lines from the MCG to the transformer. Using Eqs. 5.100 and 5.101, it can be shown that

$$\begin{aligned}(L_B + L_{1T})I_1 + MI_2 &= \text{const} \\ &= (L_B + L_{1m})I_{1m} + MI_{2m}.\end{aligned} \tag{5.108}$$

From this expression, it is clear that not all the electromagnetic energy in the primary circuit is delivered to the load; that is, when $t \to 0$, then $I_2 \to 0$ and $I_1 \to I_{1\infty} = I_{1m} + M/(L_B + L_{1T})I_{2m}$.

It was shown in Chapter 3 that it is necessary that the linear current density, along the line where the plates of the PMCG make contact, is constant to provide the maximum conversion of explosive energy into electromagnetic energy. If this condition is accepted, then $j = I_1/y = j_0$, where $y(x)$ is the width of the plates at a distance $x$ from the origin. Optimization of this problem thus reduces to that of seeking the proper form of the expression $y(x)$ and of Eqs. 5.100 and 5.101 for the condition that $I_1/y = j_0$ and when $dL_g/dt = -f_g/y$, where $f_g = 2\mu_0 bD$ and $b$ is the length of the gap between the plate and cassette (for a PMCG). The solutions of the resulting equations are

$$y = y_0 e^{\frac{t}{\tau}}, \tag{5.109}$$

$$L_g = \frac{f_g \tau}{y_0}\left(e^{-\frac{t}{\tau}} - e^{\frac{\tau_p}{\tau}}\right), \tag{5.110}$$

$$I_1 = I_{10} e^{\frac{t}{\tau}}, \tag{5.111}$$

$$I_2 = I_{20} e^{\frac{t}{\tau_p}}, \tag{5.112}$$

$$k_c^2 = 2\left(1 + \frac{L_B}{L_1} - \frac{f_g \tau}{L_1 y_0} e^{-\frac{t}{\tau_p}}\right), \tag{5.113}$$

$$I_{20} = -\frac{k_c}{2}\sqrt{\frac{L_{1T}}{L_{2T}}} I_{10}. \tag{5.114}$$

Equation 5.113 directly follows from the energy balance equation for an PMCG with an exponential profile, and Eq. 5.114 must be satisfied while the generator is operating. The following equations can be derived from Eqs. 5.103, 5.104, and 5.109–5.114:

$$W_M = \frac{f_g \tau}{2 y_0}\left(e^{\frac{\tau_p}{\tau}} - 1\right) I_{10}^2, \tag{5.115}$$

$$W_{M0} = \frac{1}{2}\left(L_{g0} + L_B + \left(1 - \frac{3}{4}k_c^2\right) L_{1T}\right) I_{10}^2, \tag{5.116}$$

$$\frac{W_M}{W_{m0}} = e^{\frac{\tau_p}{\tau}}\left(1 + \frac{L_B}{L_{g0}} + \left(1 - \frac{3}{4}k_c^2\right)\frac{L_{1T}}{L_{g0}}\right)^{-1}. \tag{5.117}$$

The most efficient mode of operation occurs when $W_M/W_{M0} >> 1$, which means that $e^{\tau_p/\tau} >> 1$, $L_N \leq L_{g0}$, and $L_1 \leq L_{g0}$. The electrical efficiency is

$$\begin{aligned}\frac{W_H}{W_{M0} + W_M} &= k_c^2 \frac{2 - \frac{k_c^2}{1 + \frac{L_B}{L_{1T}}} - e^{-\frac{2\tau_p}{\tau}}}{1\left(1 + \frac{L_B}{L_{1T}}\right) - 2k_c^2\left(1 + 0.5e^{-\frac{2\tau_p}{\tau}}\right)} \\ &= \frac{k_c^2}{1\left(1 + \frac{L_B}{L_{1T}}\right)},\end{aligned} \tag{5.118}$$

which in the most favorable case, i.e., when $k_c = 1$ and $L_B/L_{1T} << 1$, is 0.5. When $k_c = 0.9$ and $L_B/L_{1T} << 1$, the efficiency is 0.4. Thus, the requirement that the linear current density is constant implies that not more than half of the total electromagnetic energy can be coupled into the load.

### 5.2.7 *Case 7: Pulse-Shaping Transformers*

Transformers allow some shaping of the current pulse being delivered to the load. Peak power is usually achieved during the final stages of operation of the MCG. The method used to sharpen the current pulse front with the opening of the circuit is based on the following scheme: the transformer operates initially with the secondary circuit in the open circuited mode, and the switch K is closed a time $\tau_s$ after the MCG begins to operate. In this case, the pulse length will be $\tau_p - \tau_s$, provided that the current $I_2$ increases throughout the operation of the generator. After the switch K is closed and the inductive load is introduced into the circuit ($R_2 = 0$), the currents are

$$I_1 = \frac{\Phi_0 \lambda_{kS} \lambda_k L_{1S}}{\lambda_{kS} L_{ES} L_E}, \quad I_2 = \frac{\Phi_0 \lambda_{kS} M}{L_{13} L_{12}} \left( \frac{\lambda_k \lambda_{ES}}{\lambda_{ks} \lambda_E} - 1 \right), \tag{5.119}$$

where the index S refers to the value of the parameter at the moment the switch is closed. The rate of change of $I_1$, $dI_1/dt$, increases sharply (steplike) by a factor of $L_1/L_{ES}$ at the moment the switch is closed.

The parameter $\varepsilon_{Lf}$ decreases as $\tau_p - \tau_s$ decreases. Denoting the fraction of energy in the MCG by $\varepsilon_S$, which is the ratio of the energy in the load when the secondary circuit is closed, $W_{Lf}$, to the energy in the load at the moment, $\tau_S$, the switch is closed, $W_{LS}$. When $R_1 = 0$, it can be shown that $\varepsilon_S = 1 - L_{1T}/L_{1S}$. The degree to which the pulse front decreases is $\tau_p/(\tau_p - \tau_S) = (\gamma_L - 1)\left(e_S^{-1/2} - 1\right)$ for linear coupling of the inductance, $L_g$, and $\tau_p/(\tau_p - \tau_S) = -\ln \gamma_L / \ln\left(1 - \epsilon_S^{-1/2}\right)$ for exponential coupling. If $R_1 \neq 0$, then

$$\varepsilon_S = \left[ \frac{\lambda_{kS}}{L_{1S}} \left( \frac{L_{ES}}{\lambda_{ES}} - \frac{L_{Ef}}{\varphi_{Ef}} \right) \right]^2 . \tag{5.120}$$

In the case of an inductive load, this method allows one to shorten the current pulse front by a factor of 5–10 and still have an acceptable value of $\varepsilon_S$. During those periods when the transformer is idle, the values of $\Phi_0$ should be higher, because, in comparison to the operating mode when the switch is closed, the generator will operate as in an unloaded mode. If in both cases the value of the parameter $W_0$ is the same, then $\varepsilon_S$ increases by a factor of $l_L(1 + a_L)/[l_L(1 + a_L) - k_c^2]$, and if $I_{10}$ is the same for both cases, then $\varepsilon_S$ increases by a factor of $l_L^2(1 + a_L)^2 / \left[l_L(1 + a_L) - k_c^2\right]^2$. In the case of resistive loads, sharpening the front of the current pulse also sharpens the front of the power pulse.

If the length of the current pulse delivered to a resistive load needs to be longer than $\tau_p$, a storage inductor can be connected in series to $R_L$. This effectively increases $L_L$, so that during the time that the MCG is operating, the bulk of the energy is stored in $L_L$ and it is subsequently released into $R_L$ with a relaxation time of $L_L/R_L$. The transformer, which is destroyed when the MCG has completed operating is shunted by an additional switch. When the MCG is used in this manner, it operates, in practice, like an inductive store. The high powers generated by MCGs makes them different from low-power inductive stores and allow the inductive losses to be significantly lower during the charging process.

Attenuation of the current at the end of the MCG operation is described by the equations for an $RL$ circuit with initial currents of $I_{1f}$ and $I_{2f}$. If the transformer is functioning during the attenuation period, an additional amount of the energy in the MCG may be transferred to the load. If $R_2$ has a low value when the MCG is operating and if the secondary circuit of the transformer is closed, then $\gamma_{I2f} = \gamma_{I1f}$. While the current is being attenuated,

$$\gamma_{I2} = \gamma_{I2f} \left[ e^{\lambda_2 t} + \frac{\left(e^{\lambda_1 t} - e^{\lambda_2 t}\right) \lambda_1}{\lambda_1 - \lambda_2} \right], \tag{5.121}$$

where

$$\begin{aligned} \lambda_{1,2} &= \left[ -(\delta_1 + \delta_2) \mp \frac{\sqrt{(\delta_1 - \delta_2)^2 + \frac{4\delta_1 \delta_2 k_c^2}{1+a_L}}}{2 - \frac{2k_c^2}{1+a_L}} \right], \\ \delta_1 &= \frac{R_1}{L_{1T}}, \\ \delta_2 &= \frac{R_2}{L_2}. \end{aligned} \tag{5.122}$$

At the end of the MCG operation, when $t_1 = \ln(\lambda_2/\lambda_1)/(\lambda_1 - \lambda_2)$, $\gamma_{I2} = 0$, after which it changes sign. The peak value of $\gamma_{I2}$ occurs when time $t_2 = 2t_1$, after which the value of $\gamma_{I2}$ decreases. The additional energy, $\Delta W_R$, delivered to $R_2$ while the current is being attenuated is given by

$$\frac{\Delta W_R}{W_f} = \frac{k_c^2}{(1 + a_L)\left(1 + \frac{\delta_1}{\delta_2}\right)}, \tag{5.123}$$

which means that the condition $\delta_2 >> \delta_1$ must be satisfied. This condition is rather hard to achieve, when a prolonged current pulse must be provided to the load. When $a_L$ is varied, its effect on the MCG should be taken into account while it is operating. If $R_1$ is also neglected as well, then

$$\frac{W_R + \Delta W_R}{W_0} = \frac{k_c^2 \left[\gamma_L(1 + a_L) - k_c^2\right]}{(1 + a_L)(1 + a_L - k_c^2)\left(1 + \frac{\delta_1}{\delta_2}\right)}. \tag{5.124}$$

If the secondary circuit is closed at time $\tau_S$, then for small values of $R_1$ and $R_2$, $\gamma_{I2f} = \gamma_{I1f} - \gamma_{I1S}$. The parameter $\gamma_{I2f}$ decreases and the amplitude in the backward-traveling half-wave increases, along with an increase in $\tau_S$, so that the peak amplitude in the second half-wave lies between $t_1$ and $t_2$. The magnetic flux captured by the secondary circuit at the instant the switch is closed also decreases. In practice, when $\tau_S = \tau_p$, all the magnetic flux has been captured and $\gamma_{I2f} = 0$, and only the backward-traveling half-wave of the current exists with a peak amplitude at $t_1$, which determines the rise time of the current pulse. Since, in this mode of operation, the secondary circuit does not affect the operation of the MCG for any values of $R_1$ and $R_2$, it can be shown that

$$\gamma_{I2} = \frac{\gamma_{I1f}\delta_1(1+a_L)}{(1+a_L-k_c^2)(e^{\lambda_2 t} - e^{\lambda_1 t})} \tag{5.125}$$

and

$$\frac{\Delta W_2}{W_f} = \frac{k_c^2}{(1+a_L)\left(1+\frac{\delta_2}{\delta_1}\right)}. \tag{5.126}$$

If $\delta_2 << \delta_1$, then $W_L$ decreases, which is why it is desirable to increase $\delta_1$ only after the MCG has completed its operation by introducing, for example, an opening switch. The most effective way of increasing $\delta_1$ is not by changing $a_L$, but rather by decreasing $R_2$, which increases the Q-factor of the secondary circuit and, thus, the dimensions of the transformer.

Plots of $\gamma_{I2}$ versus the operating time of the MCG, calculated by assuming linear coupling of $L_L$ for $\gamma_L = 10$, $k_c = 0.9$, $a_L = 1$, $\delta_1\tau_p = 0.5$, and $\delta_2\tau_p = 0.1$ and by neglecting $R_1$ and $R_2$, are presented in Fig. 5.15. Curve 1 corresponds to the operating mode when the secondary circuit is closed, and Curve 2 to the case when $\tau_S = 0$, Curve 3 to $\tau_S = 0.4\tau_p$, and Curve 4 to $\tau_S = \tau_p$ (the flux capture mode).

During the flux capture mode, which occurs a time $t_1$, so that after the circuit is closed, the energy in $L_L$ is at its peak value

$$\frac{W_L}{W_f} = \frac{a_L k_c \delta_1^2}{(1+a_L-k_c^2)\lambda_1^2}\left(\frac{\lambda_2}{\lambda_1}\right)^{\frac{2\lambda_2}{\lambda_1-\lambda_2}}. \tag{5.127}$$

When $\delta_2 >> \delta_1$, then

$$\gamma_{I2} = -\gamma_{I1f}\left\{1 - \exp\left[-\frac{\delta_1(1+a_L)t}{1+a_L-k_c^2}\right]\right\} \tag{5.128}$$

and

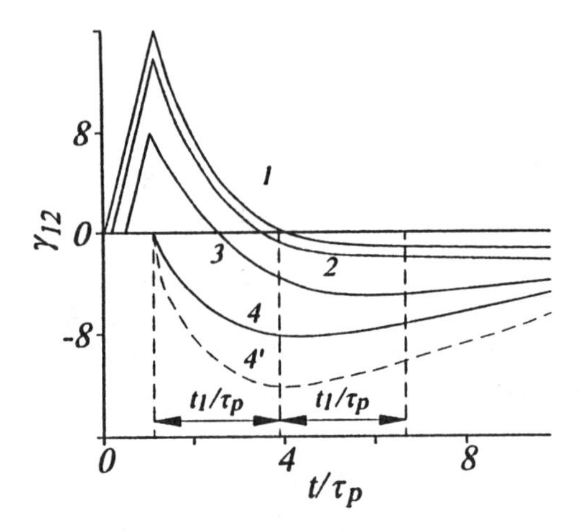


FIGURE 5.15. Plots of relative current in the load versus operating time of the MCG assuming that the MCG is linearly coupled and that $\gamma_L = 10$, $k_c = 0.9$, $a_L = 1$, $\delta_1 \tau_p = 0.5$, and $\delta_2 \tau_p = 0.1$. Curve 1 corresponds to the case when the secondary circuit is closed, Curve 2 to the case when $\tau_S = 0$, Curve 3 to $\tau_S = 0.4\tau_p$, and Curve 4 to $\tau_S = \tau_p$.

$$\frac{W_H}{W_f} \approx \frac{a_k k_c^2}{(1 + a_L)^2}. \tag{5.129}$$

If, at the end of the operation cycle of the MCG, $\delta_1$ increases sharply, then, for significantly low values of $\delta_2$, a current pulse with a fast risetime and attenuation over a long period of time may be generated.

During the flux capturing mode, it is desirable to select the value of $L_{1T}$ to be equal to the optimal value of the final inductance, while preserving the values of $\gamma_L$, $k_c$, and $a_L$. This provides the same mode of operation as for a closed secondary circuit, and the peak value of $\gamma_{I2}$ increases, as shown in Curve 4 of Fig. 5.15.

## 5.3 Connecting Through an Electroexplosive Switch

This section will consider the coupling of energy from an MCG into a load through an electroexplosive switch [5.16], used to match the impedance of the load to that of the generator. Three types of loads will be discussed: complex, active, and inductive.

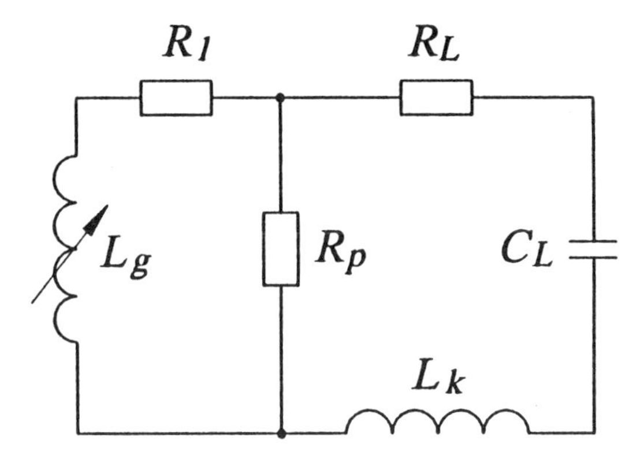


FIGURE 5.16. Equivalent circuit diagram for an MCG connected through an EEB to a complex load.

### 5.3.1 Complex Load

If the load exhibits a complex behavior, then the processes that take place in the circuit may be separated into three stages. During the first of these, the resistance of the electroexplosive switch is $R_p << |Z_L|$, where $|Z_L|$ is the absolute value of the complex impedance of the load. If it is assumed that the switch resistance during this stage of operation is equal to its resistance at the temperature at which the material vaporizes, the equation of the equivalent circuit diagram in Fig. 5.16 is

$$\frac{d(L_g I)}{dt} + R_1 t = 0, \tag{5.130}$$

where $R_1 = R_p(t) + \bar{R}_g$ is the active resistance of the circuit, $\bar{R}_g$ is the active resistance for the generator, and $L_g$ is the inductance of the generator.

In order to solve this equation for its first stage of operation, the number of conductors required in the switch must first be determined. If it is assumed that the generator inductance behaves linearly, i.e., $L_g = L_0(1-\alpha t)$, then the overall dimensions of the switch can be found by using Eq. 4.10. Knowing the geometrical dimensions of the switch conductors and the details of the materials used in their construction, the number of parallel conductors, $N$, required and their total resistance can be determined. By substituting this information into Eq. 5.25 and solving Eq. 5.130, the current flowing through the MCG-switch circuit during the first stage of operation can be calculated.

Since the resistance of the switch is small ($\sim$ 0.01 Ohm), it shunts the complex load. Therefore, during this first stage of operation, practically no current flows through the load and no energy is coupled into the load.

This stage of operation continues until the exploding conductors in the switch are vaporized, at which time the second stage of operation begins. The resistance of the switch can be found by using Eq. 4.8. Operation of the circuit during the second stage is described by the following system of equations:

$$\frac{d\left(L_g I\right)}{dt} + R_p I_p = 0, \tag{5.131}$$

$$\frac{d\left(L_g I\right)}{dt} + L_L \frac{dI_L}{dt} + R_L I_L + \frac{1}{C_L} \int_{T_s}^{T_s + t_k} I_L dt = 0, \tag{5.132}$$

$$I = I_p + I_L, \tag{5.133}$$

where $I_p$ is the current flowing through the switch and $I_L$ is that flowing through the load. This system of equations can only be solved numerically. Eqs. 5.131–5.133 can be rewritten in a form more suitable for numerical solution:

$$\frac{dI}{dx} = -t_k \frac{R_p}{L_k} I_p, \tag{5.134}$$

$$\frac{dI_L}{dx} = y, \tag{5.135}$$

$$\frac{dI_p}{dx} = -\left(t_k \frac{R_p}{L_k} I_p + y\right), \tag{5.136}$$

$$\frac{dy}{dx} = -\frac{1}{L_L}\left(t_k R_p \left(\frac{t_k R_p}{L_k} + y\right) + t_{kz} R_L y + \frac{t^2 I_L}{C_L}\right), \tag{5.137}$$

$$\frac{dR_p}{dx} = -\frac{2R_p(0)}{(1-x)^{-3}}, \tag{5.138}$$

by introducing the normalized variable $x = t/t_k$, which is the ratio of the moment in time under consideration to the time at which the conductors explode. This transition to normalized variables simplifies the calculations.

Equations 5.134–5.138 completely describes the electrical processes that occur in the circuit. When these equations are solved numerically, it can be determined how the currents depend on the variable $x$ throughout all branches of the circuit during this entire stage of operation. The currents and resistances calculated on completion of the first stage of operation serve as the initial conditions for the second stage of operation. The energy coupled into the load can be calculated by summing the products of $I_L\,|Z_L|$ for each step $\Delta x$.

During the third stage of operation, the resistance of the switch tends to infinity and the MCG operates only on the load. The equation that describes the third stage of operation of this circuit is

$$\frac{dI}{dt} + R_L I + L_L\frac{dI}{dt} + \frac{1}{C_L}\int_{\tau_s+t_k}^{t_f} I\,dt = 0, \tag{5.139}$$

which can be rewritten as

$$(L_g + L_L)\frac{d^2I}{dt^2} + \left(2\frac{dL_g}{dt} + R_L\right)\frac{dI}{dt} + \left(\frac{d^2L_g}{dt^2} + \frac{1}{C_L}\right)I = 0. \tag{5.140}$$

If it is assumed that the inductance varies linearly, this equation can be reduced to a second-order linear differential equation:

$$a_1\frac{d^2I}{dt^2} + b_1\frac{dI}{dt} - \frac{1}{C_L} = 0, \tag{5.141}$$

where $a_1 = L_L + L_0(1 - \alpha t)$, $b_1 = R_L - 2L_S$, and $L_S = \alpha L_S$. Introducing the variable

$$I = A_1 \exp\left[\int_{\tau_s+t_k}^{t_f} \bar{I}dt\right], \tag{5.142}$$

where $\bar{I} = (1/I)dI/dt$, Eq. 5.141 becomes the Ricatti equation:

$$a_1\frac{d\bar{I}}{dt} + a_1\bar{I}^2 + b_1\bar{I} + \frac{1}{C_L} = 0. \tag{5.143}$$

Assuming that $\bar{I} = z + \beta(x)$, where $\beta(x) = -b_1/2a_1$, and substituting it into this equation yields

$$a_1\left[z + \beta'(x)\right] + a_1 z^2 + \frac{b_1^2}{4a_1} + \frac{1}{C_L} = 0. \tag{5.144}$$

Solving for $\beta'(x)$ and substituting into this equation, yields the canonical equation

$$a_1 z + a_1 z^2 - \left( \frac{R_L^2 - 2\alpha L_0 R_L}{4a_1} - \frac{1}{C_L} \right) = 0. \tag{5.145}$$

Introducing the following notation

$$\xi = \frac{\alpha L_0 \left[2a_1 - C_L(a_1 + b_1)\right]}{2a_1^3 C_L}, \tag{5.146}$$

$$\eta = \frac{16a_1^2 - 8a_1 b_1 C_L - b_1^2 C_L}{16a_1^4 C_L^2}, \tag{5.147}$$

Eq. 5.145 becomes

$$\eta A_1^2 + \xi A_1 + \bar{C}(t) = 0. \tag{5.148}$$

The roots of this equation are

$$A_{1,2} = \frac{-\xi \pm \left[\xi^2 - 4\eta \bar{C}(t)\right]^{\frac{1}{2}}}{2\eta}, \tag{5.149}$$

and its solutions are

$$z_1 = A_1 \bar{C}(t), \; z_2 = A_2 \bar{C}(t). \tag{5.150}$$

Taking into account these solutions and solving the Ricatti equation, i.e., Eq. 5.143, it follows that

$$\bar{I}_1 = z_1 - \frac{b_1}{2a_1}, \; \bar{I}_2 = z_2 - \frac{b_1}{2a_1}, \tag{5.151}$$

where the general solution is

$$\bar{I} = \bar{I}_1 + \frac{\bar{I}_2 - \bar{I}_1}{1 + C_L \exp\left[\int_{T_s + t_k}^{t_f} \left(\bar{I}_2 - \bar{I}_1\right) dt\right]}. \tag{5.152}$$

The current flowing through the circuit can be found by using the following expression:

$$I = \Gamma \exp\left[\int \bar{I} dt\right]. \tag{5.153}$$

The integration constants, $\Gamma$ and $C_L$, can be found by introducing the initial conditions.

Solving the circuit for the third stage of operation begins by finding the integration constants. If one then finds the coefficients $\xi$ and $\eta$ and the general solution of the Ricatti equation, then the time variation of current flowing through the circuit can be determined. The energy transmitted to the load is

$$W_L = \int_{\tau_s + t_k}^{t_f} I^2 \left| z_L \right| dt. \tag{5.154}$$

### 5.3.2 *Active Load*

If the MCG is connected to an active load, the system of equations that describe its operation is

$$L \frac{dI_1}{dt} + R_L I_L = 0, \tag{5.155}$$

$$R_L I_L + R_P I_P = 0, \tag{5.156}$$

$$I_L + I_P = I. \tag{5.157}$$

Solving these equations for the initial conditions $I(0) = I_0$, $R_L = \text{const}$, $R_P(0) = R_k$, and $L = \text{const}$, where $R_k$ is the switch resistance at the temperature at which the conductors vaporize, $R_P$ is the resistance of the switch, and $R_L$ is the resistance of the load, the general expression for the current in the load is

$$I_L(t) = I_0 \frac{1}{g(t)} \exp \left[ -\frac{1}{\theta_L} \int_0^t \frac{dt}{g(t)} \right], \tag{5.158}$$

where $g(t) = 1 + R_L / R_P$ and $\theta_L = L / R_L$. By letting

$$g(t) = 1 + \frac{R_L (1 - \eta)^2}{R_P(0)}, \tag{5.159}$$

the time variation of the switch resistance can be determined:

$$\int_0^t \frac{dt}{g(t)} = -\tau \sqrt{\frac{R_P(0)}{R_L}} \,. \tag{5.160}$$

$$\left\{ \arctan \left[ \sqrt{\frac{R_L}{R_P(0)}} \left( 1 - \frac{t}{\tau} \right) \right] - \arctan \sqrt{\frac{R_L}{R_P(0)}} \right\}.$$

Therefore, the current flowing through the load is

$$I_L = \frac{I_0}{1 + \beta_R \left(1 - \frac{t}{\tau}\right)^2} \cdot \tag{5.161}$$

$$\exp \left\{ \frac{\tau}{\theta_L} \sqrt{\frac{1}{\beta_R}} \left[ \arctan \sqrt{\beta_R} \left( 1 - \frac{t}{\tau} \right) - \arctan \sqrt{\beta_R} \right] \right\},$$

where $\beta_R = R_L/R_P(0)$. Equation 5.161 characterizes the operation of an MCG connected to an intermediate inductive store. If the MCG is connected directly to the load, then Eqs. 5.155–5.157 must be solved by using the three stages discussed in the previous section.

If it is assumed that $W$ is the energy stored in the inductor when there is no switch present and that $\tau_P = t_0$, where $t_0$ is the effective time it takes to store the energy, then $t_0$ can be found by using

$$t_0 = \frac{1}{I_*^2} \int_0^{t_1} I^2 dt, \tag{5.162}$$

where $I_*$ is the peak current of the MCG and $t_1$ is the point in time at which the load is connected to the MCG.

Nearly 15% of the available energy is coupled into the load. Based on this, the following relationships can be written

$$P = \frac{6W}{t_0} \approx \frac{40 W_L}{t_0}, \tag{5.163}$$

where $W_L$ is the energy delivered to the load and $P$ is a limit on the amount of pulsed power in the active load. The relationship between the pulse length in the load, $\tau_L$, and the time $t_0$ to store the energy is

$$\tau_L \approx 0.045 t_0. \tag{5.164}$$

For example, estimating the accumulation time and energy required to generate $10^{14}$ W of power in $\tau_L = 10^{-7}$ s, it follows that $t_0 = 2$ $\mu$s and that $W = 30$ MJ.

### 5.3.3 *Effects of Switch Inductance on Energy Coupling Coefficient for an Inductive Load*

It was stated in the previous section that a peak energy coupling coefficient of 15% is achieved when the inductance of the storage device equals the inductance of the load. Consider now the effects of the switch inductance on the energy coupling coefficient when an MCG is connected to an inductive load.

In Section 3.4, the exponent $F$ (the "perfectness" factor), Eq. 3.36, was introduced. It was found that as the factor $F$ increases, the system becomes more efficient. Energy amplification by an MCG occurs when $F > 0.5$. To calculate the energy coupling coefficient when an MCG is connected to an inductive load with a coefficient $F$ and with spurious inductance in the switch, consider the equivalent electrical circuit in Fig. 5.17. Here the inductance $L_1(t)$ is the decreasing inductance of the generator, $L_2$ is the inductance of the load, and $L_3$ is the inductance of the switch. Assuming that the switch resistance increases sharply from 0 to infinity at the moment the circuit opens, then at the moment the flux compression begins ($t = 0$), the current flowing in the generator with an initial inductance of $L_0$ is $I_0$. If the value of the initial magnetic flux is $\Phi_0$, then the flux $\Phi_1$ at the moment the switch opens, $\tau_3$, during which the inductance decreases from $L_0$ to $L_1(\tau_3)$, is

$$\Phi_1 = \Phi_0 \left( \frac{L_0 + L_3}{L_1 + L_3} \right)^{F-1}, \tag{5.165}$$

and the currents in the generator and the load are

$$I_1 = \frac{\Phi_1}{L_1 + L_2} = I_0 \left( \frac{L_0 + L_3}{L_1 + L_3} \right)^{F}, \tag{5.166}$$

$$I_2 = \frac{I_1 L_1}{L_1 + L_3} = I_0 \left( \frac{L_0 + L_3}{L_1 + L_2} \right)^{F} \frac{L_1}{L_1 + L_2}, \tag{5.167}$$

respectively. Since the initial energy in the generator and in the load are equal to

$$W_0 = \frac{I_0^2 (L_0 + L_3)}{2}, \tag{5.168}$$

$$W_2 = \frac{I_0^2}{2} \left( \frac{L_0 + L_3}{L_1 + L_3} \right)^{2F} \frac{L_1^2 L_2}{(L_1 + L_2)^2}, \tag{5.169}$$

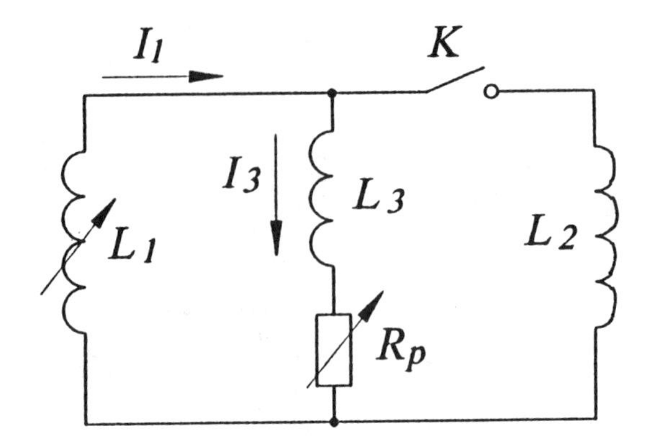


FIGURE 5.17. Equivalent circuit diagram for an MCG connected through a switch with spurious inductance to a load.

respectively, the energy coupling factor for an MCG connected to an inductive load is

$$\begin{aligned} \varepsilon_L &= \frac{W_2}{W_0} \\ &= (L_0 + L_3)^{2F-1} \frac{L_1^2 L_2}{(L_1 + L_3)^{2F} (L_1 + L_2)^2} \\ &= \frac{(d_L + \beta_L)^{2F-1}}{(\alpha_L + \beta)^{2F} (1 + \alpha_L^{-1})^2}, \end{aligned} \tag{5.170}$$

where $\alpha_L = L_1/L_2$, $\beta_L = L_3/L_2$, and $d_L = L_0/L_2$ are normalized parameters.

If the perfectness factor $F$ equals 0.5, the generator does not amplify energy and Eq. 5.170 may be reduced to

$$\begin{aligned} \varepsilon_L &= \frac{1}{\left(1 + \frac{L_2}{L_1}\right)^2 \left(\frac{L_1}{L_2} + \frac{L_3}{L_2}\right)} \\ &= \frac{1}{\left(1 + \frac{1}{\alpha_L}\right)^2 (\alpha_L + \beta_L)}, \end{aligned} \tag{5.171}$$

which characterizes the energy coupled into a load through a fixed inductive store. The family of calculated curves for $\varepsilon_L(\alpha_L)$ for different values of $F$ are presented in Fig. 5.18. In calculating these curves, it was assumed that the initial values of the parameters $\alpha_L$ and $\beta_L$ are 100 and 0.25, respectively. The plots of $\varepsilon_L(\alpha_L)$ clearly have peaks, which decrease as $F$ decreases and shifts toward higher values of $\alpha_L$. The higher the efficiency of the MCG, the less the final inductance has to be to deliver peak energy to the load. The value of the energy coupling factor can now be optimized with respect to the parameter $\alpha_L$, by taking the derivative of Eq. 5.170 with respect to $\alpha_L$, setting it equal to zero, and solving the resulting second-order equation:

$$\alpha_L^2 - \alpha_L \left( \frac{1}{F} - 1 \right) - \frac{\beta_L}{F} = 0, \tag{5.172}$$

which yields

$$\alpha_{L_*} = \frac{1-F}{2F} + \sqrt{\left( \frac{1-F}{2f} \right)^2 + \frac{\beta_L}{F}}. \tag{5.173}$$

Substituting this equation into Eq. 5.170 gives

$$\varepsilon_{L_*} = \frac{(\alpha_L + \beta_L)^{2F-1}}{[\beta_L + F^*]^{2F} \left[1 + \frac{1}{F^*}\right]^2}, \tag{5.174}$$

where

$$F^* = \frac{1-F}{2F} + \sqrt{\left( \frac{1-F}{2F} \right)^2 + \frac{\beta_L}{F}}. \tag{5.175}$$

The family of calculated curves of $\varepsilon_{L_*}$ for different values of $F$ is presented in Fig. 5.19. The results of the calculations clearly show the importance of the spurious inductance in the switch during the coupling of current into the load. When $F = 1$ and $\beta_L = 0$, the energy coupling coefficient is $\varepsilon_{L_*} = 100$, but when $F = 1$ and $\beta_L = 0.25$, it decreases by a factor of 5.

Usually, the MCG couples energy into the load in the following way: initially, the main circuit of the generator is deformed and it attains a small final inductance, and then, by means of a switch, the circuit is opened and the generator circuit is connected to the external load. As a result, a portion of the energy stored in the final inductance of the MCG is transferred into the load. There is, however, another method for coupling energy from the MCG into the load. This consists of opening the circuit at an earlier moment in time, i.e., when the inductance in the generator is still fairly high. The deformation of the MCG circuit and the compression of the magnetic field continues even after the switch is opened. As opposed to the previous scheme, the energy in the load increases due to compression after the switch has opened. It is worth noting that the energy gain due to

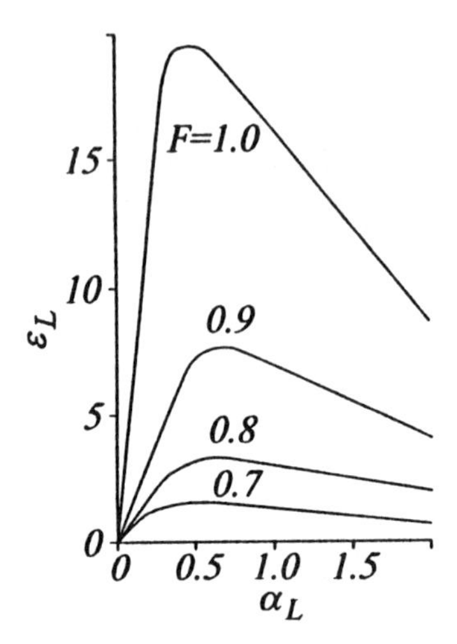


FIGURE 5.18. Family of calculated curves for $\varepsilon_L(\alpha_L)$ versus the perfectness factor ($F$) of an MCG, assuming $\alpha_L = 100$ and $\beta_L = 0.25$. The amplitude of the peaks decrease as $F$ decreases, which implies that the higher the efficiency of the MCG, the less the final inductance has to be to deliver peak energy to the load.

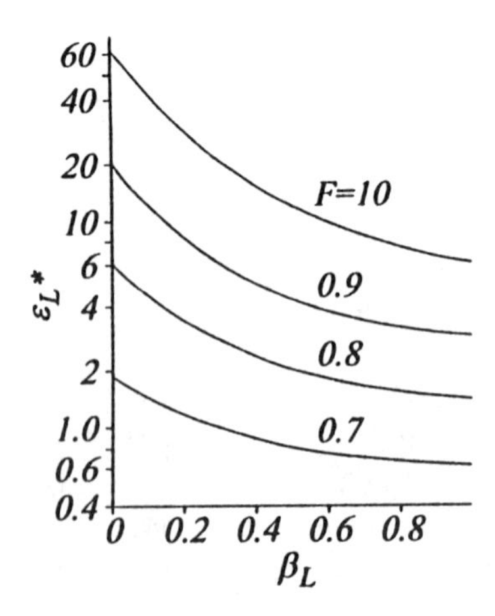


FIGURE 5.19. Family of plots of $\varepsilon_L$ versus perfectness factor ($F$). It can be seen that the spurious inductance in the switch impacts the current coupled into the load.

flux compression is feasible only when the value of the integral $\int_0^t \dot{L}(t)\, dt$ is comparable in value to that of $L_1(t) + L_2$, where $\dot{L}_1$ is the rate at which the inductance of the MCG changes after the switch is opened and $\tau$ is the specified pulse length of the current in the load. It is informative to estimate the amount of energy delivered to the load for the second scheme and to compare the result with that for the previous scheme. In Fig. 5.20, let the main circuit of the MCG, with a given initial inductance, be deformed over a time period of $t_0 - t_1$, where the final inductance is $L_1(t_1)$. At time $t_2$, the load $L_2$ is connected and the switch opens. The Ohmic resistance of the circuit, when it is opened, increases sharply from zero to infinity. After the circuit is opened, flux compression continues and the inductance decreases to $L_1(t_2)$ in a time $t_1 - t_2$. If the initial flux is $\Phi_0$ at time $t_0$, then the flux, taking into account losses, at time $t_1$ is

$$\Phi_1 = \Phi_0 \left( \frac{L_0 + L_3}{L_1(t_1) + L_3} \right)^{F-1} \tag{5.176}$$

and the currents in the generator and load are

$$I_1(t) = \frac{\Phi_1}{L_1(t_1) + L_3} = I_0 \left( \frac{L_0 + L_3}{L_1(t_1) + L_3} \right)^F, \tag{5.177}$$

$$\begin{aligned} I_2(t_1) &= \frac{I_1(t_1) L_1(t_1)}{L_1(t_1) + L_2} \\ &= I_0 \left( \frac{L_0 + L_3}{L_1(t_1) + L_3} \right)^F \frac{L_1(t_1)}{L_1(t_1) + L_2}, \end{aligned} \tag{5.178}$$

respectively. After the circuit has been opened, the current in the load at $t_2 > t_1$ is

$$I_2(t_2) = I_0 \left( \frac{L_0 + L_3}{L_1(t_1) + L_3} \right)^F \frac{[L_1(t_1) + L_2]^{F-1} L_1(t_1)}{[L_1(t_2) + L_2]^F}. \tag{5.179}$$

The energy coupling coefficient for this circuit is

$$\begin{aligned} \varepsilon_L' &= \frac{W_2}{W_0} \\ &= \frac{(L_0 + L_3)^{2F-1} L_1^2(t_1) \left[L_1(t_1) + L_2\right]^{2(F-1)} L_2}{\left[L_1(t_1) + L_3\right]^{2F} \left[L_1(t_2) + L_2\right]^{2F}} \\ &= \frac{(\alpha_L + \beta_L)^{2F-1} \alpha_L^2 (\alpha_L + 1)^{2(F-1)}}{(\alpha_L + \beta_L)^{2F} (\mu_L + 1)^{2F}}, \end{aligned} \tag{5.180}$$

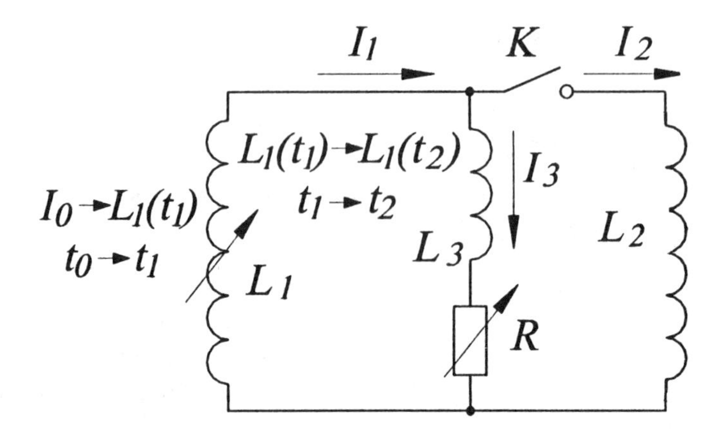


FIGURE 5.20. Equivalent circuit diagram for an MCG connected through a switch with spurious inductance to a load. In this case, the circuit is openned earlier than the circuit in Fig. 5.17.

where $d_L = L_0/L_2$, $\alpha_L = L_1(t_1)/L_2$, $\beta_L = L_3/L_2$, and $\mu_L = L_1(t)/L_2$ are normalized parameters.

To determine the advantages of the second scheme over the first, compare the respective energy coupling factors. Taking the ratio of Eqs. 5.170 and 5.180 yields

$$\frac{\varepsilon'_L}{\varepsilon_L} = \left(\frac{\alpha_L + 1}{\mu_L + 1}\right)^{2F}. \tag{5.181}$$

Since $\alpha_L > \mu_L$, when $F > 0$, then $\varepsilon'_L/\varepsilon_L > 1$. As can be seen by examining Eq. 5.181, the advantages of the second scheme over the first in terms of energy occurs when the values of $\alpha_L$ are high and the values of $\mu_L$ are low. In practice, this means that the earlier the circuit is opened and the smaller the final inductance of the generator, the greater is the amount of energy delivered to the load.

## 5.4 Pulsed Transformer and Electroexplosive Switch

In this section, the case will be considered of the MCG connected to a load through a transformer and an electroexplosive switch [5.17], where the switch is connected to the secondary winding of the transformer. The transformer and the switch are used to match the impedances of the MCG and the load. In this scheme, the pulse transformer is used to decrease the current flowing through the switch, as well as adjust impedances. The switch is used to shorten the length and to increase the power of the pulse delivered to the load. Both complex and active loads are considered.

### *5.4.1 Complex Load*

The circuit diagram of an MCG connected through a pulsed transformer and switch to a complex load is shown in Fig. 5.21. The operation of this circuit can be divided into three stages. In the first, the resistance of the switch, $R_P$, is much less than the resistance of the load, since the electroexplosive switch does not receive enough energy to vaporize its conductors. This stage of operation of the circuit is described by the equations

$$\frac{d(LI_1)}{dt} + R_1 I_1 - M\frac{dI_2}{dt} = 0, \tag{5.182}$$

$$L_T\frac{dI_2}{dt} + R_P I_2 - M\frac{dI_1}{dt} = 0, \tag{5.183}$$

where $L_T$ is the inductance of the secondary winding of the transformer, $M = k_c\sqrt{L_k L_T}$ is the mutual inductance of the transformer windings, and $L$ is the inductance of the MCG and primary circuit of the transformer, which changes during magnetic flux compression from an initial value of $L_0$ to $L_k$. If it is assumed that the resistance of the switch, $R_P$, goes to zero during this stage of operation, than Eq. 5.183 becomes

$$\frac{M}{L_T}\frac{dI_1}{dt} = \frac{dI_2}{dt}. \tag{5.184}$$

Using this equation to eliminate $I_2$ from Eq. 184 results in

$$L\frac{dI_1}{dt} + I_1\frac{dL}{dt} - R_1 I_1 - \frac{M^2}{L_T}\frac{dI_1}{dt} = 0. \tag{5.185}$$

If the normalized variable $x = t/t_k$ is introduced, where $t$ is any moment in time and $t_k$ is the time at which the switch opens, Eq. 5.185 can be rewritten as

$$I_1 = I_0\frac{L_0 - \frac{M^2}{L_T}}{L(x) - \frac{M^2}{L_T}}\exp\left[-\int_0^t \frac{R_1 t_k}{L(x) - \frac{M^2}{L_T}}dx\right]. \tag{5.186}$$

Assuming that the resistance of the generator is zero and that the inductance of the generator varies linearly, i.e., $L = L_0 - L_S$, then the current, in terms of the normalized parameter $x$, is

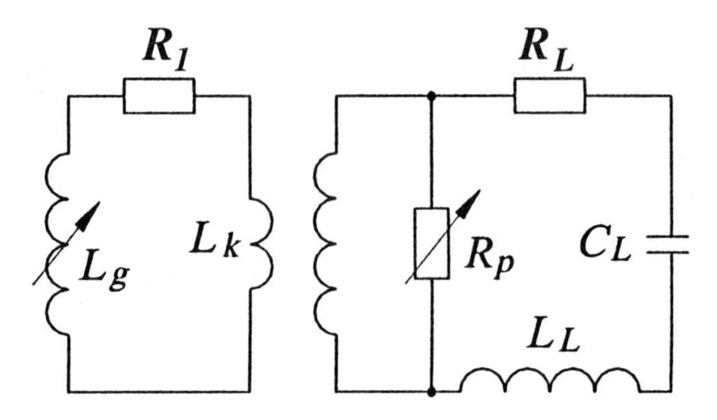


FIGURE 5.21. Equivalent circuit diagram for an MCG connected through an EEB and pulsed transformer to a complex load.

$$I_1 = I_0 \left[ \frac{L_0 - \frac{M^2}{L_T}}{L(x) - \frac{M^2}{L_T}} \right]^{1 - \frac{R_1 t_k}{L_S}} \tag{5.187}$$

when $x \leq 1$. The current in the switch circuit can be found from Eq. 184 by taking into account the initial conditions

$$I_2 = I_1 \frac{M}{L_T}. \tag{5.188}$$

Solving the circuit equations begins with determining the number of conductors required by the switch. If the inductance varies nonlinearly, then the geometry of the switch, i.e., the number of conductors needed, can be found by using Eq. 4.11:

$$N = \frac{4S_P}{\pi d^2}, \tag{5.189}$$

where $d$ is the diameter of the wire conductors. Knowing the value of $N$, the resistance of the switch, $R_P$, can be calculated. Knowing how the current varies with changes in $x$ (Eqs. 5.187–5.188), the energy required to vaporize the wires is

$$W_{R_P} = t_k \int_0^t I_2^2 R_P dx. \tag{5.190}$$

Since the resistance of the switch during this stage of operation shunts the load, practically none of the energy is coupled into the load.

During the second stage of operation, the resistance of the switch changes from the resistance at the temperature at which the wire vaporizes to infinity. The system of equations that describes this is

$$\frac{dI_1}{dt} = \frac{M}{L_k}\frac{dI_2}{dt}, \tag{5.191}$$

$$L_T\frac{dI_2}{dt} + R_P I_P - M\frac{dI_1}{dt} = 0, \tag{5.192}$$

$$-R_P I_P + L_L\frac{dI_L}{dt} + R_L I_L + \frac{1}{C_L}\int_{\tau_P}^{\tau_P+t_k} I_L dt = 0, \tag{5.193}$$

$$I_P = I_2 - I_L, \tag{5.194}$$

where $I_1$ is the current flowing through the primary winding of the transformer, $I_2$ is the current in the secondary winding, $I_L$ is the current flowing through the load, and $I_P$ is the current flowing through the switch. Equations 5.191–5.194 can only be solved numerically, so it is best to rewrite them in a more convenient form:

$$\frac{dI_1}{dx_2} = \frac{M}{L_k}\frac{t_k R_P I_P}{\left(L_T - \frac{M^2}{L_k}\right)}, \tag{5.195}$$

$$\frac{dI_2}{dx_2} = -\frac{t_k R_P I_P}{L_T - \frac{M^2}{L_k}}, \tag{5.196}$$

$$\frac{dI_L}{dx_2} = y, \tag{5.197}$$

$$\frac{dI_P}{dx_2} = -\left[\frac{t_k R_P I_P}{L_T - \frac{M^2}{L_k}} + y\right], \tag{5.198}$$

$$\frac{dy}{dx_2} = \frac{1}{L_L}\left[\frac{t_p^2 R_P^2 I_P}{L_T - \frac{M^2}{L_k}} + t_k R_P y - t_k R_L y + t_k I_P \frac{2R_P(0)}{(1-x)^3} - \frac{t_k^2}{C_L} I_L\right], \tag{5.199}$$

$$\frac{dR_P}{dx_2} = \frac{2R_P(0)}{(1-x)^3}, \tag{5.200}$$

where $y$ is an intermediate value, $x_2 = t_k/t$, and $t_k$ is the time at which the wires in the switch are vaporized.

The amount of energy coupled into the load during the second stage of operation is

$$W_L = t_k \int_0^t I_L \, |z_L| \, dx. \tag{5.201}$$

The currents generated at the completion of the second stage serve as the initial values for the calculations performed for the third stage of operation. If it is assumed that $R_P = \infty$, the system of equations describing this stage are

$$\frac{d(LI_1)}{dt} + R_1 I_1 - M\frac{dI_2}{dt} = 0, \tag{5.202}$$

$$L_L \frac{dI_2}{dt} + R_2 I + \frac{1}{C_L}\int_{\tau_P + t_k}^{t_f} I_2 dt - M\frac{dI_1}{dt} = 0. \tag{5.203}$$

Rewriting the equations to facilitate their solution numerically yields

$$\frac{dI_1}{dt} = \frac{1}{L}\left[M\frac{dI_2}{dx_3} + I_1(R_1 t_3 - L_S)\right], \tag{5.204}$$

$$\frac{dI_2}{dx_3} = y_2, \tag{5.205}$$

$$\begin{aligned} \frac{dy_2}{dx_3} &= \left(L_2 - \frac{M^2}{L}\right)\left\{y_2\left(\frac{ML_S}{L} - R_2 t_3\right)\right. \\ &\quad + \frac{M}{L^2}\left(R_1 t_3 - L_S\right)\left[My_2 + I_1\left(R_1 t_3 - L_S\right)\right] \\ &\quad \left. + \frac{ML_S}{L}(R_1 t_3 - L_S)I_1 - \frac{t_3^2}{C_L} I_2\right\}, \end{aligned} \tag{5.206}$$

where $y_2$ is an intermediate value, $t_3$ is the time it takes to complete the third stage, and $x_3 = t/(t_k + t_3 + \tau_P)$.

If the explosion of the conductors occurs at the instant the generator completes its operation, then, assuming that $R_1 = 0$, the calculations for the third stage are simplified and Eqs. 5.196–5.201 reduce to

$$\left(L_2 - \frac{M^2}{L}\right)\frac{d^2 I_2}{dx_3^2} + t_3 R_2 \frac{dI_2}{dx_3} + t_3^2 \frac{I_2}{C_L} = 0, \tag{5.207}$$

$$I_1 = \frac{M}{L} I_2, \tag{5.208}$$

where $L_2 = L_{1T} + L_L$.

Equation 5.204 is a linear homogeneous second-order differential equation with constant coefficients. Solving it yields

$$I_2 = C_1 e^{S_1 x_3} + C_2 e^{S_2 x_3}, \tag{5.209}$$

where $C_1$ and $C_2$ are the integration constants, which have the following values:

$$C_{1,2} = \frac{-t_3 R_2 \pm \left[(t_3 R_2)^2 - 4\left(L_2 - \frac{M^2}{L}\right)\frac{t_3^2}{C_L}\right]^{\frac{1}{2}}}{2\left(L_2 - \frac{M^2}{L}\right)}. \tag{5.210}$$

To simplify calculations in the third stage of operation, assume that the point $x_3 = 0$ is the origin and that the current in the load at this time is $I_m$. Therefore,

$$I_2 = C_1\left(e^{S_1 x_3} + e^{S_2 x_3}\right) + I_m e^{S_2 x_3}. \tag{5.211}$$

The integration constant $C_1$ can be found by using the conditions at the moment the second stage has completed its operation, since the current in the load equals zero at this time, i.e.,

$$C_1 = \frac{I_m e^{S_2 t_3}}{e^{S_1 t_3} + e^{S_2 t_3}} = -\frac{I_m e^{S_2}}{e^{S_1} + e^{S_2}}. \tag{5.212}$$

If the condition $R_2 < 2\sqrt{L_2 C_L}$ is satisfied during the last stage of operation, high-frequency oscillations are formed in the secondary circuit at a frequency

$$\omega = \sqrt{\frac{t_3}{\left(L_2 - \frac{M^2}{L}\right) C_L} - \frac{(t_3 R_2)^2}{4\left(L_2 - \frac{M^2}{L}\right)^2}}. \tag{5.213}$$

The current is

$$I_2 = \frac{U_0}{\omega\left(L_2 - \frac{M^2}{L}\right)} \exp\left[-\frac{R_L t_3 x_3}{2\left(L_2 - \frac{M^2}{L}\right)}\right] \cdot \sin\left(\omega x_3 + \frac{\pi}{2}\right). \tag{5.214}$$

Depending on the ratio of the reactive and active parameters of the circuit, the current during this stage can be calculated by using Eqs. 5.204–5.206 or 5.214. The energy coupled into the load can be found by substituting the operating time of the generator for $t_k$ and Eq. 5.214 for the current into Eqs. 5.202 and 5.203:

$$W_L(x_3) = \frac{2\omega_S |z_L| I_0^2}{R_L^2 + 4\omega_S\omega^2}\left\{R_L - \omega_S \exp\left(-\frac{R_L}{2\omega_S}x_3\right) \times \left[\frac{R_L}{2\omega_S}\sin\left(\omega t + \frac{\pi}{2}\right) - \omega\cos\left(\omega t + \frac{\pi}{2}\right)\right]\right\}. \tag{5.215}$$

Since the inductance of the generator during the final stage of compression of the magnetic flux is constant and equal to $L_k$, this equation becomes

$$W_L = \frac{2\omega_s I_0 |z_L|^2}{R_L^2 + 4\omega_s\omega^2}, \tag{5.216}$$

where $\omega_s = L_2 - M^2/L_g$. This last equation can be used to calculate the amount of energy coupled into an active load during the third stage of operation.

### 5.4.2 Active Load

Consider the operation of an MCG connected to an active load as shown in Fig. 5.22. Analyzing this situation, it is assumed that the EEB opens the circuit at the moment the current in the MCG achieves its peak value. The parameters $I_1$, $I_2$, $I_P$, and $I_L$ are the currents in the primary and secondary circuits of the transformer, in the switch, and in the load, respectively. The inductance of the primary circuit is $L_1$, the inductance of the secondary circuit is $L_2$, the resistance of the switch is $R_P$, and the resistance of the load is $R_L$.

If an equivalent inductance, $L_E$, defined by

$$L_E = \frac{1}{L_1}\left(L_1 L_2 - M^2\right) = L_2(1 - k_c^2), \tag{5.217}$$

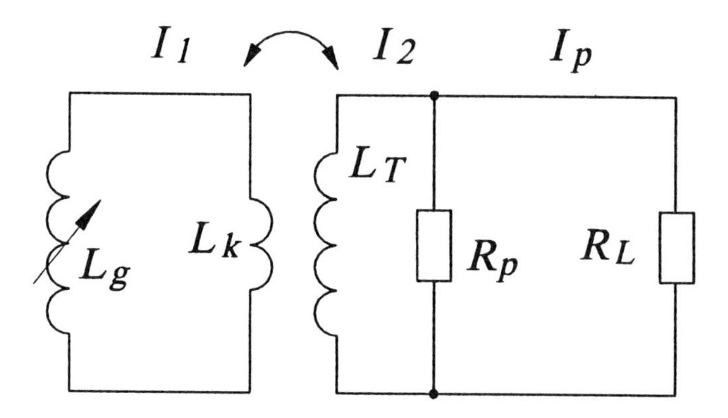


FIGURE 5.22. Equivalent circuit diagram for an MCG connected through an EEB and pulsed transformer to an active load.

is introduced, the system of equations that describe the circuit is

$$L_E \frac{dI}{dt} + R_L I_L = 0, \tag{5.218}$$

$$I_L R_L = I_P R_P, \tag{5.219}$$

$$I = I_P + I_L, \tag{5.220}$$

where $I$ is the current flowing through the equivalent inductance. In this case, the current flowing through the load is

$$\begin{aligned} I_L &= \frac{I_0}{1 + \beta_R \left(1 - \frac{t}{\tau}\right)} \\ &\quad \cdot \exp\left\{ \frac{\tau}{\tau_L} \sqrt{\frac{1}{\beta_R}} \left[ \arctan \sqrt{\beta_R} \left(1 - \frac{t}{\tau}\right) - \arctan \sqrt{\beta_R} \right] \right\}, \end{aligned} \tag{5.221}$$

where $\tau_L = L_E/R_L$, $\tau_R = L_E/R_P$, and $\beta_R = R_L/R_P(0) = \tau_R/\tau_L$. The initial current in the load is

$$I_L(0) = \frac{I_0 k_c \sqrt{L_1 L_2}}{L_2}. \tag{5.222}$$

Taking the derivative of of Eq. 5.221 with respect to time and setting it equal to zero, gives the time at which the current in the load reaches its peak value as

$$t_E = \left(1 - \frac{\tau}{2\tau_L \beta_R}\right)\tau = \left(1 - \frac{\tau}{2\tau_P}\right)\tau. \tag{5.223}$$

and the peak current is therefore

$$I_L = \frac{I_0}{1 + \frac{\tau^2}{4\tau_L \tau_P}} \exp\left\{\frac{\tau}{\tau_L}\sqrt{\frac{1}{\beta_R}}\left[\arctan \frac{\tau}{2\tau_P}\sqrt{\beta_R}\right]\right\}. \tag{5.224}$$

If a pulsed transformer is added to the circuit, then the parameters $\tau_L$, $\tau_P$, and $I_0$ in Eqs. 5.223 and 5.224 are defined by Eq. 5.222.

The peak current in the load for a given load resistance, according to Eq. 5.224, depends only on the resistance of the load. Therefore, to deliver maximum power to the load, the ratio of $\tau_P$ to $\tau_L$ must be optimized. Taking the derivative of Eq. 5.224 with respect to $\tau_P$ and setting it equal to zero and taking into account that $\tau_P > \tau_L$ yields

$$\tau_L = \frac{\tau^2}{4\tau_P}. \tag{5.225}$$

Substituting $\tau_L$ from this expression into Eq. 5.224 leads to an expression for the maximum possible peak current in the load as

$$I_{L_{\max}} = I_0 \exp\left[-\arctan 2\sqrt{\frac{\tau_P}{\tau}}\right] \exp\left[\arctan 1\right] = 2.25 I_0 \exp\left[-\arctan 2\sqrt{\frac{\tau_P}{\tau_L}}\right], \tag{5.226}$$

which expression is in good agreement with experimental data.

The methods presented in this chapter for getting good impedance matching between various MCGs and various loads, as well as the analytical expressions and computer simulation models developed, enables one to calculate the processes that occur in various experimental arrangements. To aid in making these calculations, the physical properties of various explosives and materials used to construct MCGs are provided in the Appendices A and B of [5.18]. In addition, the formulas used to calculate the magnetic fields of various geometries are presented in Appendix C.

# References

[5.1] S.N. Golosov, Y.V. Vlasov, V.A. Demidov, and S.A. Kazakov, "Investigation of Magneto-Plasma Compressor Operation and its Powering from a Helical Flux Compression Generator," to be published in the Proceedings of the Megagauss Conference, Tallahassee, FL (1998).

[5.2] I.I. Divnov, N.I. Zutov, and O.P. Karpov, "Magnetocumulative Generator with Plasma Load," PMTF, No. 6, pp. 46–51 (1979).

[5.3] C.R. Jones, C.M. Fowler, and K.D. Ware, "High-Energy Atomic Iodine Laser Driven by Magnetic Flux–Compression Generator," *Megagauss Technology and Pulsed Power Applications* (eds. C.M. Fowler, R.S. Caird, and D.J. Erickson), Plenum Press, New York, pp. 747–755 (1987).

[5.4] A.I. Pavlovskii, R.Z. Lyudaev, L.N. Plyashkevich, N.B. Romanenko, G.M. Spirov, and L.B. Sukhanov, "MCG Application for Powered Channeling Neodim Lasers," *Megagauss Magnetic Field Generation and Pulsed Power Applications* (eds. M. Cowan and R.B. Spielman), Nova Science Publ., New York, pp. 969–976 (1994).

[5.5] A.I. Pavlovskii, A.Ya. Brodskii, B.P. Giterman, V.Ye., Gurin, N.V. Deulin, D.I. Zenkov, A.S. Krachenko, B.V. Lazhintsev, R.Z. Lyudaev, N.N. Petrov, L.N. Plyashkevich, G.M. Spirov, and L.V. Sukhanov, "Magnetic Cumulation Generator Application for Photodissociation Laser Powering," *Megagauss Magnetic Field Genera-*

*tion and Pulsed Power Applications* (eds. M. Cowan and R.B. Spielman), Nova Science Publ., New York, pp. 977–986 (1994).

[5.6] A.I. Pavlovskii, N.F. Popkov, V.I. Kargin, A.S. Picar, and E.A. Ryaskov, "Magnetic Cumulation Generator as a Power Source to Accelerate Intense Electron Fluxes," *Megagauss Fields and Pulsed Power Applications* (eds. V.M. Titov and G.A. Shvestov), Nova Science Publ., New York, pp. 449–452 (1990).

[5.7] A.I. Pavlovskii, A.S. Kravchenko, V.D. Selemir, A.Ya. Brodskii, Yu.B. Bragin, V.V. Ivanov, I.V. Konovalov, V.G. Suvorov, K.V. Shibalko, V.V. Chernyshev, V.A. Cherepenin, V.A. Vdovin, A.V. Korzhenevskii, and S.A. Sokolov, "EMG Magnetic Energy for Superpower Electromagnetic Microwave Pulse Generation," *Megagauss Magnetic Field Generation and Pulsed Power Applications* (eds. M. Cowan and R.B. Spielman), Nova Science Publ., New York, pp. 969–976 (1994).

[5.8] R.S. Hawke, W.J. Nellis, G.H. Newman, J. Rego, and A.R. Susoeff, "Summary of Railgun Development for Ultrahigh-Pressure Research," *Megagauss Technology and Pulsed Power Applications* (eds. C.M. Fowler, R.S. Caird, and D.J. Erickson), Plenum Press, New York, pp. 803–810 (1987).

[5.9] V.D. Selemir, V.A. Demidov, L.N. Plyashkevich, A.S. Kravchenko, S.A. Kazskov, A.M. Shuvalov, A.S. Boriskin, V.A. Zolotov, G.M. Spirov, and M.M. Kharlamov, "High-Current (30 MA or More) Energy Pulses for Powering Inductive and Active Loads," *Megagauss and Megaampere Pulse Technology and Applications* (eds. V.K.Chernyshev, V.D. Selemir, and L.N. Plyashkevich), VNIIEF, Sarov, pp. 241–247 (1997).

[5.10] A.B. Prishchepenko and M.V. Shchelkachev, "The Work of the Implosive Generator with Capacitive Load," *Megagauss and Megaampere Pulse Technology and Applications* (eds. V.K.Chernyshev, V.D. Selemir, and L.N. Plyashkevich), VNIIEF, Sarov, pp. 304–307 (1997)

[5.11] I.I. Divnov, Iu.A. Guskov, and N.I. Zutov, "Magnetocumulative Generator with Inductive Load," FGV, **12** (6), pp. 959–962 (1976).

[5.12] V.F. Buharov, V.A. Vasiukov, and V.E. Gurin, "Magnetocumulative Generators with Transformer Energy Coupling," PMTF, No. 1, pp. 4–10 (1982).

[5.13] V.A. Demidov, E.I. Zharikov, S.A. Kazakov, and V.K. Chernyshev, "High-inductive Magnetoexplosive Generators with High Energy Gain," PMTF, No. 6, pp. 106–111 (1981).

[5.14] L.S. Gerasimov, "Adjusting of Magnetoexplosive Generators with Active Loads by Means of a Transformer," PMTF, No. 4, pp. 50–54 (1978).

[5.15] A.S. Kravchenko, R.Z. Ludaiev, and A.I. Pavlovskii, "Feeding of the Inductive and Ohmic Loads of the MCG by Means of a Transformer," PMTF, No. 5, pp. 116–121 (1981).

[5.16] V.A. Demidov, E.I. Zharikov, S.A. Kazakov, and V.K. Chernyshev, "Energy Coupling from Inductive Stores and Magnetoexplosive Generators into Inductive Loads by Means of Circuit Breakers," PMTF, No. 4, pp. 54–60 (1978).

[5.17] V.P. Isahov, "Operation of the Inductive Store through a Discharger into an Active Load," *Continuous Medium Dynamics*, Nauka, Novosibirsk, **49**, pp. 118–124 (1981).

[5.18] G. Knoepfel, *Pulsed High Magnetic Fields*, North-Holland Publishing Company, Amsterdam (1970).

# 6 Design, Construction, and Testing

In this chapter, two simple models that can be used in designing and predicting the performance of an MCG are discussed. Owing to the many different arrangements of such generators, attention is focused on one particular type, i.e., a helical generator developed by the authors at Loughborough University in the United Kingdom. The generator has been named **FLEXY I**, and a description of how it was built and tested is presented. This chapter is intended to put into perspective the theoretical concepts presented in Chapters 1–5, by considering in detail the design, construction, and testing of MCGs, and at the same time addressing their limitations. This chapter is based primarily on a series of papers written by two of the authors (I.R. Smith and B.M. Novac) [6.1–6.7].

## 6.1 A Brief Description of FLEXY I

In this section, a brief description of FLEXY I is presented, and Fig. 6.1a is a schematic drawing of the major components of an end-initiated helical generator such as FLEXY I. A capacitor bank is used to create the initial magnetic field within the helical stator winding. The generator converts the chemical energy stored in a high explosive into kinetic energy, which, in turn, is converted into electromagnetic energy. The detonation process accelerates an armature (liner) to a velocity that is a significant proportion of the detonation velocity of the explosive. At the same time, there is a corresponding decrease in the inductance of the generator. The coaxial

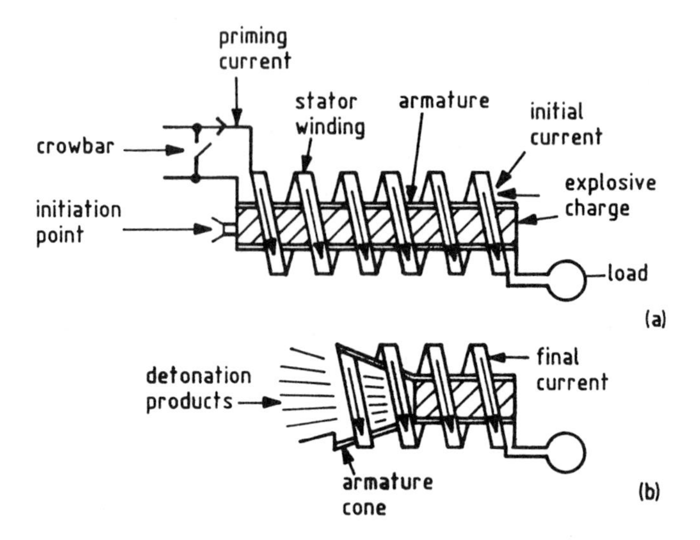


FIGURE 6.1. An end-initiated explosively driven helical generator (a) prior to detonation and (b) during detonation.

armature is filled with the high explosive, which is initiated at one end of the armature. As a result of the detonation, the armature expands into the conical form shown in Fig. 6.1b and, as it expands, it makes contact with the stator. The point of contact moves progressively along the length of the generator. When the armature makes contact with the crowbar, it shorts out the input to the stator winding and subsequently does work against the magnetic field. As a result of this process, the current in the load increases, with a substantial portion of the kinetic energy of the accelerating armature being converted into electromagnetic energy in the load [6.1]. To begin, two simple computer models developed at Loughborough University, which were used in the design and construction of FLEXY I, are introduced.

## 6.2 Computer Models

In this section, a simple zero-dimensional (0D) approach to the design of a simple 1 MJ generator is discussed. This is followed by a description of a simple, but complete 2-dimensional (2D) model for a helical generator that overcomes many of the limitations of the 0D model. The reason it was decided to present the material in this manner is that the authors were able to verify the 0D model experimentally for medium to large generators, which gave them many important insights into developing the 2D model and into understanding the physics of small (mini and micro) helical generators.

### 6.2.1 *Simple Zero-Order Model for a Helical MCG*

A Simple General Model

When designing an MCG, it is important that a mathematical model of adequate accuracy is available to determine the initial conditions and to predict the performance of the generator. Unfortunately, most existing models are quite complex [6.8,6.9] and require long run times on large, fast computers to generate results of limited accuracy. The much simpler model presented here provides computational results that are of comparable accuracy to those obtained from the more complicated models. With this simpler model, both the computing power and the run time are much less, and it provides a good qualitative guide to the various physical phenomenon involved in the operation of the generator and the methods used to reduce the various losses.

Both high-intensity magnetic and electric fields are generated during a flux compression process, but it is possible to neglect the effects of the magnetic field on the armature dynamics and the conductor assembly for both small and medium-energy generators. In these latter devices, the magnetic field intensity is kept low by suitably choosing the turn-splitting [6.10,6.11], so that its effects can be neglected. The effects of the electric field, however, cannot similarly be neglected. It has been found that ignoring the high voltages that develop between the helical coil and the armature leads either to extremely inaccurate results [6.12] or to results that could not be reproduced on a regular basis [6.13].

If the effects of the electric field are included in the analysis [6.11,6.14], then an upper limit is imposed on the maximum voltage, $V_{\mathrm{max}}$, that can be allowed to form between the coil and the armature. This imposed limitation leads to an expression that describes how the inductance, $L$, of the generator varies in time following crowbarring at $t = t_0 = 0$ [6.14] as

$$L = L_0 \exp\left\{ V_{\mathrm{max}} \frac{1 - \exp(\gamma t)}{L_0 I_0 \gamma} - \gamma t \right\}, \tag{6.1}$$

where $I$ is the load current, the subscript 0 denotes the initial conditions, and $\gamma$ is the ratio of the resistance $R$ to the inductance $L$, which is assumed to be constant. Although Eq. 6.1 has been used in generator designs, it is in general not adequate, since, in practice, the value of $\gamma$ varies widely with time. A better model for determining the parameters of the generator is provided by the system of equations

$$L\frac{dI}{dt} + I\frac{dL}{dt} + IR = 0, \tag{6.2}$$

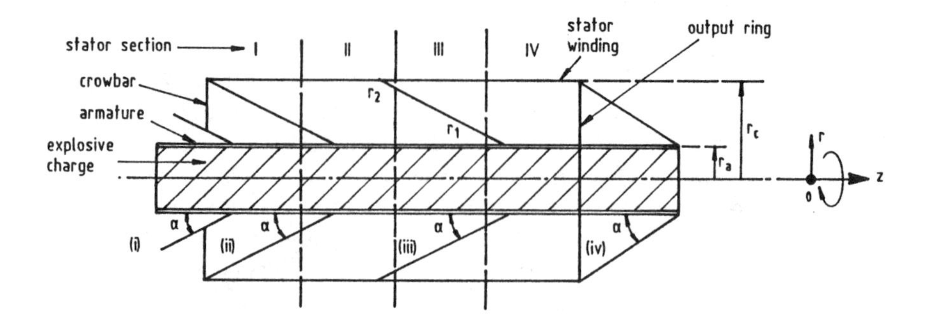


FIGURE 6.2. A four-section generator: (i) the armature cone at the crowbar position, (ii) the armature cone at the final position of period (1), (iii) the armature cone at an intermediate position of period (2), and (iv) the armature cone at the final position of period (2).

$$R = R\left(t, I, \frac{dI}{dt}\right), \tag{6.3}$$

$$L = L(t), \tag{6.4}$$

where the functions $R$ and $L$ are described later in this section. The second term in Eq. 6.2 is the energy conversion term, which describes the maximum voltage that is induced in the generator. The energy multiplication factor for the generator is therefore directly related to the maximum permitted voltage.

The model presented in this section is for a generator in which the stator is divided into a number of equal length sections, where each section has a constant winding pitch and the same number of turns in any parallel current paths. The axial dimensions of the crowbar and the output ring in a typical four-section generator shown in Fig. 6.2 are neglected. In developing this model, it was assumed that the behavior of the circuit can be separated into two distinct time periods. The first of these is the time taken for the armature cone to expand from the crowbar position (i) to position (ii), where it makes contact with the helical coil, as shown in Fig. 6.2. The second period is the time taken for the point of the contact between the armature cone and the helical coil to move along the length of the coil to its final position (iv).

Another important parameter in developing this model is the length of the explosive charge. If the end of the explosive charge is not beyond the output ring of the stator, any additional compression beyond this point is not considered.

Ohmic Resistance

Calculation of the Ohmic resistance of a generator is usually based on a one-dimensional model for diffusion of the magnetic field into both the coil conductors and the armature wall. The skin depth $\delta$ for the diffusion process can be found by using the following equation [6.15]:

$$\delta = \left( \frac{I}{\mu_0 \sigma_0 \frac{dI}{dt}} \right)^{\frac{1}{2}}, \tag{6.5}$$

where $\mu_0$ is the magnetic permeability of free space and $\sigma_0$ is the conductivity of the conductors at room temperature. Equation 6.5 is only valid for an exponentially increasing current, which is only approximately true in practice. Equation 6.5 yields somewhat pessimistic estimates during the final moments of compression, because the rate of the current increase may be far from exponential.

To determine the initial resistance when a capacitor bank generates the initial magnetic field, a skin depth is used given by [6.15]

$$\delta = \left( \frac{2 \left( L_0 C \right)^{\frac{1}{2}}}{\mu_0 \sigma_0} \right)^{\frac{1}{2}}, \tag{6.6}$$

where $C$ is the bank capacitance.

The resistance $R_c$ of one section of the helical stator coil can be found by using both a *skin effect factor* $f_\delta$ and a *proximity effect factor* $f_p$ [6.16], thus,

$$R_c = R_{DC} f_\delta f_p, \tag{6.7}$$

where $R_{DC}$ is the DC coil resistance,

$$f_\delta = \frac{\Phi}{4\delta}, \tag{6.8}$$

and

$$f_p = \begin{cases} n \left[ 1 = 2 \left( \frac{n\Phi}{\delta} \right)^2 \right], & \Phi \geq 2\delta \\ n \left[ 1 + \frac{n^5 \Phi^2}{4p^2 \delta^3} \right], & \Phi < 2\delta \end{cases} \tag{6.9}$$

and where $\Phi$ is the diameter of the wire, $n$ is the number of parallel paths in a section, and $p$ is the pitch of each section. Both the skin and proximity effects are described in [6.16].

The current in the generator armature can be found by summing the axial current to the circular currents induced in the stator. The total current then forms a helix having the same pitch as the corresponding stator section. The armature resistance can be calculated by using the above equations, except that in this case of course $f_p = 1$.

A significant phenomenon that occurs during the first period of the flux compression process is a rapid increase in the generator resistance. As the armature expands, the length of the path followed by the helical current increases. To calculate the length of the armature cone, the truncated cone section of the armature with radii $r_1$ and $r_2$, where $r_2 > r_1$ [see Fig. 6.2, intermediate position (iii)], will be considered. Introducing cylindrical coordinates $(r, \theta, z)$, a helix on this surface can be described by the equations

$$z = \frac{p\theta}{2\pi}, \tag{6.10}$$

$$r = r_2 - \frac{p\theta \tan\alpha}{2\pi}. \tag{6.11}$$

Using these equations, the elemental length

$$dl = \left[(r d\theta)^2 + (dz)^2 + (dr)^2\right]^{\frac{1}{2}} \tag{6.12}$$

can be rewritten as

$$dl = \left[\frac{p\tan\alpha}{2\pi}\right]\left\{\left[\frac{2\pi r_2}{p\tan\alpha} - \theta\right]^2 + \left(1 + \cot^2\alpha\right)\right\}^{\frac{1}{2}} d\theta. \tag{6.13}$$

The total length $l_h$ of the helix in this section is given by the following equation:

$$l_h = \int_{\theta_0}^{\theta_1} \frac{dl}{d\theta} d\theta, \tag{6.14}$$

where $\theta_0 = 0$ and $\theta_1 = 2\pi(r_2 - r_1)/p\tan\alpha$. Integrating Eq. 6.14 leads to

$$
\begin{aligned}
l_h \;=\; & \frac{1}{2}\left\{ r_2 \left[ 1+\cot^2\alpha + \left(\frac{2\pi r_2 \cot\alpha}{p}\right)^2 \right]^{\frac{1}{2}} \right. \\
& - r_1 \left[ 1+\cot^2\alpha + \left(\frac{2\pi r_1 \cot\alpha}{p}\right)^2 \right]^{\frac{1}{2}} \qquad (6.15) \\
& \left. + \left[\frac{p\tan\alpha}{2\pi}\right] \left(1+\cot^2\alpha\right) \ln\frac{R_2}{R_1} \right\},
\end{aligned}
$$

where

$$R_1 = r_1 + \left\{ r_1^2 + (1+\cot^2\alpha)\left[\frac{p\tan\alpha}{2\pi}\right]^2 \right\}^{\frac{1}{2}}, \tag{6.16}$$

$$R_2 = r_2 + \left\{ r_2^2 + (1+\cot^2\alpha)\left[\frac{p\tan\alpha}{2\pi}\right]^2 \right\}^{\frac{1}{2}}. \tag{6.17}$$

Inductance

Several methods have been developed for calculating the inductance of individual stator sections of a helical generator, and these are summarized in [6.10,6.17–6.21]. The method used in [6.21] has the advantage of being relatively simple, and has an error of less than 5% provided that the ratio of the diameter of the helical coil to the length of the section is less than 1.5. Based on this premise, the inductance $L_s$ of a stator section of length $l$ is given by

$$L_s = \frac{k_1 l^2 (r_c^2 - r_a^2)}{p^2[l + k_2(r_c - r_a)]}, \tag{6.18}$$

where $r_c$ and $r_a$ are the radii of the coil and armature, respectively, and the constants are $k_1 = 0.003948$ and $k_2 = 0.45$.

The mutual inductance between adjacent sections of the stator, denoted as $x$ and $y$, is much less than the self-inductance of either section, and may be calculated from [6.22]

$$M_{xy} = \frac{k_2 L_x p_x (r_c - r_a)}{p_y[2l + k_2(r_c - r_a)]}. \tag{6.19}$$

For coil sections that have an expanded armature within their axial length, and possibly a point where the armature and coil make contact, the reduced inductance can be derived by subtracting the following expression [6.22] from Eq. 6.18:

$$\Delta L_s = \frac{k_1[r_a(r_1^2 - r_2^2) + \frac{1}{3}(r_1^3 - r_2^3)]}{p^2 \tan\alpha\{l + k_2[r_c - \frac{1}{2}(r_1 + r_2)]\}}, \tag{6.20}$$

where $r_1$ and $r_2$ are the radii of the armature cone at each end of the section, and $\alpha$ is the angle of the cone [see Fig. 6.2, position (iii)]. All the lengths in Eqs. 6.18–6.20 are measured in millimeters and the inductances are obtained in microhenrys. When the contact point between the coil and the armature lies within a section, the length $l$ of the section is decreased accordingly. Equations 6.18–6.20 were derived under the assumption that the value of the flux density is constant throughout the volume between the armature and the coil. Further approximations from [6.21] were introduced in deriving Eqs. 6.18–6.20, to improve the accuracy of prediction. The most important of these was to replace $(l^2 + 4r_c^2)^{1/2}$ by $(l + 0.9r_c)$.

Because the self-inductance of the stator winding is much greater than the mutual inductance between any two sections, it is not necessary that variations in the latter, due to the expansion of the armature, be taken into account. However, this omission, together with the approximations used in deriving Eqs. 6.18–6.20, leads to small discontinuities in the calculated values of the inductance each time the contact point moves from one section to another.

### Verification of Zero-Order Model

In general, the experimental data available in the open literature are inadequate to verify satisfactorily the simple model presented above. Nevertheless, a few important results do exist that can be used for this purpose. In this section, results provided by the 0D model results are compared to published data on both the Mark IX and the EF-3 generators.

#### *The MARK IX Generator*

The MARK IX is a helical generator developed at the Los Alamos National Laboratories (LANL) [6.20]. It was designed to deliver high currents into a low-impedance load, and it is a low-gain generator that requires a large input energy.

The active coil has a length of 1118 mm and consists of four sections with lengths of 217, 217, 223, and 461 mm, as shown in Fig. 6.3. But since the computer model that was developed only accepts equal-length sections, it was assumed that the helical coil has a length of 1085 mm and that it consists of five equal-length sections each of which is 217 mm in length. This will affect the computed results during the final stage of operation of the generator, since the operation time is reduced by 3.6 $\mu$s and there is a corresponding reduction in inductance. The inside diameter of the coil is

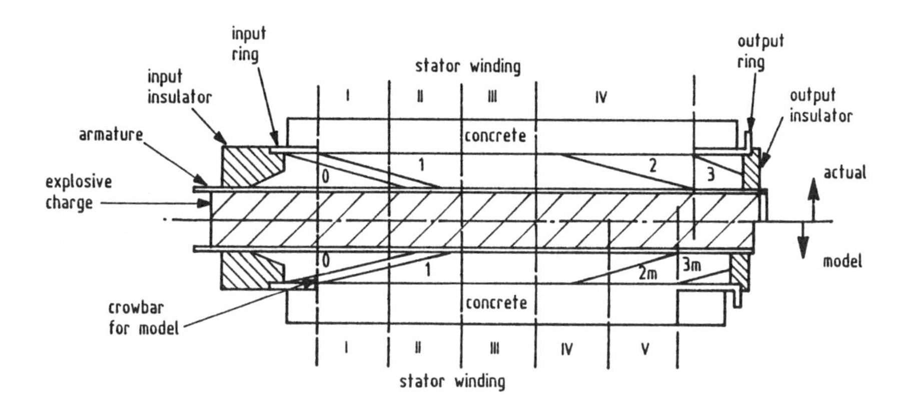


FIGURE 6.3. The MARK IX generator. The upper half is the actual arrangement (0, 1, 2, and 3 are the armature positions at $t_0, t_1, t_2$, and $t_3$), the lower half is the model arrangement (0, 1m, 2m, and 3m are the armature positions at $t_0, t_{1m}, t_{2m}$, and $t_{3m}$). Here $t_0$ is the crowbar time, $t_1$ is the end of period (1), $t_2$ and $t_{2m}$ are the end times of the purely helical phase (the armature cone enters the output ring), and $t_3$ and $t_{3m}$ are the end times of period (2).

| Section | Pitch (mm) | Number of Parallel Cables | Cable Diameter (mm) |
|---|---|---|---|
| I | 54.25 | 5 | 9.3 |
| II | 108.5 | 10 | 9.3 |
| III | 223.0 | 20 | 9.3 |
| IV | 461.0 | 40 | 9.3 |
| V | 461.0 | 40 | 9.3 |

TABLE 6.1. Data for five sections of the MARK IX helical coil.

356 mm. The ratio of the diameter to the section length exceeds 1.6, which implies that a 10% error is expected in the values of the inductance. The pitch, number of cables, and cable diameter of each section are given in Table 6.1.

The copper armature had an outside diameter of 173 mm and a wall thickness of 9 mm. The cone angle was $14^o$. The detonation velocity was not reported in the literature, and was estimated therefore from the time taken for it to expand and make contact with the coil by considering the radial velocity to be $v_{\text{det}} \tan \alpha$. Thereby, the detonation velocity was obtained as approximately 9 km/s.

According to [6.20], the MARK IX has no crowbar, which means that the input ring has the same diameter as the coil. Nevertheless, a crowbar time was introduced, based on $dI/dt$ and resistance data, and the calculation begins at the instant the armature cone enters the coil. Therefore, it is expected that there will be some differences between the experimental and the computed data, even for the first sections of the stator.

Because this model considers only the helical part of the generator, the inductance of the output coaxial ring, which is 26 nH, was added to the load inductance. The ring inductance was calculated from the cross-sectional area of the generator [6.20], and was used for the first $t_2 = 102$ $\mu$s of the generator operation. It was assumed, therefore, that the load inductance was 60 nH.

To account for the peculiar design of the MARK IX a coaxial coil was added, so that the currents during the final stage of flux compressions could be compared, i.e., after 102 $\mu$s. This coaxial coil had an active length of 153 mm and a passive length of 26 mm, where the latter accounts for the output insulator. The 3.9 nH inductance of the passive length was added to the inductance of the load until the compression ended. The calculations end when the contact point reaches the output end of the helical coil.

The priming current required to create the initial magnetic field in the MCG was 413 kA, which was supplied by a 1500 $\mu$F, 40 kV capacitor bank.

When the variations of inductance and rate of change of inductance with time, calculated by the more complex LANL model [6.20], were compared with those calculated by the simpler model presented here, it can be seen from Figs. 6.4a and 6.4b that there is good agreement between the values of the inductances and its rate of change. In Fig. 6.5a, it can be seen that there is good agreement between the measured and calculated load currents for the first 100 $\mu$s following crowbar action at time $t_0$. At time $t_2$ for the actual generator and $t_{2m}$ for the model being used, the armature cone enters the output ring and the subsequent error in the predictions increases slightly. Nevertheless, when the detonation process is complete at time $t_3$, the error in the predictions is still only about 20%. The shape of the experimental and calculated $dI/dt$ curves in Fig. 6.5b are in agreement. This indicates that the model tracks the normal operation of the generator, without the need to include any unusual loss mechanisms.

The very high resistances observed experimentally, and presented in Fig. 6.5c, result from the very high voltages that are generated within the generator, while the armature is expanding from the crowbar position. These high voltages are evident in Fig. 6.5d. There are substantial energy losses when the armature cone makes contact with the uninsulated input ring, which gradually disappear as the contact point moves along the coils of the helical stator winding.

### *The EF-3 Generator*

The EF-3 generator delivers medium-level currents to loads with fairly high inductances and is primed with a low-energy capacitor bank. It was

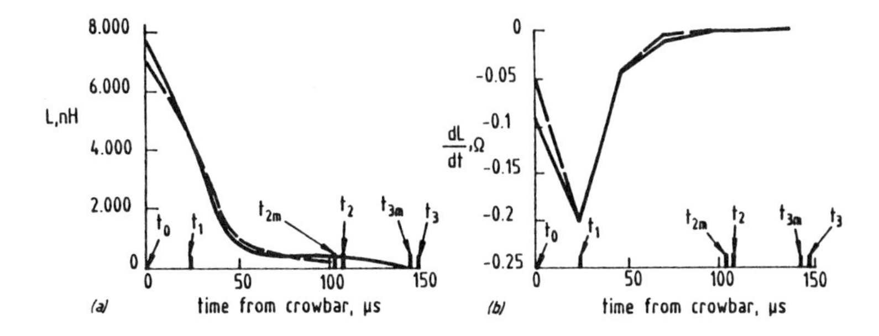


FIGURE 6.4. Calculated parameters of the MARK IX generator: (a) inductance variation with respect to time and (b) rate-of-change of inductance variation with respect to time. The lines (—) are calculated and (- - -) are from [6.20]

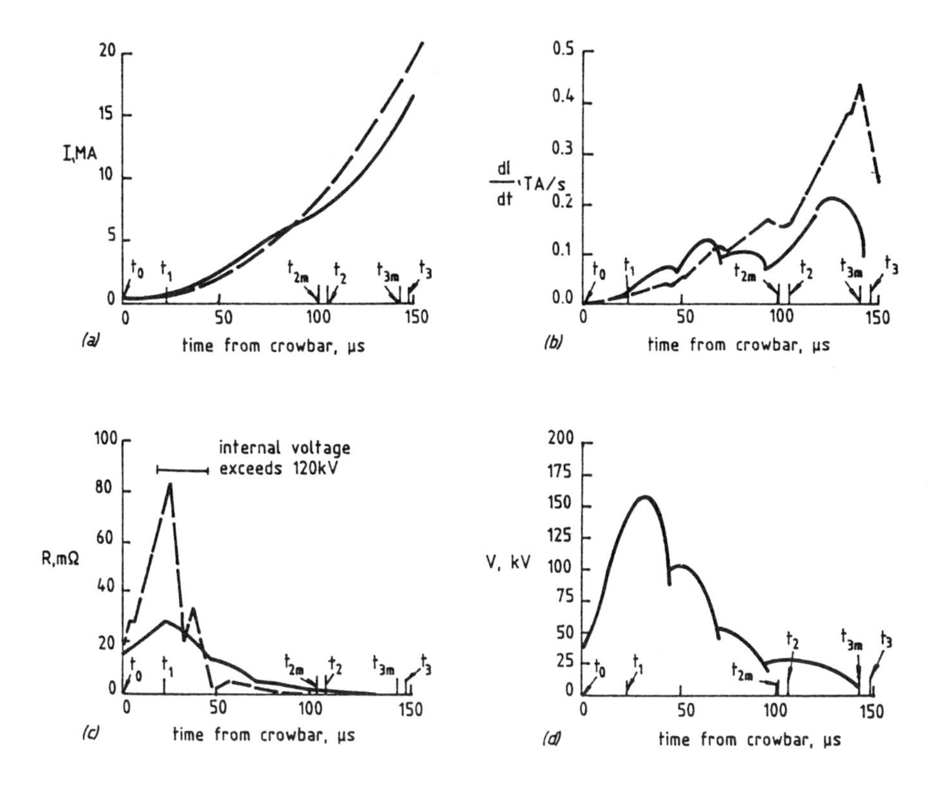


FIGURE 6.5. Characteristics of the MARK IX generator: (a) load current variation with respect to time, (b) rate-of-change of load current variation with respect to time, (c) resistance variation with respect to time, and (d) maximum internally generated voltage variation with respect to time. The lines (—) are calculated and (- - -) are measured [6.20].

| Section | Pitch (mm) | Number of Parallel Cables | Cable Diameter (mm) |
|---|---|---|---|
| I | 3.2 | 1 | 2.5 |
| II | 3.5 | 1 | 3.0 |
| III | 4.0 | 1 | 3.0 |
| IV | 4.7 | 1 | 4.0 |
| V | 5.5 | 1 | 4.5 |
| VI | 6.3 | 1 | 4.5 |
| VII | 7.9 | 2 | 3.0 |
| VIII | 9.5 | 2 | 4.0 |
| IX | 11.1 | 2 | 4.5 |
| X | 14.3 | 3 | 4.0 |
| XI | 19.0 | 3 | 4.5 |
| XII | 25.4 | 4 | 4.5 |
| XIII | 35.0 | 6 | 4.5 |
| XIV | 51.0 | 8 | 4.5 |
| XV | 76.0 | 12 | 4.5 |

TABLE 6.2. Parameters of the 15-section EF-3 helical coil.

developed by one of the authors (B.M. Novac) at the Institute of Atomic Physics in Romania [6.22–6.24].

The helical coil of the EF-3 generator had a length of 1500 mm and an inside diameter of 160 mm. It consisted of 15 equal-length sections, with the pitch, number of cables, and cable diameter for each section being presented in Table 6.2.

The copper armature had an outside diameter of 80 mm and a wall thickness of 3 mm. The cone angle was $13^o$, the detonation velocity 7.7 km/s, the crowbar diameter 107 mm, the load inductance 100 nH, and a priming current of 5 kA was provided from a 1650 $\mu$F, 4.5 kV capacitor bank.

The calculated time dependencies of the inductance and of its rate of change with time are presented in Figs. 6.6a and 6.6b, respectively. In Fig. 6.6b, the presence of the 15-stator sections are evident in the discontinuities in the plot. The calculated and the experimental variations of the current and the rate of change of the current with time can be seen in Figs. 6.7a and 6.7 b, respectively. It is clear that there is good agreement between the calculated and experimental results until the detonation process has been completed, and that beyond this point, there is considerable disagreement. As can be seen, the experimental result reaches a peak current and then declines, while the computed current continues to increase. It is also clear that, in order to obtain better agreement, account must be taken of losses in the generator, in addition to Ohmic losses. This is done in the next

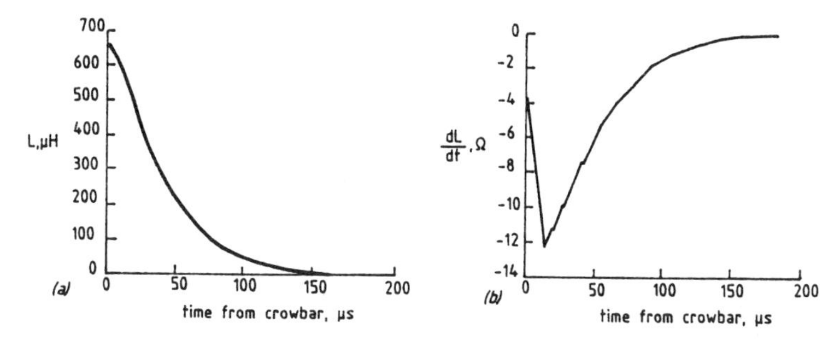


FIGURE 6.6. Calculated parameters for the EF-3 generator: (a) the inductance variation with respect to time and (b) the rate-of-change of inductance variation with respect to time.

section. Although these effects are not significant in 1 MJ generators, such as FLEXY I, they are included here for completeness. The variations in the resistance and internal voltage of the generator with time are shown in Figs. 6.7c and 6.7d, respectively. Again, the presence of the various sections of the stator winding can clearly be seen. Toward the end of the flux compression stage, the voltage reaches 110 kV, which was too high to be sustained by the insulation in the cables, even when a 1.2 mm layer of polyethylene foil was added to the interior of the coil [6.24].

### Non-Ohmic Losses

Calculated results, based on the 0D model that is being considered in this section, are in fairly good agreement with the corresponding measured results for at least two generators, i.e., the MARK IX and the EF-3. Nevertheless, if non-Ohmic losses are included in the analysis, even better agreement can be obtained. The non-Ohmic losses to be considered include nonlinear diffusion of the magnetic field, $2\pi$-clocking, geometric defects in the armature, and voltage breakdown.

#### *Nonlinear Diffusion of the Magnetic Field*

There are two ways of calculating the resistances and the inductances in helical generator circuits. The first includes diffusion losses in the definition of the inductance and assumes these to be a permanent energy loss as the stator turns are removed from the circuit. The second method, which is more commonly adopted, is to neglect the diffusion losses when calculating the inductances, but to include them in the resistance calculations. In this case, there is no energy loss at the contact point, where the diffusion losses are linear due to the current density being low [6.25]. However, additional losses occur at the point of contact between the armature cone and the coil, where a high-magnetic field is formed, as the wires, which are heated due to nonlinear diffusion, are removed from the circuit [6.25]. Under these conditions, a modified skin depth $\delta^*$ is used, which can be calculated from

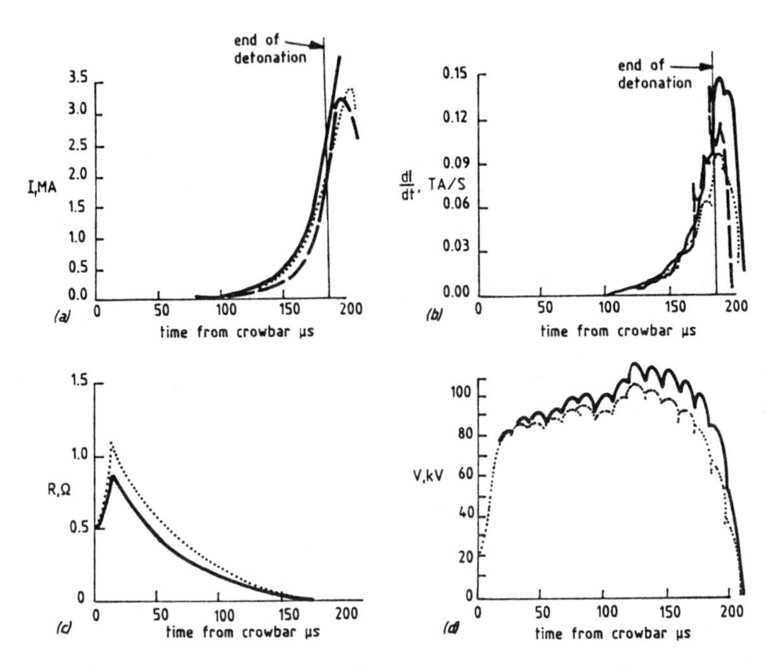


FIGURE 6.7. Characteristics of the EF-3 generator: (a) the load current variation with respect to time, (b) the rate-of-change of load current variation with respect to time, (c) the resistance variation with respect to time, and (d) the maximum internally voltage variation with respect to time. The lines (—) are calculated, (- - -) are measured [6.23], and (...) are calculated using modified codes.

$$\delta^* = \delta \left[ \frac{1}{T_0} \left( T_0 + \frac{B^2}{\mu_0 \rho c_V} \right) \right]^{\frac{1}{2}}, \tag{6.21}$$

where $T_0$ is the initial temperature of the coil, $B$ is the magnetic flux density arising from the coil current, $\rho$ is the density of the coil conductors, and $c_V$ is the specific heat of the coil conductors. The additional loss is then calculated in terms of the magnetic field confined to the volume between $\delta$ and $\delta^*$. To derive an expression for the equivalent resistance in this case, the following analysis is carried out. If a volume $\Delta V$ is removed from the internal generator volume, either by movement or by an unexpected jump of the contact point, the energy contained in this volume is lost. If the reduction in inductance is $\Delta L$, then the change in the inductance can be related to the change in volume by

$$\frac{1}{2} I^2 \Delta L = \frac{B^2}{2\mu_0} \Delta V. \tag{6.22}$$

For nonlinear diffusion,

$$\Delta V = 2\Delta l(S^* - S), \tag{6.23}$$

where $\Delta l$ is the cable length removed in time $\Delta t$ and $S^*$ and $S$ are circular cross-sectional areas inside the cable defined by skin depths of $\delta^*$ and $\delta$. The factor 2 is included to account for the effects of the armature, that is, diffusion occurs at the surface of the armature as well as at the surface of the coils.

The approximate velocity of the contact point of

$$v_{cp} = \frac{2\pi r_c v_{\text{det}}}{p} \tag{6.24}$$

can be extremely high. In the case of the EF–3 generator, the contact point velocity in the first section is 880 km/s. Therefore, the equivalent non-Ohmic resistance is

$$R_{nd} = \frac{\Delta L}{\Delta t} = 2B^2 v_{cp} \frac{S^* - S}{\mu_0 I^2}. \tag{6.25}$$

If the magnetic flux density is calculated from

$$B = \frac{\mu_0 I}{\pi n \Phi} \tag{6.26}$$

and $\cos\beta$ is approximated by $p/2\pi r_c$, then the non-Ohmic diffusion for $n$ contact points of the armature with the $n$ cables gives the equivalent resistance as

$$R_{nd} = \frac{2\mu_0 v_{\text{det}}[\delta^*(\Phi - \delta^*) - \delta(\Phi - \delta)]}{\pi\Phi^2 n \cos\beta}, \tag{6.27}$$

where $\beta$ is the angle made by the coil with a plane normal to the coil and $v_{\text{det}}$ is the detonation velocity.

For a generator in which the linear current density flowing along the coil axis is kept below about 0.2 MA/cm, both calculations and experiments show that there is little loss due to magnetic diffusion [6.11,6.25].

*$2\pi$-Clocking*

This loss occurs whenever a coil section is not coaxial with the armature. If the eccentricity exceeds $p\tan\alpha/2\pi$, where $p$ is the pitch of the coil, the contact point jumps unexpectedly ahead by one turn. When the coil pitch is small, as may be the case in the earlier sections of the stator winding, this mechanism is often the only one that needs to be considered [6.26].

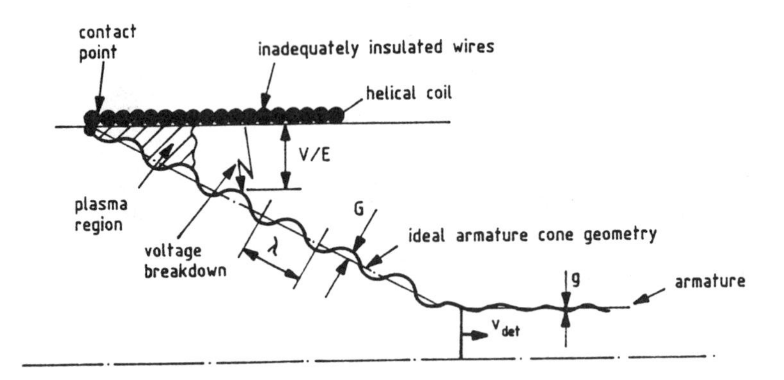


FIGURE 6.8. Mechanism for voltage breakdown losses using a sinusoidal model for armature cone defects.

*Geometric Defects*

If the generator armature is machined from a thick tube, vibrations of the cutting tool may lead to a sinusoidal modulation in the diameter of the armature along the length of the axis of the generator. It has been shown [6.25], both experimentally and theoretically, that these defects may be amplified by a factor of between 7 and 9 as the armature expands. In Fig. 6.8, if $\lambda$ is the wavelength of the sinusoidal defect, the initial amplitude of the defect $g$ is amplified to $G$ owing to expansion of the armature. This process can generate multiple contact points and as a consequence lead to additional nonlinear losses [6.25]. Similar defects can be created by using an armature made of an inhomogeneous material or an armature with too large a ratio of coil diameter to armature diameter.

*Voltage Breakdown*

Overall, the most important of the various non-Ohmic loss mechanisms in a high-inductance generator is the voltage breakdown that occurs between the coil and the armature cone. Near the contact point, the electric field is sufficiently intense to form a plasma. If the geometry is ideal, there will be good electrical contact, which will prevent the lines of force from leaving the region [6.11]. However, if armature defects are present and the generator geometry is not coaxial, electrical breakdown is likely to occur ahead of the contact point and the plasma region. In practice, the high voltage effectively amplifies the loss mechanisms that were described earlier. For instance, the effective amplitude of the sinusoidal armature defects becomes the sum of the actual amplitude and the breakdown distance $V/E$, where $V$ is the voltage between the coil and the armature at the point under consideration and $E$ is the characteristic breakdown electric field of the working gas under these particular conditions. (See Fig. 6.8.) For a detonation velocity of $v_{\text{det}}$, the equivalent non-Ohmic resistance is

$$R_{vb} = \frac{2\pi\mu_0 v_{\text{det}} A \cos\alpha\ (2r_c - A)}{p^2}, \tag{6.28}$$

where $A$ is the effective amplitude of the armature expansion defects in the presence of a voltage breakdown. For each voltage breakdown between the coil and a peak in the sinusoidal armature surface, the magnetic energy in the volume between the plasma and the breakdown point is lost. The new effective amplitude is $A = G + V/E$. The volume $\Delta V$, which can be approximated by $\pi\lambda A(2r_c - A)\cos\alpha$, is lost for the time interval $\Delta t = \lambda/v_{\text{det}}$. If the magnetic field is assumed to be generated by a copper sheet of width $p$, the magnetic field can be approximated by $B = \mu_0 I/p$. This expression was used in deriving Eq. 6.28.

The total non-Ohmic resistance for the generator is the sum of the components due to the effects described above:

$$R_{no} = R_{nd} + R_{2\pi} + R_{vb}. \tag{6.29}$$

This expression is effective only after the armature has made contact with the coil, for only then does it have any significant effect on the calculated current. The results of calculations conducted using more complicated computer models are added to Fig. 6.7. It is clear that although the accuracy of the computations are improved, the net improvement is somewhat less than expected. The maximum error in predicting the current along the compression is 35%. However, the error in predicting the peak current is less than 10%, while in practice, the generator, *even* with improved stator winding insulation, had a ±15% variation in the peak current reproducibility.

### 6.2.2 Simple 2D Model for a Helical MCG

In this section, a simple but more complete 2D model for a helical MCG that overcomes many of the limitations of the 0D model discussed above is presented. This code can be easily adapted to model high-energy and high-current generators having variable geometry for both the helical coil and the armature. It provides valuable information on both the magnetic and the electric field distributions within the generator and the likely radial and axial movement of the stator turns. The model can yield important insights into the various phenomena that differentiate the performance of small generators primed with either a capacitor, a battery, or an externally produced magnetic field. In addition, it opens the way for an understanding of the behavior of cascaded systems of MCGs inductively coupled through dynamic transformers, by the so-called *flux-trapping* technique.

As noted earlier, the 0D model was successfully used to design and produce a simple 1 MJ generator. Although the experimental performance of the generator was in overall agreement with the theoretical predictions of the model, and subsequent improvements led to an enhanced 2 MJ version, a number of detailed questions related to the multidimensional geometry of the generator could not be answered. An example of this arises in the 3D region near the armature-coil contact point, where extremely high magnetic and electric fields are generated and where most of the kinetic energy of the expanding armature is transformed into electrical energy as work is done by the armature to compress the magnetic flux. Any loss mechanisms such as magnetic diffusion, $2\pi$-clocking, electrical breakdown, and those related to the axial or radial movement of the coil, are dependent on the distribution of the magnetic and electric field intensities within this region. Accurate information about these processes is clearly required to optimize the performance of high-performance generators, in which the coil and the armature could have a variable diameter along the length of the axis of the generator.

Other questions that require multidimensional analysis are related to the different ways in which the generator can be excited, e.g., by either a capacitor bank or a battery, and to the manner in which the generator behaves when it is excited by an external magnetic field source. Finally, accurate modeling is required in the design of cascaded systems, which consist of two or more MCGs inductively coupled through dynamic transformers by flux-trapping.

While a number of 2D generator models have been described in the literature [6.8,6.27], these are complex and require long run times on large computers, and attention will be focused on a novel 2D model that is much easier to use and for which the basic concepts are simple. The resulting computer code is extremely fast.

## The 2D Model

In this section, the basic working equations for the major components of the helical generator are presented.

### *The Working Equations*

The equations used to model a helical generator can be derived by superimposing the magnetic flux densities $B_\theta$ and $B_z$ generated by currents flowing in the $z$- and $\theta$-directions [6.19]. This technique is used since it allows expressions for the inductance [6.20] to be derived. The helical generator shown in Fig. 6.9 can be decomposed into equivalent $z$- and $\theta$-current-carrying circuits. The $z$-current circuit consists of the helical coil and the armature, with the load used to complete the circuit. The coil can be further divided into $N$ rings, through which the same load current $I_z$ flows azimuthally to generate the $B_z^z$ field. On the other hand, these rings, to-

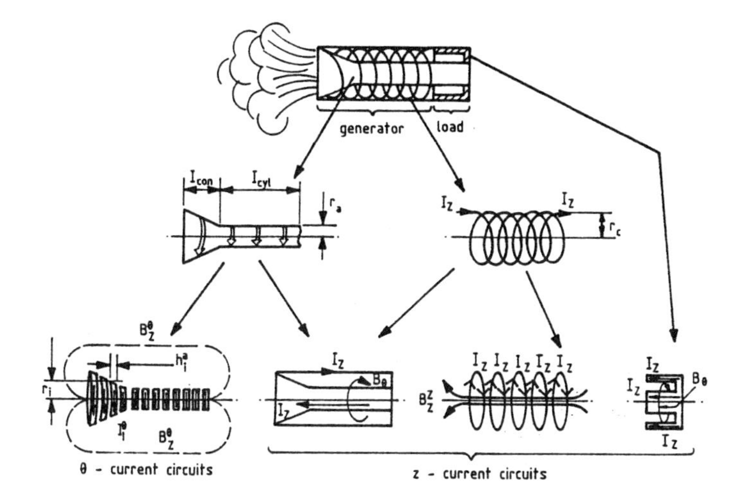


FIGURE 6.9. Decomposition of a helical generator circuit into a single $z$-circuit and $N$ $\theta$-current circuits.

gether with the armature, form a coaxial structure that generates the $B_\theta$ field. Although Fig. 6.9 shows a constant pitch, single-section coil with a constant diameter along the full axial length, an extension to multisection, variable diameter coils is straightforward.

The armature current in the $\theta$-direction induced by the time-dependent helical coil current $I_z$ has a more or less pronounced axial distribution in intensity. In order to model adequately the armature current, the armature is divided into a number of separate $\theta$-circuits (rings), with a different current flowing through each ring, as shown in Fig. 6.9. Plots of these ring currents showing their axial distribution are presented in Fig. 6.10 for the case of a generator powered by a capacitor bank (no explosive present). This approach differs from others in which the armature is decomposed into separate rings [6.10], because in these models the magnetic field diffusion through the armature wall is neglected so that only one circuit is needed to describe the generator (all ring currents had the same value, i.e., $I_z$ value). To simplify further this analysis, it is assumed that the number of armature circuits (or rings) is equal to the number of helical rings, i.e., the number of helical turns. However, the number of turns or rings can be varied to provide a more precise description of the magnetic field distribution (as in Fig. 6.10) or to model special cases where the generator is primed by an outer coil. As with the helical coil, the initial armature geometry is assumed to be cylindrical, although alternative geometries can readily be implemented.

Based on the above discussion, it can be seen that there are $N + 1$ equivalent circuits in the armature as shown in Fig. 6.11. On this basis, the following $N$ equations for the $z$-circuit, i.e., the load circuit, can be written

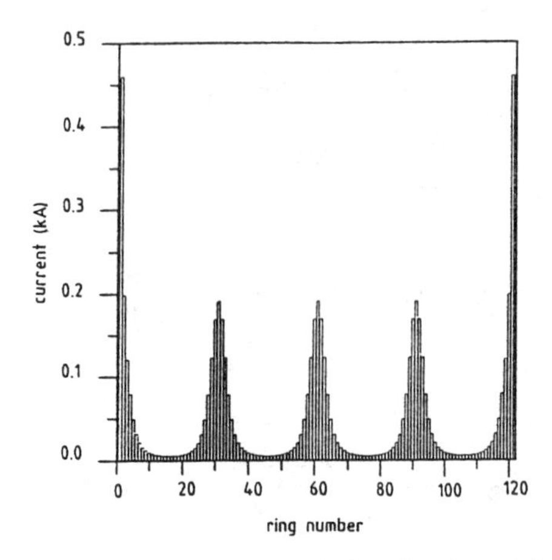


FIGURE 6.10. Example of a continuous axial distribution of the armature current described by $\theta$-circuit rings. The armature current is produced by a sinusoidal $I_z$ current in the helical coil (not shown).

$$L_z \frac{dI_z}{dt} + \sum_{i=1}^{N} \left( M_{zi} \frac{dI_i^\theta}{dt} + \frac{dM_{zi}}{dt} I_i^\theta \right) + \left( R_z + \frac{dL_z}{dt} \right) I_z = 0 \tag{6.30}$$

and for the $\theta$-circuit

$$L_i^\theta \frac{dI_i^\theta}{dt} + M_{zi} \frac{dI_z}{dt} + \frac{dM_{zi}}{dt} I_z \tag{6.31}$$

$$+ \sum_{\substack{j=1 \\ j \neq i}}^{N} \left( M_{ij}^\theta \frac{dI_j^\theta}{dt} + \frac{dM_{ij}^\theta}{dt} I \right) + \left( \frac{dL_i^\theta}{dt} + R_i^\theta \right) I_i^\theta = 0,$$

where $i = 1 \cdots N$. In the following sections, expressions for the inductance and resistance are derived.

*Inductance Expressions*

Based on the decomposed generator model presented in Fig. 6.9, the total inductance in the $z$-circuit is $L_z = L_z^z + L_z^\theta + L_{\text{load}}$. The components $L_z^z$ and $L_z^\theta$ are related to the $B_z^z$ and $B_z^\theta$ fields, respectively. For the $B_\theta$ field, the inductance $L_z^\theta$ can be decomposed into two terms:

$$L_z^\theta = L_{\text{cyl}} + L_{\text{con}}, \tag{6.32}$$

where

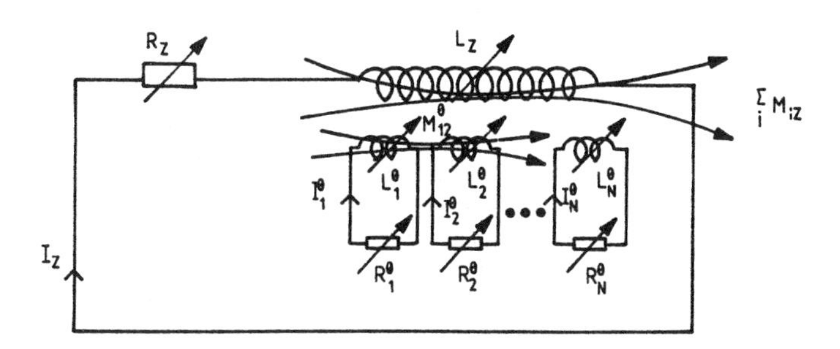


FIGURE 6.11. Equivalent circuit diagram with the armature separated into $\theta$-rings.

$$L_{\text{cyl}} = \frac{\mu_0 l_{cyl}}{2\pi} \ln \frac{r_c}{r_a} \tag{6.33}$$

and

$$L_{\text{con}} = \frac{\mu_0}{2\pi} K \tag{6.34}$$

are the inductances of the corresponding cylindrical and conical parts of the coaxial structure, respectively, and

$$K = l_{\text{con}} + \frac{1}{\tan\alpha}\left[r_a \ln \frac{r_a}{r_c} - (r_a + l_{\text{con}} \tan\alpha) \ln\left(\frac{r_a + l_{\text{con}} \tan\alpha}{r_a}\right)\right]. \tag{6.35}$$

To calculate the inductance $L$ of an arbitrary coaxial structure for which the inner and outer radii are expressed in terms of cylindrical coordinates $(r, z, \theta)$ by $r = r(z)$ and $R = R(z)$, respectively, the energy definition of inductance is used:

$$\frac{LI^2}{2} = \int \frac{B^2}{2\mu_0} dv, \tag{6.36}$$

where the current $I$ flowing axially through the structure generates an azimuthal magnetic flux density $B = \mu_0 I / 2\pi r$. The inductance in cylindrical coordinates can be derived from

$$L = \frac{\mu_0}{2\pi} \int_0^l \int_{r(z)}^{R(z)} \frac{1}{r} dr dz, \tag{6.37}$$

where $l$ is the length of the structure. Letting $r(z) = r_a$ and $R(z) = r_c$ for a cylindrical geometry, leads to Eq. 6.33, and letting $r(z) = r_a + z \tan\alpha$ and $R(z) = r_c$ for a conical geometry, leads to Eq. 6.34. For the geometries used in most generator models, in which $r$ and $R$ change from $r_1$ and $R_1$ to $r_2$ and $R_2$, respectively, it is found that $r(z) = z(r_2 - r_1)/l$ and $R(z) = R_1 + z(r_2 - r_1)/l$. For this case, the inductance is given by

$$L = \frac{\mu_0 l}{2\pi} \ln \left[ \left( r_2^{r_2} r_1^{-r_1} \right)^{r_2 - r_1} \left( R_2^{R_2} R_1^{-R_1} \right)^{R_1 - R_2} \right]. \tag{6.38}$$

For the $B_z^z$ field, the inductance $L_z^z$ is given by

$$L_z^z = \sum_{i=1}^{N} \left( L_i^z + \sum_{\substack{j=1 \\ j \neq i}}^{N} M_{ij}^z \right), \tag{6.39}$$

where the inductance of a very thin ring is

$$L_i(r_i, h_i) = \mu_0 r_i \left( \ln \frac{8 r_i}{h_i} - 0.5 \right) \tag{6.40}$$

in which $r_i$ is the radius of the ring and $h_i$ is its dimension in the $z$-direction as shown in Fig. 6.9. Equation 6.40 may be slightly modified depending on the cross-section of the ring [6.28,6.29].

The mutual inductance between two coaxial rings having radii of $r_i$ and $r_j$ is

$$M_{ij}(r_i, r_j, d_{ij}) = \mu_0 \left[ -\{(r_i + r_j)^2 + d_{ij}^2\}^{\frac{1}{2}} E(x_{ij}) + \frac{(r_i^2 + r_j^2 + d_{ij}^2) K(x_{ij})}{\{(r_i + r_j)^2 + d_{ij}^2\}^{\frac{1}{2}}} \right], \tag{6.41}$$

where $d_{ij}$ is the distance between the rings and $E$ and $K$ are, respectively, complete elliptic integrals of the first and second kind of modulus:

$$x_{ij} = \frac{4 r_i r_j}{(r_i + r_j)^2 + d_{ij}^2}. \tag{6.42}$$

If the mutual inductance between the $i^{th}$ armature ring having radius $r_i^a$ and the $j^{th}$ coil ring having radius $r_j^c$ is taken to be $M_{ij}(r_i^a, r_j^c, d_{ij})$, then

$$M_{zi} = \sum_{j=1}^{N} M_{ij}(r_i^a, r_j^c, d_{ij}). \tag{6.43}$$

The inductances associated with the $\theta$–circuits are described by equations similar to Eqs. 6.40–6.42.

*Resistance Expressions*

In this section, the basic principles used to calculate the resistance are discussed. A complete description of the equivalent resistance required to model all the various losses in MCGs can be found in [6.2] for high-energy generators and in [6.17] for small generators.

Adopting the same notation that was used above, the total resistance of the generator circuit is

$$R_z = R_z^\theta + R_z^z + R_z^P + R_{load}, \tag{6.44}$$

where:

$$R_z^\theta = R_{cyl} + R_{con}, \tag{6.45}$$

$$R_{cyl} = \frac{l_a}{\pi \delta_a} \cdot \frac{l_{cyl}}{2r_a - \delta_a}, \tag{6.46}$$

$$R_{con} = \frac{\rho_a}{2\pi \delta_a \tan\alpha} \ln\left[\frac{2l_{con}\tan\alpha}{2r_a - \delta_a} + 1\right], \tag{6.47}$$

and

$$R_z^z = \sum_{i=1}^{N} R_i^z(r_i^c, h_i^c). \tag{6.48}$$

In these equations, $\rho$ is the electrical conductivity and $\delta$ is the skin depth. The resistance of the $i^{th}$ ring, in which the superscript $\theta$ is used to represent the armature and the superscript $z$ to represent the helical coil, is

$$R_i(r_i, h_i) = \frac{2\pi \rho r_i}{\delta h_i}. \tag{6.49}$$

The proximity effect term $R_z^P$ in Eq. 6.44 is calculated only for the helical coil, where the axial length of the ring is less than the coil pitch because of the insulation between the turns of the coil.

### Ring Dynamics

In this section, the dynamics of the rings introduced above is examined.

#### *Armature Ring Dynamics*

If the detonation front makes contact with the first ring at $t = 0$, then the $n^{th}$ ring begins to move at time

$$\Delta t_n = \sum_{i=1}^{n-1} \frac{h_i^a}{D}, \tag{6.50}$$

where $D$ is the detonation velocity. In this analysis, it is assumed that the radial velocity of expansion, $v_r = D \tan \alpha$, is constant, but it will be necessary in more accurate models to include the initial impulse expansion phase, which could be determined from x-ray or photographic data [6.30]. The ring acceleration is given by the following expression:

$$a_n(t) = \theta(t - \Delta t_n) \left[ V_0 \dot{\theta}(t - \Delta t_n) + \frac{V_1}{T} e^{-\frac{t}{T}} \right], \tag{6.51}$$

where $a_n$ is the acceleration of the $n^{th}$ ring, $\dot{\theta}$ is the time derivative of the step function $\theta$, and $V_0$, $V_1$, and $T$ are parameters that must be fitted to the experimental data.

For very-high-current generators, the effects of the magnetic field on the ring dynamics can be important, and a more general treatment including the corresponding magnetic deceleration may be necessary, which means including the term $2\pi P_M^n r_n^a h_n^a / M_n^a$, where $P_M^n = (B_z^n)^2/2\mu_0$ is the magnetic pressure due to the local field $B_z^n$ and $M_n^a$ is the initial mass of the $n^{th}$ ring of the armature.

#### *Helical Coil Ring Dynamics*

It will be assumed that the ring only moves as the result of magnetic forces. Any radial movement, which is extremely important in the design of MCGs, is kept to a minimum by a supplementary mass. A model for assessing this motion is presented in [6.2]. Axial ring movement is also a very important limiting factor in designing high-efficiency, very-high-energy generators [6.31]. To model this axial movement, the local field $B_\theta$ was used as the accelerating term, which is similar to the decelerating term used to calculate the radial deceleration of the armature.

### The Switching Problem

Equations 6.30 and 6.31 can be solved to determine the generator currents and the magnetic field distribution produced by the ring currents can be calculated by using established methods [6.32]. However, although the solution of the problem appears to be straightforward, in fact it is not so. The radial movement of the armature rings was initially assumed to be a continuous process, but it was found that this is not the case when the first ring comes into contact with the helical coil. When this happens, the ring is effectively removed from the circuit and the process becomes discontinuous.

In one approach to the general problem of solving circuits that are undergoing an inductance change due to a switching process [6.32], it was assumed that the switching process was continuous. This assumption is correct for the turns of a helical coil carrying the same current, when the finite differences can be replaced with differentials, since the coil inductance is being reduced continuously. However, this approach cannot be applied to the case under consideration, as the different currents being carried by the independent circular closed rings prevent changes in the ring inductance, Eq. 6.40, from being infinitesimally small, even if the axial dimension $h$ is made infinitesimally small. Unlike the helical coil, which can be considered to have a fractional number of turns, a fraction of a ring has no meaning. Thus, in the model being presented, it is not possible to replace the finite differences with differentials.

To solve this problem, the generator currents can be calculated by using the two-step method illustrated in Fig. 6.12. The currents following movement of the armature ring from the original position of Fig. 6.12a, during the first time interval, can be calculated using Eqs. 6.30 and 6.31. At the end of this time interval, the "first" two armature rings are both touching the coil as shown in Fig. 6.12b. The first of these two rings is removed from the calculations in the second time interval, but the magnetic flux linkage is preserved [6.33]. This process results in $N$ algebraic equations, which can be solved using the LU decomposition method [6.34] to yield values for the $N$ currents flowing in the circuit. The algebraic equations are

$$L_z \Delta I_z + \sum_{i=2}^{N} M_{zi} \Delta I_i^0 + M_{z1} I_1^0(0) = 0, \tag{6.52}$$

$$L_i^0 \Delta I_i^0 + M_{zi} \Delta I_z + \sum_{\substack{j=2 \\ j \neq i}}^{N} M_{ij}^0 \Delta I_j^0 + M_{i1} I_1^0(0) = 0, \tag{6.53}$$

where $\Delta I_z = I_z(0) - I_z(1)$ and $\Delta I_i^0 = I_i^0(0) - I_i^0(i)$ in which 0 and 1 denote the values immediately before and after ring $i = 1$ is removed from the circuit, as shown in Fig. 6.12c. This two step process is repeated for every time step, as more armature rings are removed from the circuit.

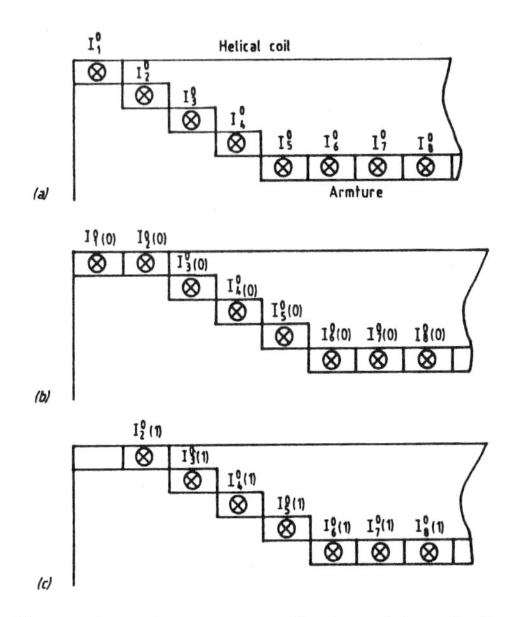


FIGURE 6.12. Illustration of twp-step method used to calculate generator currents: (a) initial position, (b) first step, and (c) second step.

## SOME TYPICAL RESULTS

Some typical results of this analysis are presented in Figs. 6.13–6.20. The input data for these calculations are

- a 70-turn helical insulated copper coil with an inner diameter of 6 cm, a length of 50 cm, and a conductor diameter of 0.45 cm,
- an aluminum armature with an outer diameter of 3 cm, divided into 70 rings,
- a detonation velocity of 8 km/s, which expands the armature to a cone angle of $12^o$,
- a coaxial inductive load with an inductance of 100 nH.

The time dependence of the typical self- and mutual inductances, calculated by using Eqs. 6.30 and 6.31, is shown in Fig. 6.13. In these calculations, a pair of supplementary rings was included to account for the azimuthal currents induced in the coaxial load at the ends of the inner and outer electrodes.

Different mechanisms for generating the initial magnetic field in the active volume of the generator, i.e., between the armature and the helical coil, were also considered: a capacitor bank (case I) and a battery (case II). In case I, the magnetic field was generated in tens of microseconds,

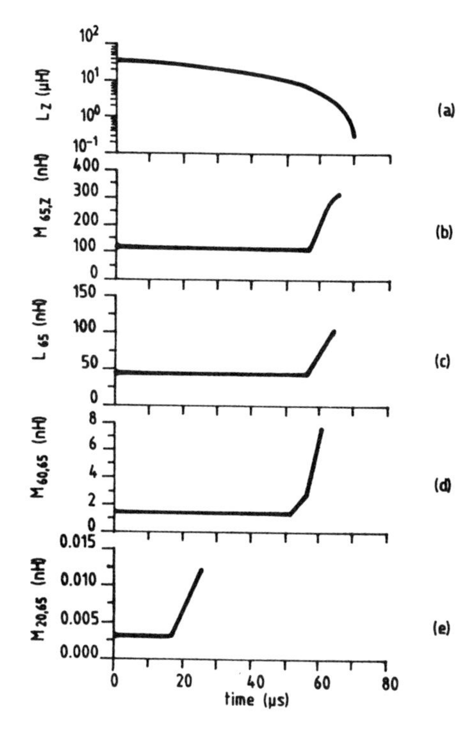


FIGURE 6.13. Plots of typical self- and mutual inductances.

which was insufficient for the magnetic field to penetrate into the armature wall and means that the entire magnetic flux in the active volume is compressed. Movement of the first ring was synchronized with the discharge of the capacitor, so that the circuit was closed at a peak current of 1.5 kA. In case II, part of the magnetic flux penetrated into the armature, which means the initial current from the battery had to be increased to 1.57 kA to generate the same magnetic flux as in case I.

The time histories of the generator currents and resistances for cases I and II are presented in Figs. 6.14–6.16. The very different initial conditions are evident, for in case I there is an axial armature current distribution and in case II there are no $\theta$-currents. In addition, the difference can also be seen in the differing behaviors as soon as a ring has began its expansion. It is clear from Fig. 6.16 that the final load current, and therefore the magnetic flux, is greater for case II, which agrees with experimental evidence [6.21]. Obviously, the mechanism responsible for this higher current is related to the existence of initially induced $\theta$-currents when the capacitor is used to provide the priming current and to their absence when a battery is used. This is illustrated in Fig. 6.15. An interesting inductive phenomenon that can be observed in Fig. 6.15, and later in Fig. 6.19, is that once the ring begins to move, the current in its stationary neighbor decreases.

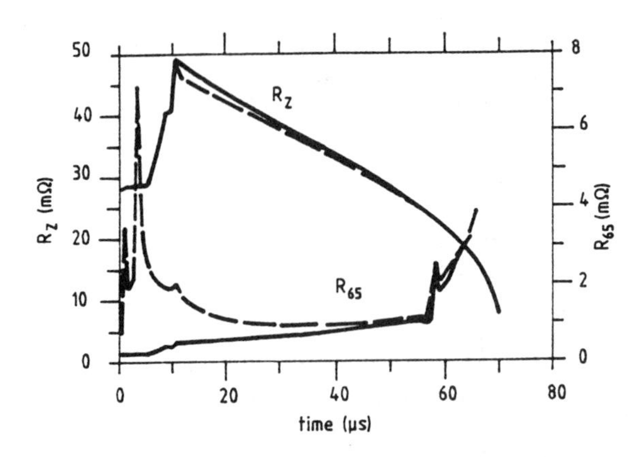


FIGURE 6.14. Time history of generator resistances: total $z$-circuit restance $R_z$ and a typical armature by resistance $R_{65}$. Line (—) are for Case I ( capacitor) (Fig. a) and (- - -) for Case II (battery) (Fig. b).

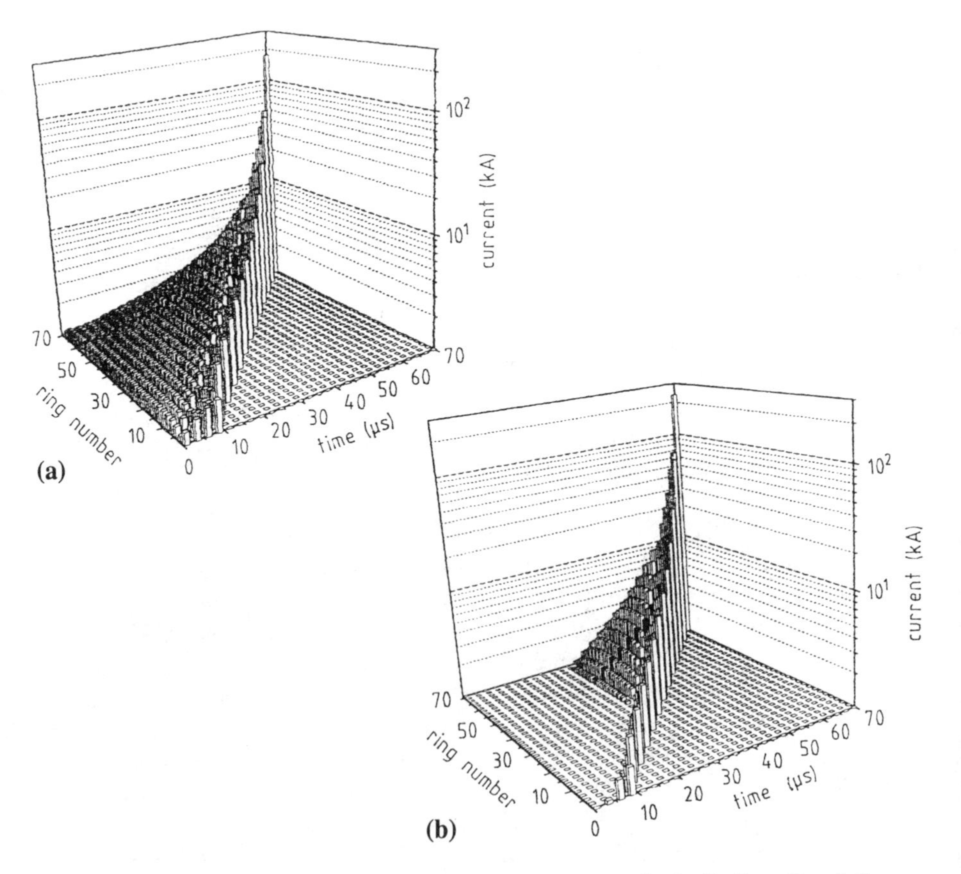


FIGURE 6.15. Time history of armature ring currents for both Case I and Case II.

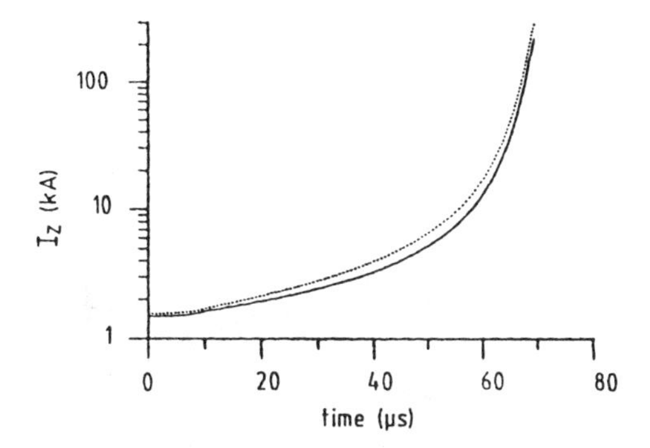


FIGURE 6.16. Time history of current $I_z$. Lines are (—) for Case I and (- - -) for Case II.

The time history of the total inductance of the generator and its time derivative, which were calculated by using the 0D model in [6.2] and the 2D model with the energy definition of inductance, are presented in Fig. 6.17. Figure 6.18 shows the 2D axial component of the magnetic field distribution at a typical moment in time during the operation of the generator and Fig. 6.19 shows the corresponding ring radii and currents. The initial penetration of the magnetic field into the armature walls is evident. These figures confirm the fact that the maximum magnetic field is produced near the armature–coil contact point. The time history of the radial electric field distribution inside the active volume of the generator is presented in Fig. 6.20.

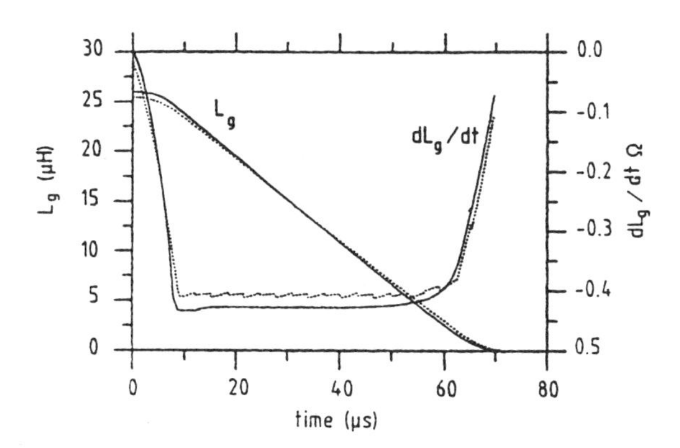


FIGURE 6.17. Time history of the inductance of the generator and the rate-of-change for the inductance using the 0D (- - -) and 2D (—) models.

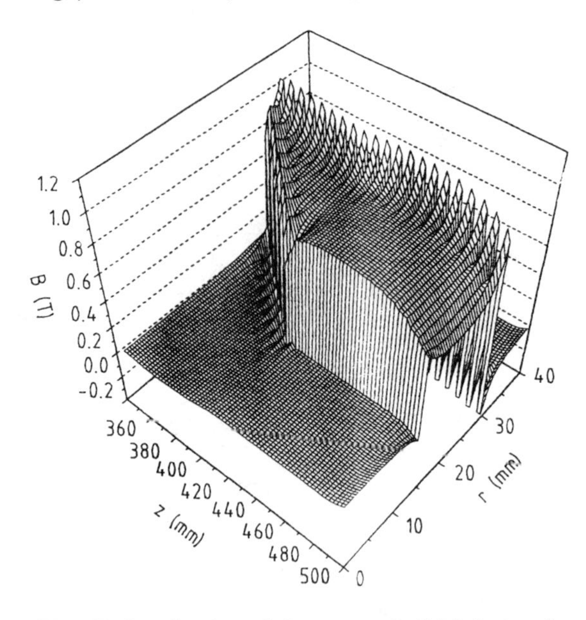


FIGURE 6.18. The 2D distribution of the magnetic field during flux compression (Case I).

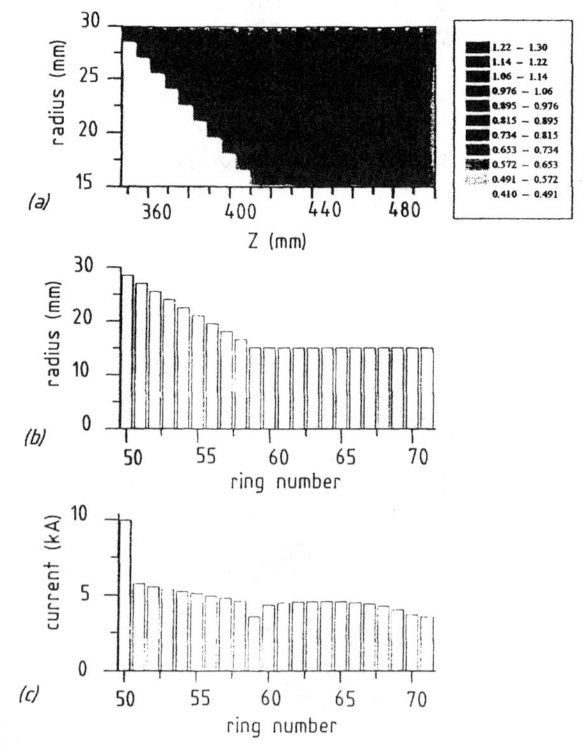


FIGURE 6.19. Illustration of the internal action of a generator (Case I): a) magnetic field intensity in the active volume, b) corresponding armature ring positions, and c) $\theta$-current distribution.

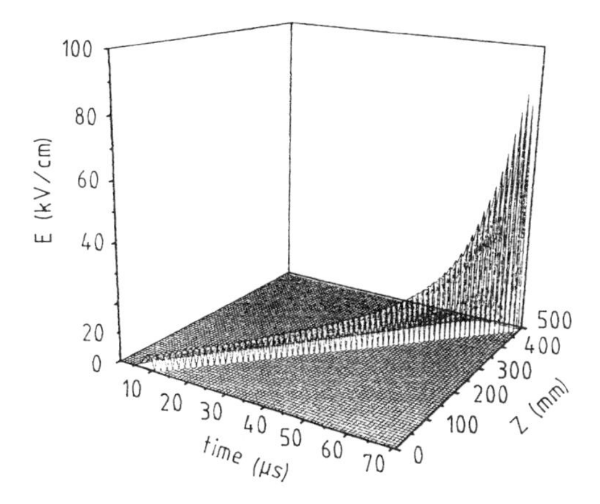


FIGURE 6.20. Time history of the radial electric field distribution inside the active volume of the generator (Case I).

### 6.2.3 *Comparison to Other Codes*

The main reason for developing the 2D numerical code described above is to model the mini to micro size generators normally found in dynamic transformer chains. Two sample problems from the literature are therefore examined, and results provided by the 2D code described above are compared to results obtained from more detailed and lengthy 2D calculations.

A 2D code developed by Sandia National Laboratories [6.35–6.37] for the micro MCGs solves the complete set of Maxwell's equations for the magnetic stream function $\Psi = rA_\theta$, where $A_\theta$ is the magnetic vector potential and $r$ is the radius. The magnetic field configuration 6 $\mu$sec after the explosive charge is detonated is presented in Fig. 6.21a. The field configuration calculated by using the code described in this section is presented in Fig. 6.21b. The load was connected in parallel with the helical coil, and the armature compresses the magnetic field *without* being connected to the load ($Z$) circuit. In calculating the inductance, it was necessary to omit the $L_z^\theta$ term, which is related to the $B_\theta$ field. Although the ring geometry is fully accounted for in calculating the magnetic field intensity, the total magnetic vector potential was determined from filamentary ring calculations. This simplification is undoubtedly responsible for the small irregularities observed in the field lines. In the Sandia model, the armature extended beyond the ends of the coil, whereas it was assumed to have the same length in the calculations performed as with the simpler model. This also probably contributes to the minor differences in the results presented in Figs. 6.21a and 6.21b. Calculations of the inductance, resistance, and time history of the load current were also in very good agreement with the Sandia results.

Results of an analysis of a mini helical generator, developed as a compact pulsed power source [6.38], using the 2D code presented in this section are compared to the experimental results in Fig. 6.22. In this generator, the

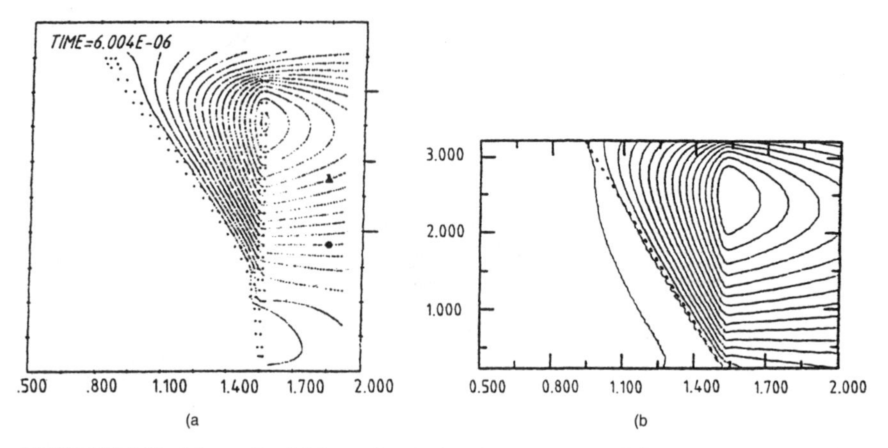


FIGURE 6.21. Magnetic field configuration for a very small helical generator: (a) reproduced from Figure 6 in Reference [6.9] and (b) calculated from the present model and covering only the length of the helical coil. Both the radial (horizontal) and axial (vertical) coordinates are in cm.

helical coil was uninsulated, and the high voltage between the coil and the armature led to electrical breakdown during the compression process. The 0D model was used to model this phenomenon [6.39]. The electric field between the coil and armature ring pairs is calculated, and when this exceeds a specified breakdown value, the corresponding pair of rings is removed from the circuit. A first estimate of the breakdown field (in kV/cm) is obtained from [6.40]:

$$E = 24.5p + 6.7\left(\frac{p}{R_{eff}}\right)^{\frac{1}{2}}, \tag{6.54}$$

where $p$ is the air pressure in atmospheres and $R_{eff}$ is the effective radius, which is 23% of the armature radius in centimeters. This equation yields a breakdown field of 34 kV/cm, which was used in the calculations presented in Fig. 6.22. As can be seen, the calculated results are in good agreement with the experimental results. This may however be somewhat fortuitous, since the conditions inside the generator are strongly affected by shock waves, ionization, and high temperatures, which means that, in other generators, it may be necessary to regard the breakdown fields calculated using the above equation as an adjusting factor [6.37]. In any event, this example illustrates that the models presented in this section can easily incorporate various loss mechanisms.

In conclusion, the simple and efficient 2D model presented in this section is a valuable tool in the design of helical flux compression generators with almost any initial geometry, and it provides important information on the magnetic and electric field distributions, which is needed to optimize the

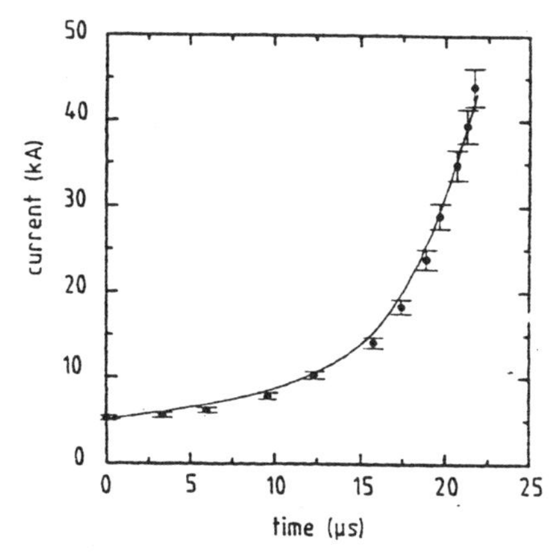


FIGURE 6.22. Load Current: (—) calculated and experimental points (Φ) from [6.38].

generator. In addition, it can serve as the cornerstone for developing a model for cascaded systems that use dynamic transformers. The accuracy of the model can be increased by going to a full 3D description of the current distribution in the generator, but this would necessitate the armature rings being divided further into a number of concentric radial rings and the coil rings into elements for accurately describing the 3D proximity effects.

## 6.3 Helical Generator Design

In the previous sections of this chapter, the important features of a helical generator were discussed. In addition, two models of differing degrees of complexity for calculating the performance parameters of the generator were presented, as well as a comparison of the results of these models with experimental data. Now, the formal design of a generator, with particular focus on the design of a 1 MJ generator (Fig. 6.23), will be considered.

### *6.3.1 Basic Input Data*

The basic data needed to design a helical generator is summarized below. The specific values of each parameter are those required to construct the particular 1 MJ generator.

The capacitor bank used to create the seed field in the MCG has a capacitance of 250 $\mu$F and a voltage $V_{cb}$ of 20–30 kV. The usual practice of connecting the input cables to the armature and coil, as shown in Fig. 6.1a, is followed in this design. However, it should be pointed out that systems exist in which the current is returned directly to the capacitor bank and

FIGURE 6.23. FLEXY I ready to be fired.

not through the armature [6.11,6.17], thereby reducing the resistance of the generator.

A load inductance $L_{\text{load}}$ of 40 nH and a maximum current $I_{\max}$ of about 7 MA were selected as appropriate for a particular application. The need to keep the linear current density to an acceptable level of about 0.2 MA/cm imposes a minimum limit on the outer diameter of the armature, which is 106 mm in this case. The normal practice is to produce an armature from a tube that only requires machining on the outside and to make it from copper rather than aluminum.

The high explosive that was used has a detonation velocity of 8.2 km/s, which is determined by specifying the inner diameter of the armature and the initial mass density of the charge, i.e., $\rho_{ex} = 1.71$ g/cm$^3$. For helical generators, the magnetic energy output directly depends on the product $\rho_{ex} v_{\text{det}} \Delta Q$, where $\Delta Q$ is the characteristic heat of detonation of the high explosive [6.22].

The armature dynamics are preferably determined from field experiments by using high-speed photography. If these data are not available, a simple numerical code or calculations based on the Gurney model [6.40] can be used to estimate the armature dynamics. From the armature dynamics, the optimal wall thickness, which is 9 mm, and cone angle, which is $12^o$, of the actual aluminum armature were determined. For such an armature, the inner coil diameter should be about twice the outside diameter of the armature, which in this case is 212 mm [6.14,6.20]. For a copper armature, a higher ratio of inside to outside diameters can be used [6.11,6.14,6.25]. The crowbar diameter was chosen as 146 mm in order to withstand the reflected input voltages of about twice the charging voltage $V_{cb}$ (no steps were taken to match the impedance of the input cables).

The amount of high-explosive charge that could be used was restricted to about 15 kg, which limited the length of the armature to 1350 mm.

The distance from the initiation point of the high-explosive charge to the crowbar was 100 mm. The requirement that the end of the explosive charge should be one cone length, which is 250 mm, beyond the output ring of the coil, so as to avoid inertial movement, sets the maximum coil length at 1000 mm.

The above information fully defines the geometry of the generator and provides sufficient information to design a helical coil.

### *6.3.2 Helical Coil Design Rules*

A number of rules should be observed in designing the coil:

(i) *Constant voltage* or *constant electric field rule* [6.11,6.14]. The requirement that the generated voltage $IdL/dt$ must not exceed a certain maximum predetermined voltage $V_{\max}$ is a major design issue, since the higher this value, the greater is the energy multiplication ratio. Although values exceeding 150 kV have been used, it was shown earlier in this chapter that voltages above 125 kV are accompanied by large increases in the generator resistance if inadequate insulation is provided to prevent premature breakdown. A working gas, such as Freon or $SF_6$, at high pressure can be used to reduce the chance of electrical breakdown [6.13,6.41]. To meet the requirement of a simple design, a maximum voltage of 100 kV was selected.

(ii) *Constant linear current density* or *constant magnetic intensity* rule [6.11]. It is common practice in helical generators to restrict the current density to less than 0.2 MA/cm to avoid nonlinear diffusion losses.

(iii) *Containment* rule. The radial movement of the cables within any section of the coil during flux compression, which is assumed to be asymmetric for conservation reasons, should be less than the $2\pi$-clocking eccentricity allowed for that section.

To simplify construction of the coil, the various section pitches were selected as integer multiples of the cable diameter. In addition, the number of parallel coils was doubled (a process referred to as *bifurcation*) between adjacent sections, whenever possible, although for the final section only 20 (not 24) of the largest diameter cables could be used due to space limitations. Furthermore, the overall winding was divided into a number of equal-length sections. The maximum possible number of sections, which was subject to the constraint that the ratio of the inside diameter of the coil to the length of the section should not exceed 1.5, was used to ensure a smooth inductance–time characteristic. In order to provide a reasonable load current margin of 9 MA, designs for different values of $V_{cb}$ were investigated. No designs were found for $V_{cb} = 20$ kV, and $V_{cb} = 30$ kV was regarded as being too near the primary source limitations. Thus, the final design was for $V_{cb} = 25$ kV, which as the code indicated to obtain the necessary output parameters, the stator coil had to have a length of 1120 mm. The coil therefore consisted of eight equal-length sections, each having a length of 140 mm. The resulting optimized pitch, number of parallel cables,

| Section | Pitch (mm) | Number of Parallel Cables | Strands/Diameter (mm) |
|---|---|---|---|
| I | 12 | 3 | 7/0.85 |
| II | 18 | 3 | 7/1.35 |
| III | 27 | 3 | 7/2.14 |
| IV | 36 | 6 | 7/1.35 |
| V | 54 | 6 | 7/2.14 |
| VI | 84 | 12 | 7/1.70 |
| VII | 140 | 20 | 7/1.70 |
| VIII | 180 | 20 | 7/2.14 |

TABLE 6.3. The sections of the helical coil of FLEXY I.

and conductor diameters (and number of strands) for each section of this generator are given in Table 6.3.

### Design Limitations

Plots of the calculated time variations of the parameters of FLEXY I are presented in Figs. 6.24 and 6.25. Calculations were performed both with and without non-Ohmic resistances added into the equations. It is evident from Figs. 6.25a and 6.25d that it is not possible to meet both the constant voltage rule and the constant current density rule. This arises from the fact that the conducting cross-sectional areas, which depend on the coil pitches required to satisfy the constant voltage rule as limited by the skin depths in both the armature and the coil, are insufficient to meet the constant current density rule. The required high currents can only be obtained at the expense of a progressively decreasing internally generated voltage, which decreases the energy multiplication. Experimental data confirm in fact that simple high-current, high-input-energy generators have low energy multiplication factors [6.11,6.20]. To prevent the decrease in generated voltage and energy requires the adoption of complex and costly techniques, such as tilted pitches or variable armature and coil diameters to be employed, so as to increase the $IdL/dt$ term in Eq. 6.2, while at the same time satisfying the current density rule [6.11]. To satisfy both the constant voltage rule and the constant current rule requires either the current to be maintained at a reduced total energy gain or the energy gain to be achieved at a reduced current. If only high-energy multiplication is required, then it can be achieved, if the current is kept low, by using a high-inductance load [6.14].

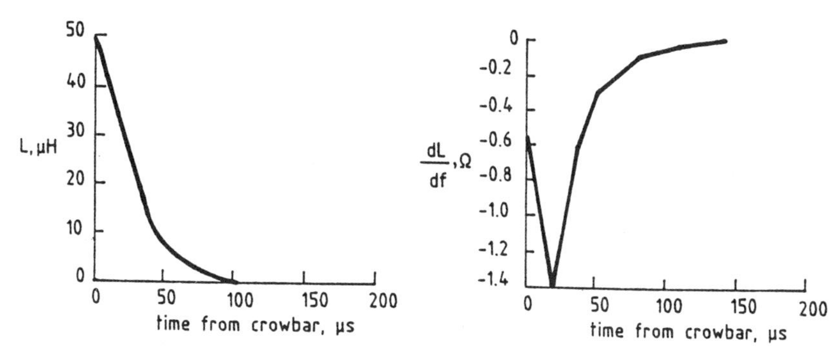


FIGURE 6.24. Calculated parameters of the first design FLEXY I generator: (a) inductance variation with respect to time and (b) rate-of-change of inductance variation with respect to time.

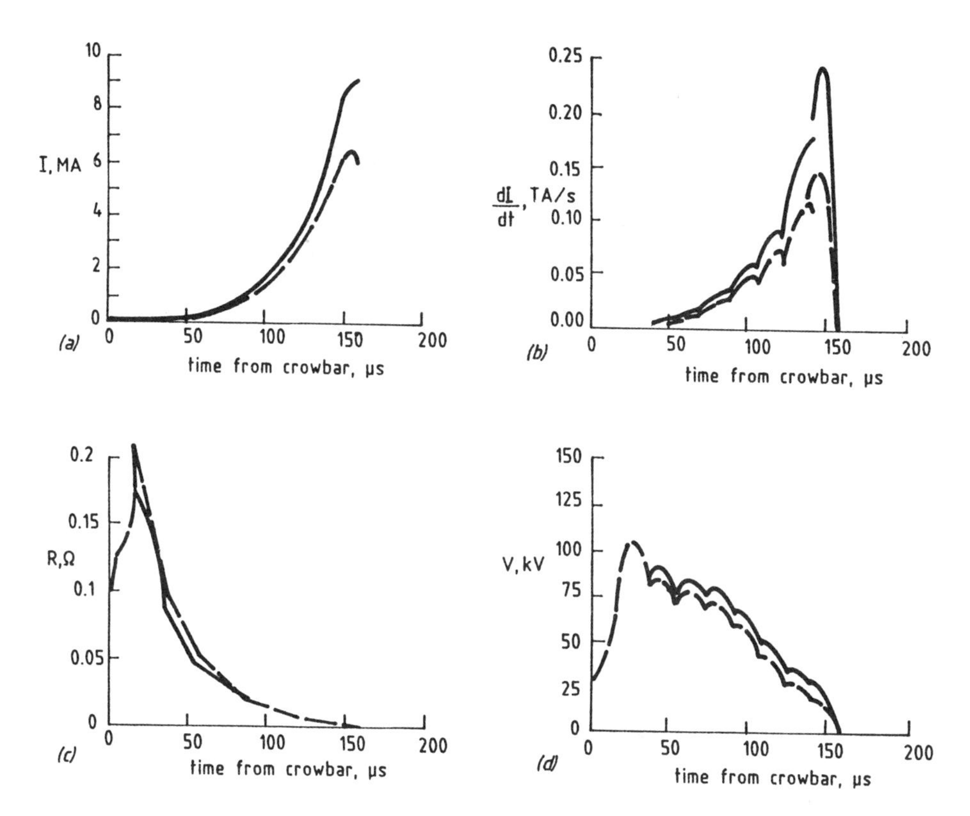


FIGURE 6.25. Predicted characteristics of the first design of FLEXY I generator: (a) load current variation with respect to time, (b) rate-of-change in load current variation with respect to time, (c) resistance variation with respect to time, and (d) maximum internally generated voltage variation with respect to time. The lines are (—) calculated and (- - -) calculated using modified design.

## 6.4 Construction of the FLEXY I

A schematic drawing of the FLEXY I generator is provided in Fig. 6.26. The stator coil was wound on a special-purpose cylindrical mandrel with a diameter that tapered slightly along the length of its axis. The surface of the mandrel was liberally coated with wax to facilitate removal of the coil. To further assist removal of the coil, the mandrel was cooled with liquid nitrogen. After positioning the start (7) and end rings (12) of the generator at the correct distance from each other on the mandrel, the eight sections of the coil were wound in sequence. Single-core, PVC-insulated, nonsheathed, stranded cables were used to construct the coil. The number of strands and the size of the cables used in each section are given in Table 6.3. Figure 6.27 shows the interconnections being made between the cables of adjacent sections by using special soldering techniques. Supplementary insulation is provided at the joints. The input and output cables of the first and last section of the coil are terminated on lugs and bolted to the appropriate ring. It was found to be more convenient to commence the winding at the load end where the cables are bigger. It was also found better to connect the cables of the last two sections prior to winding on the mandrel. To reduce the likelihood of voids, several interconnections were made between sections having different numbers of turns in several different planes normal to the coil axis. After completing the winding, each section was coated with layers of fiberglass glued with epoxy resin. Additional reinforcement was provided at sites adjacent to the start and end rings. Figure 6.28 shows the completed coil on the mandrel, prior to applying the concrete casing.

The armature was made from unmachined aluminum tubing having a diameter of 106 mm and a sag of less than 3 mm over its 1500 mm length. However, in order to meet the length requirements of the design, a welded extension had to be added, as shown in Fig. 6.26. The plastic explosive, which is PE4, was hand-shaped into balls and packed into the armature. To simplify the design, the crowbar consisted of three insulated and symmetrically mounted bolts, positioned at a effective diameter of 146 mm, rather than the more usual ring.

When the armature cone made contact with the crowbar at time $t = t_0 = 0$ in Fig. 6.26, the combined coil and load inductance is 37.5 $\mu$H. When the contact point entered the ring at time $t = t_2$, the load inductance by itself is 38 nH. The generator current and its rate of change in time were measured with calibrated Rogowski coils mounted inside the load. The total error in the current measurements was estimated to be $\pm 5\%$. Inlet and outlet valves were provided, so that the generator and load could be filled with $SF_6$ gas. The energy required to create the initial magnetic field in the generator was provided by a mobile capacitor bank via 12 parallel-connected coaxial cables.

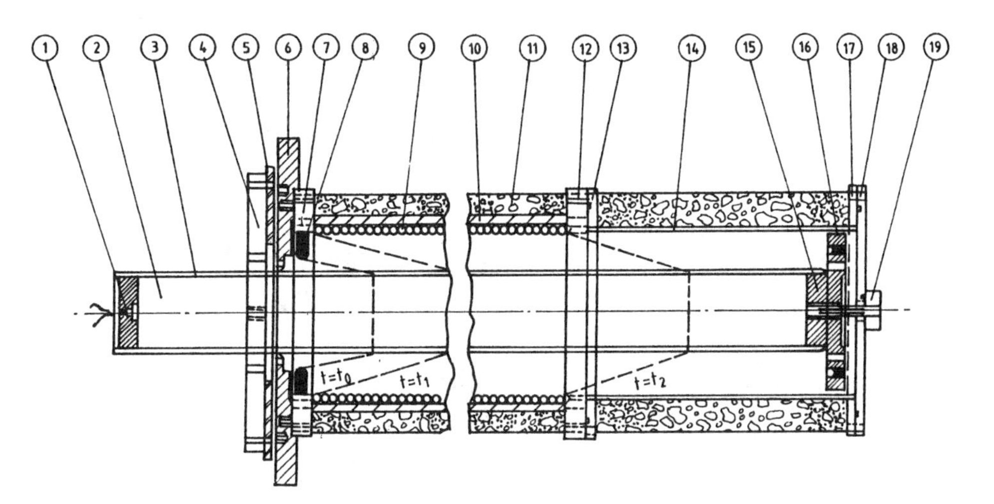


FIGURE 6.26. Schematic drawing of the FLEXY I: (1) detonator, pellet holder, and pellet, (2) high explosives, (3) aluminum armature with extensions welded at each end, (4) aluminum start plate, (5) polyvinylchloride separator plate, (6) polyvinylchloride insulator, (7) aluminum starting ring, (8) three insulated crowbar bolts, (9) helical coil, (10) fiber glass reinforcement, (11) concrete inertial mass, (12) aluminum end ring, (13) and (17) load attachment rings, (14) aluminum coaxial load, (15) aluminum armature plug, (16) polyvinylchloride armature end plug, (18) aluminum end plate, and (19) end screw. Broken lines are the positions of the armature cone at times $t = t_0$, $t_1$, and $t_2$.

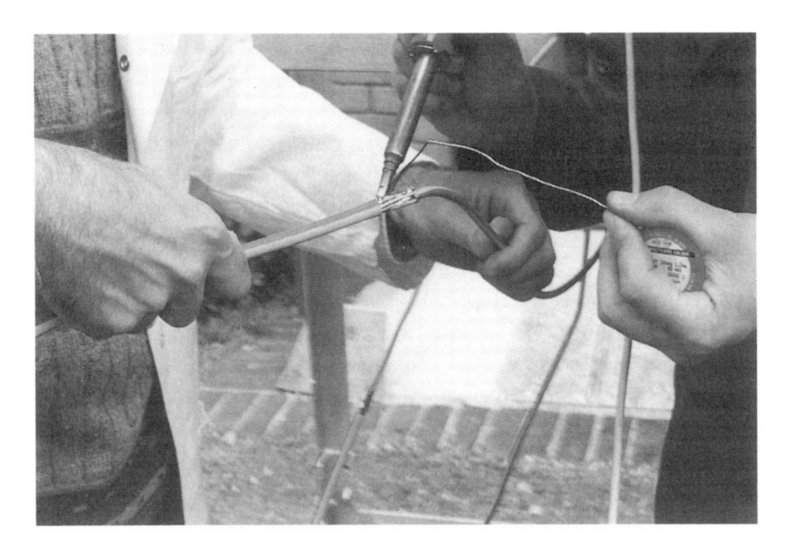

FIGURE 6.27. The joining of cables.

FIGURE 6.28. The completed coil on its mandel.

Prior to loading the armature with high explosive, a series of preliminary tests were conducted on the generator. During one of these tests, which was intended to calibrate the Rogowski coils, the generator was excited at 25 kV from the capacitor bank. Although rough calculations predicted that the coil should survive these tests without adding supplementary inertial mass, the first section was destroyed. A similar problem was encountered by others [6.41] during a full-energy firing test. To solve this latter problem, other researchers replaced the capacitor bank with a Marx bank so that the risetime of the injected current was reduced. The approach used in FLEXY I was to leave the capacitor bank configuration unchanged, but to apply the inertial containment rule by modifying the generator design so as to avoid this problem.

The containment rule must be followed, if the coil geometry is to be preserved both during generator pretesting and ahead of the contact point during testing. Although axial and radial forces are both present, the former are important only during the final stages of operation of a complex variable-geometry multimegajoule generator [6.31] and will be neglected here. Since the tensile forces produced in the coil following discharge of the capacitor bank are greater than the yield point of copper, the elastic forces can be neglected and conservative calculations can be made based on the assumption that only the inertial movement of the coil be considered [6.42]. Thus, if $B_i$ is the flux density within the coils of the $i^{th}$ section, which has an inner surface area $S_i$, inner radius $r_i$, and mass $m_i$, then the radial force and the corresponding radial acceleration of this section are

$$F_r^i = \frac{B_i^2 S_i}{2\mu_0} \tag{6.55}$$

and

$$\ddot{r}_i = \frac{F_r^i}{m_i}, \tag{6.56}$$

respectively. While the capacitor bank is discharging, the variation of flux density with time is described by

$$B_i^0(t) = B_{0i} \sin \frac{\pi t}{2\tau}, \tag{6.57}$$

where $B_{0i}$ is the magnetic flux at the moment the discharge current is at its peak value and $\tau$ is a quarter period of the discharge. Integrating the radial acceleration equation gives the coil displacement as [6.42]

$$\Delta r_i^0(t) = \frac{B_{0i}^2 S_i \tau^2}{4\mu_0 m_i \pi^2} \left[ \frac{\pi^2 t^2}{2\tau^2} + \cos \frac{\pi t}{\tau} - 1 \right]. \tag{6.58}$$

Between the time that the cone makes contact with the crowbar and the time it makes contact with the end ring, the flux density can be assumed to be increasing exponentially from a value of $B_i^0(\tau)$, which is the magnetic flux at peak discharge current. Thus, during this period, the magnetic flux can be approximated by

$$B_i(t) = B_i^0(\tau) e^{\Lambda t}. \tag{6.59}$$

From the above equations, it can be shown that the total radial displacement of the coil is

$$\Delta r_i(t) = \Delta r_i^0(\tau) + v_0^i t + \frac{(B_i^0(\tau))^2 S_i}{8\mu_0 m_i \Lambda^2} e^{2\Lambda t}, \tag{6.60}$$

where $v_0^i$ is the velocity at the end of the capacitor discharge. The containment rule requires that the final radial movement of each section should be less than the permitted eccentricity $\epsilon_i$ of that section, that is,

$$\Delta r_i \leq \epsilon_i, \tag{6.61}$$

where $\epsilon_i$ is determined by the $2\pi$-clocking expression in Section 6.2.

Using the total radial displacement expression above, calculations show that it was impossible for the first coil section to meet the containment rule, even when it is covered with concrete to provide additional inertial mass. Therefore, the generator design had to be modified. Cables of the first section were made identical to those of the second section, and a 40 mm

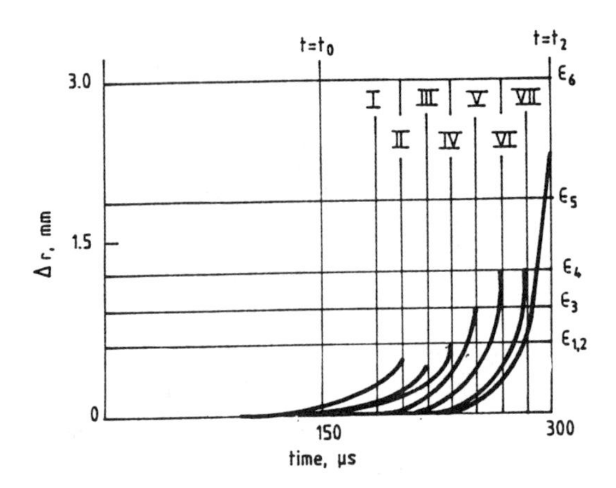


FIGURE 6.29. Radial movement of coil sections for the final design during a firing test as predicted by Eq. 6.60. The maximum predicted capacitor current was 53 kA, at an initial capacitor voltage of 25 kV.

concrete sheath was added to increase the inertial mass to 10 kg for each section. The completed, but partially concreted FLEXY I coil is shown in Fig. 6.29. This modification reduced the rate of change in the inductance of the generator, which, in turn, lowered the interior voltage and the overall performance of the generator.

Plots of the radial movement of the eight coil sections throughout the operation of the new generator design are presented in Fig. 6.29. These plots were calculated for the modified coil design using the total radial displacement expression (Eq. 6.60) for a 25 kV capacitor voltage. The first two plots are for the two identical first coil sections, and the remaining six apply in sequence for the remaining six sections. The origin of the time axis coincides with the time at which the capacitor bank switch is closed, the times $t_0(=\tau)$ and $t_2$ are identified in Fig. 6.26, and the Roman numerals correspond to times at which the armature cone leaves the corresponding coil section. The eccentricities of the different sections are also included in Fig. 6.29, which shows that, for the new design, all the coil sections satisfy the above inequality for the maximum calculated capacitor current that is represented by the broken line in Fig. 6.30.

Further calculations show that a current of 30 kA produces a maximum tensile stress in the coils of 30 kg/mm$^2$, which is sufficiently less than the yield point of 40 kg/mm$^2$ and which is acceptable for the pretests. This limited the capacitor voltage used in the pretests to about 14 kV.

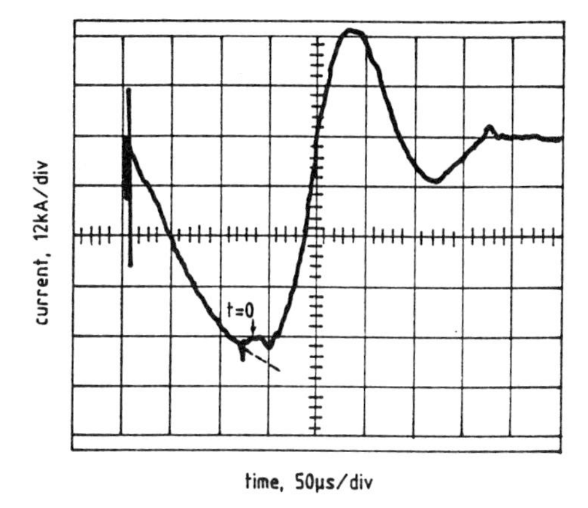


FIGURE 6.30. Test results: the current measured by self-integrating Pearson–Rogowski coil at the capacitor bank. The broken line is predicted and does not account for any generator resistance or inductance variation due to armature movement after firing. After cone–crowbar contact ($t = 0$), the waveform corresponds to a 48 mΩ, 1.4 $\mu H$, 250 $\mu F$ discharge.

## 6.5 Testing the FLEXY I

The FLEXY I was carried in a wooden frame, as shown in Fig. 6.23, that was mounted vertically above a double-layer table of 50-mm-thick steel plate with rubber sandwiched between the plates. Prior to firing the generator, its interior was pressurized with 3 atm of $SF_6$ gas. Synchronizing the closure of the switch connecting the generator to the capacitor bank and the firing of the detonator of the FLEXY I high-explosive charge was based on data provided by the design code, which did not take into account either the breakdown time of the crowbar insulation or any variations in the detonation velocity of the high explosive. Nevertheless, it is evident from the capacitor current waveform in Fig. 6.30 that the armature cone makes contact with the crowbar very near the predicted time $t = 0$, which is the time that the initial current in the generator was calculated to attain its peak value (see end of broken line in Fig. 6.30). The effects of the armature motion on the current profile of the generator can be seen in Fig. 6.30. At the moment the armature and the crowbar come into contact, the capacitor delivers 48 kA of current to the generator, which corresponds to a priming energy delivered to the generator of 43.2 kJ and initial magnetic flux of 1.8 Wb.

Rogowski coil measurements of the load current and the rate of change of the current in time are presented in Fig. 6.31. Note that there is a small irregularity in the current waveform presented in Fig. 6.31a, which is due to the interconnection discontinuities. The disruption that appears in the rate of change of current waveform after the peak current was generated (see

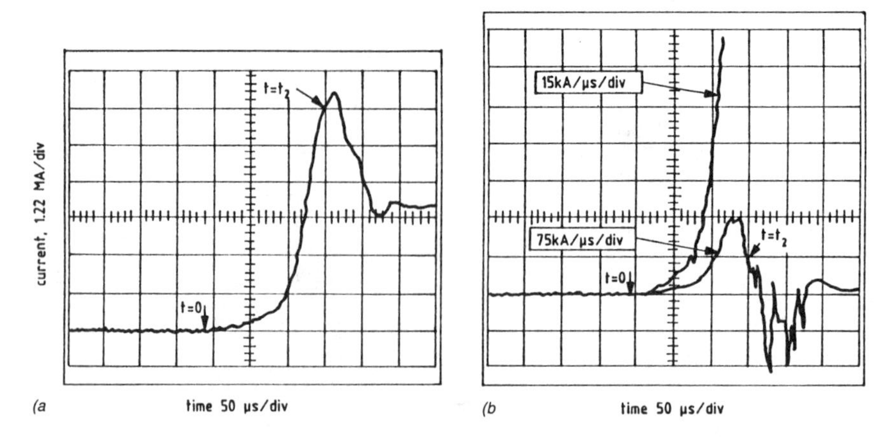


FIGURE 6.31. Test results: measurements from a Rogowski coil positioned within the load. The (a) current waveform and (b) the rate-of-change of current waveform, where $t_0 = 0$ and $t = t_2$ as shown in Fig. 6.26.

Fig. 6.31b) is difficult to explain, and was not observed in later experiments, and plays no role in the applications envisaged for FLEXY I.

As Fig. 6.31a shows, the load current at the moment the armature cone comes into contact with the end ring is 7.3 MA, which means that 1 MJ of energy is delivered to the load. The final magnetic flux is 0.28 Wb, which corresponds to a magnetic flux conservation efficiency of 15.4% and a chemical-to-magnetic energy conversion efficiency of 1.5%. The FLEXY I had an overall energy gain of 23 and a current multiplication factor of 153. Calculations show that about 1 MJ of energy is dissipated in the various resistances in the system. The main parameters are presented in Table 6.4. Further improvements to the generator design have raised the energy delivered to the load to 2 MJ.

## 6.6 Comparison of Theoretical and Experimental Results

The principal theoretically predicted parameters of the final designed FLEXY I generator, including ohmic losses, are presented in Fig. 6.32. The experimental results are added to the plots, with the appropriate error bars. The equivalent resistance in Fig. 6.32e was calculated from

$$R = -\frac{L\frac{dI}{dt} + I\frac{dL}{dt}}{I}, \tag{6.62}$$

| Parameter | Value |
|---|---|
| Overall Length | 1.8 m |
| Armature and Coil Mass | 65 kg |
| Explosive Mass | 15 kg |
| Initial Inductance | 37.5 $\mu$H |
| Final Inductance (No Load) | 38 $\mu$H |
| Final Energy | 1 MJ |
| Final Current | 7 MA |
| Operating Time | 155 $\mu$s |
| Concrete Mass | 80 kg |
| Initial Energy | 40 kJ |
| Initial Current | 50 kA |
| Maximum Diameter | 0.45 m |

TABLE 6.4. Main parameters of FLEXY I.

where the values of $I$ and $dI/dt$ were taken from Fig. 6.31 and the values of $L$ and $dL/dt$ were taken from Fig. 6.32. A number of the experimental resistance data points exceeded their corresponding predicted values, possibly as a consequence of 3D proximity effects, since electrical breakdown, severe $2\pi$-clocking, and so on are unlikely due to the correct application of the rules for generator design. There are also some differences in the predicted and measured current profiles of the generator, although, as with other generator designs that have been investigated, the final values are predicted accurately.

In Fig. 6.32f, it can be seen that the maximum internal voltage produced during the experiment was less than 70 kV, which is a rather low value imposed by the containment rule. Higher voltages are required to produce greater output. Therefore, to generate these higher outputs, without changing the design of the FLEXY I, cables with increased insulation must be used. Other, more complicated generator designs [6.43] have used cables with a breakdown voltage of 60 kV, rather than the 30 – 40 kV breakdown voltages of the cables used in the FLEXY I generator. In addition, the energy output of the FLEXY I can be easily increased by changing the coil design, without exceeding the peak current of the FLEXY I and using an increased load inductance.

## 6.7 Summary

This Chapter has shown how a practical multisection helical generator can be produced on the basis of a simple theoretical model and following a number of important design rules. No sophisticated manufacturing techniques, such as titled pitch or variable diameter coils, were used in the construction of the generator. The armature was not machined and was filled manually

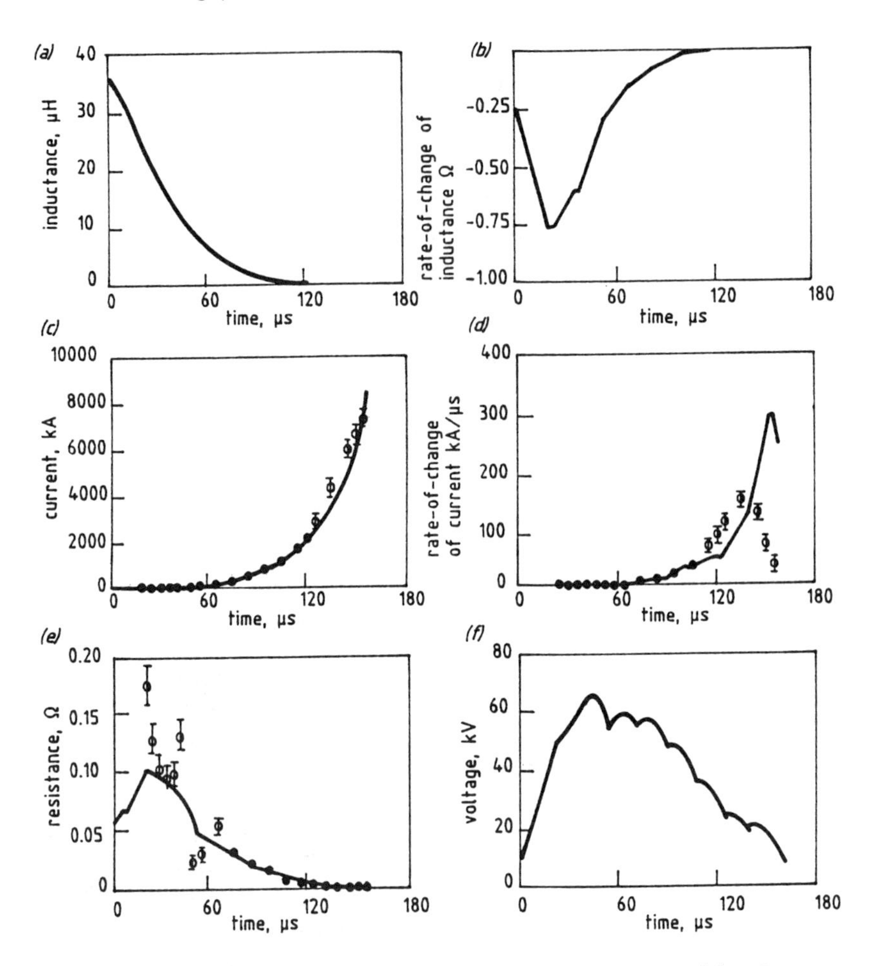


FIGURE 6.32. Predicted and measured FLEXY I performance: (a) inductance, (b) rate of change of inductance, (c) load current, (d) rate of change of load current, (e) equivalent resistance, and (f) maximum internally generated voltage. Time is measured from the armature cone–crowbar contact. The error bars on the symbols indicate the possible range of results.

with explosive. However, very good reproducibility was obtained during active load testing as is presented in Chapter 7.

**Acknowledgements:** Tables 6.1-6.4 and Figs. 6.1-6.8 and 6.23-6.32 are reproduced from *J Phys D*, **28**, pp.807-823 (1995), with permission of IOP Publishing Ltd., Bristol, U.K. Figures 6.9-6.22 are reproduced from *Laser and Particle Beams*, **15**, pp. 379-395 (1997), with permission of Cambridge University Press.

# References


[6.1] I.R. Smith, B.M. Novac, and H.R. Stewardson, "Flux Compression Generators," Power Engineering Journal, April, pp. 97-101 (1995).

[6.2] B.M. Novac, I.R. Smith, H.R. Stewardson, P. Senior, V.V. Vadher, and M.C. Enache, "Design, Construction, and Testing of Explosive-driven Helical Generators," J. Phys. D: Applied Phys, **28**, pp. 807–823.

[6.3] B.M. Novac, I.R. Smith, H.R. Stewardson, and P. Senior, "Experimental Methods with Flux-compression Generators," Engineering Science & Technology, October, pp. 211–222 (1996).

[6.4] B.M. Novac, I.R. Smith, M.C. Enache, and H.R. Stewardson, "Simple 2-Dimensional Model for Helical Flux-compression Generators," Lasers and Particle Beams, (1997).

[6.5] B.M. Novac, H.R. Stewardson, I.R. Smith, and P. Senior, "Analysis of Helical Generator Driven Exploding Foil Opening Switch Experiments," *$10^{th}$ IEEE Pulsed Power Conference* (eds. W. Baker and G. Cooperstein), **2**, pp. 1182–1187.

[6.6] H.R. Stewardson, B.M. Novac, and I.R. Smith, "Fast Exploding-foil Switch Techniques for Capacitor Bank and Flux Compressor Output Conditioning," J. Phys. D: Applied Phys., No. 28, pp. 2619–2630 (1995).

[6.7] B.M. Novac, M. Ganciu, M.C. Enache, I.R. Smith, H.R. Stewardson, and V.V. Vadher, "A Fast Electro-optic High-Voltage Sensor," Meas. Sci. Technol. **6**, pp. 241–242 (1995).

[6.8] J.M. McGlaun, S.L. Thomson, and J.R. Freeman, "COMAG–III: a 2-D MHD Code for Helical CMF Generators," *Megagauss Physics and Technology* (ed. P.J. Turchi), Plenum, New York, pp. 193–203 (1980).

[6.9] J.R. Freeman, J.M. McGlaun, S.L. Thomson, and E.C. Cnare, "Numerical Studies of Helical CMF Generators," *Megagauss Physics and Technology* (ed. P.J. Turchi), Plenum, New York, pp. 205–218 (1980).

[6.10] T.J. Tucker, "A Finite-element Model of Compressed Magnetic Field Current Generators," *Megagauss Physics and Technology* (ed. P.J. Turchi), Plenum, New York, pp. 265–273 (1980).

[6.11] A.I. Pavlovskii, R.Z. Lyudaev, V.A. Zolotov, A.S. Seryoghin, A.S. Yuryzhev, M.M. Kharlamov, A.M. Shuvalov, V. Yegurin, G.M. Spirov, and B.S. Makaev, "Magnetic Cumulation Generator Parameters and Means To Improve Them," *Megagauss Physics and Technology* (ed. P.J. Turchi), Plenum, New York, pp. 557–583 (1980).

[6.12] D.B. Cummings and M.J. Morley, "Electrical Pulses from Helical and Coaxial Explosive Generators," *Megagauss Magnetic Field Generation by Explosives and Related Experiments* (eds. H Knoepfel and F. Herlach), pp. 451–471 (1966).

[6.13] J.W. Shearer, et al., "Explosive-driven Magnetic-field Compression Generators," J. Appl. Phys., **39**, pp. 2102–2116 (1968).

[6.14] V.K. Chernyshev, E.J. Zharionov, V.A. Demidov, and S.A. Kazakov, "High Inductance Explosive Magnetic Generators with High Energy Multiplication," *Megagauss Physics and Technology* (ed. P.J. Turchi), Plenum, New York, pp. 641–649 (1980).

[6.15] H. Knoepfel, *Pulsed High Magnetic Fields*, North-Holland Pub., Amsterdam (1970).

[6.16] V.G. Welsby, *The Theory and Design of Inductance Coils*, MacDonald, London (1964).

[6.17] J.E. Grover, O.M. Stuetzer, and J.L. Johnson, "Small Helical Flux Compression Amplifiers," *Megagauss Physics and Technology* (ed. P.J. Turchi), Plenum, New York, pp. 163–180 (1980).

[6.18] M. Jones, "An Equivalent Circuit Model of a Solenoidal Compressed Magnetic Generator," *Megagauss Physics and Technology* (ed. P.J. Turchi), Plenum, New York, pp. 249–264 (1980).

[6.19] M. Cowan and R.J. Kaye, "Finite-element Circuit Model of Helical Explosive Generators," *Ultrahigh Magnetic Fields - Physics, Techniques, and Applications* (eds. V.M. Titov and G.A. Shvetsov), Nauka, Moscow, pp. 240–246 (1984).

[6.20] C.M. Fowler and R.S. Caird, "The Mark IX Generator," *Seventh Institute of Electronic and Electrical Engineers Pulsed Power Conference*, New York, pp. 475–488 (1989).

[6.21] C.M. Fowler, R.S. Caird, and W.B. Garn, "An Introduction to Explosive Magnetic Flux Compression Generators," Report LA–5890–MS, Los Alamos National Laboratories (1975).

[6.22] B.M. Novac, "Production of Ultrahigh Magnetic Fields," PhD Thesis, University of Bucharest (1989).

[6.23] I. Ursu, M. Ivascu, B.M. Novac, V. Zoita, V. Zambreanu, D. Preotescu, M. Butuman, and A. Radu, "Pulsed Power from Helical Generators," *Megagauss Fields and Pulsed Power Systems* (eds. V.M. Titov and G.A. Shvetsov), Nova, New York, pp. 403–410 (1990).

[6.24] B.M. Novac, V. Zambreanu, and V. Zoita, "A Pulsed Power Source for Plasma Focus Single-shot Experiments," *Proc. 8th IEEE Pulsed Power Conf.*, San Diego, pp. 434–437 (1991).

[6.25] B. Antoni and C. Nazet, "Experimental and Theoretical Study of Helical Explosive Current Generators with Magnetic Field Compression," Report CEA–R–4662, Commission of Atomic Energy, France (1975).

[6.26] F. Herlach, "Explosive-Driven Energy Generators with Transformer Coupling," J. Physics. E: Sci. Instrum., **12**, pp. 421–429 (1979).

[6.27] R.E. Tipton, "A 2-D Lagrange MHD Code," *Megagauss Physics and Technology* (eds. C.M. Fowler, R.J. Caird, and D.J. Erickson), Plenum, New York, pp. 299–306 (1987).

[6.28] F.W. Grover, *Inductance Calculations*, Dover, New York (1973).

[6.29] P.L. Kalantarov and L. A. Teitlin, *Inductance Calculations*, Tehnica, Bucharest (1958).

[6.30] F.S. Felber, R.S. Caird, C.M. Fowler, D.J. Erickson, B.L. Freeman, and J.H. Goforth, "Design of a 20 MJ Coaxial Generator," *Ultra High Magnetic Fields: Physics, Techniques, and Applications* (eds. V.M. Titov and G.A. Shvetsov), Nauka, Moscow, pp. 321–329 (1984).

[6.31] V.K. Chernyshev, E.I. Volkov, S.V. Pao, A.N. Skobelov, and V.P. Strekin, "Factors Limiting Energy Characteristics of a Helical Generator Axial Turns Shift and Armature Deceleration Initiated by a Magnetic Field," *Megagauss Fields and Pulsed Power Systems* (eds. V.M. Titov and G.A. Shvetsov), Nova Science, Inc., New York, pp. 367–370 (1990).

[6.32] N. Miura and S. Chikazumi, "Computer Simulation of Megagauss Field Generation by Electromagnetic Flux Compression," Jap. Journal of Applied Physics, **18**(3), pp. 553–564 (1979).

[6.33] J. Long, K. Lindner, and O. Zucker, "Analysis and Comparison of Circuits Undergoing a Change of Inductance via Continuous Sequential Switching and/or Geometrical Change," *Megagauss Technology and Pulsed Power Applications* (eds. C.M. Fowler, R.S. Caird, and D.J. Erickson), Plenum, New York, pp. 593–607 (1987).

[6.34] W.H. Press, *Numerical Receipts in Pascal: The Art of Scientific Computing*, Cambridge University Press, Cambridge (1989).

[6.35] J.E. Grover, et al., *Megagauss Physics and Technology* (ed. P.J. Turchi), Plenum Press, New York, pp. 163–180 (1980).

[6.36] J.R. Freeman and S.L. Thompson, J. Computational Phys., **25**(4), pp. 332–352 (1977).

[6.37] J.M. McGlaun, et al., *Megagauss Physics and Technology* (ed. P.J. Turchi), Plenum Press, New York, pp. 193–203 (1980).

[6.38] C.J. Brooker, et al., *Megagauss Magnetic Field Generation and Pulsed Power Applications* (eds. M. Cowan and R.B. Spielman), Nova Science Pub., New York, pp. 511–517 (1994).

[6.39] R.J. Adler, *Pulse Power Formulary*, North Star Res. Corp. (1989).

[6.40] J.E. Kennedy, "Explosive Output for Driving Metal," *Behaviour and Utilization of Explosives in Engineering Design* (eds. L. Davison et al.), ASME, New York (1972).

[6.41] J.H. Goforth, R.S. Caird, C.M. Fowler, A.E. Greene, H.W. Kruse, I.R. Lindermuth, H. Oona, and R.E. Reinovsky, "Performance of the Laguna Pulsed-power System," *6th Pulsed-Power Conference*, June 29–July 1, pp. 445–448 (1987).

[6.42] F. Herlach, "Megagauss Magnetic Fields," Rep. Prog. Phys., **25**, pp. 391–417 (1968).

[6.43] V.K. Chernyshev, et al., "Explosive Magnetic Generators of the 'Potok' Family," *8th IEEE Pulsed-Power Conference*, San Diego, pp. 419–433 (1991).

# 7
# Experimental Methods and Techniques

The first part of this chapter presents a general description of various experimental methods relating to MCG physics and technology, and concludes with a general outline of an MCG laboratory. The second part of the chapter presents important practical aspects of the various technologies that need to be associated with an MCG, either in order to match its output to the load in a particular application or to construct an autonomous very high energy multiplication system.

As with Chapter 6, much of the material that is presented is derived from the ongoing MCG activity at Loughborough University.

## 7.1 Experimental Methods

The physics of MCGs are essentially a blend of detonics and electromagnetism. **Detonics** is the science of detonation that provides information about the performance of the explosive and the energy transformation processes in the generator. **Electromagnetism** complements detonics by providing a description of the ultrahigh magnetic fields and the electric currents generated during the energy transformation process. This implies that a mix of diagnostic tools need to be used during the experimental phase of development. A number of diverse, as well as novel experimental techniques, had to be developed during the research on the FLEXY I presented in Chapter 6. However, for completeness, other more sophisticated, but related diagnostics are also described in this chapter.

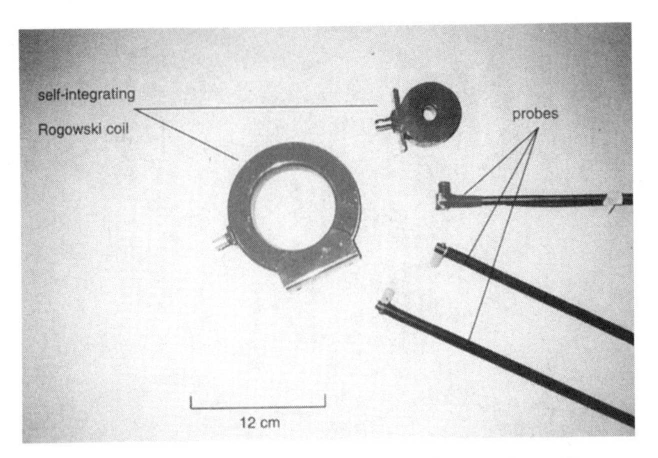


FIGURE 7.1. Pick up probes and Rogowski coil..

### 7.1.1 Electromagnetic Techniques

In this section, a number of electromagnetic and optical devices used to measure the various parameters of the MCG are discussed.

*Pick Up Probes*

The pick up or magnetic induction probes, more commonly called B-dot probes, in Fig. 7.1, are the simplest and most frequently used devices for measuring pulsed magnetic fields. While it appears to be only a simple coil of thin wire, the number of turns must be properly chosen to increase its sensitivity. The probe responds to the rate of change in magnetic flux density in time. To measure the flux density, a passive high-impedance $RC$-integrator is connected at the oscilloscope end of the coaxial cable to which the probe is attached. Although the frequency bandwidth of the probes can be hundreds of megahertz, on occasion it is difficult or impossible to use. The main difficulties that arise are related to the pick up of electromagnetic noise, electrostatic coupling, and electrical breakdown when there are unexpectedly high rates of change in the magnetic fields and when excessive voltages are induced in the probe. Another problem with using these probes is associated with the length of the cable that connects the probe to the measuring device, which could adversely affect the overall frequency response. Finally, in ultrahigh magnetic field compression experiments, the electromagnetic forces acting on current-carrying conductors of the probe can change their geometry, thus affecting the initial calibration of the probe, or on occasions inducing currents that can melt and even vaporize the wires from which the probe is made.

Although pick up probes are normally used to measure the magnetic fields, they can also be used to measure currents by either calibrating with a known current source or calculating the current for simple probe geometries from the measured field strength.

FIGURE 7.2. Rogowski coil used for voltage measurements.

*Rogowski Coils*

Rogowski coils are used to measure electrical currents flowing through conductors. The typical Rogowski coil depicted in Fig. 7.1 has a toroidal shape that surrounds the conductor in which the current is to be measured. Applying Ampere's law, it can be shown that the voltage induced in the coil is independent of its position relative to the conductor.

The construction of the Rogowski coil is not as simple as that of the pick up probe, but improvements made over the years enable experimentalists to design the probe to make measurements with the required bandwidths and even to self-integrate the signal to be measured. Although calibrated Rogowski coils are available commercially, the particular features of an experiment may require that a special purpose coil be designed and constructed. When using a Rogowski coil, the experimentalist must take into account the drawbacks of using the coil including such things as electromagnetic noise and the effects of the length of the connecting cable. As with the pick up probe, the Rogowski coil can be used, when properly calibrated, to measure quantities other than currents. For example, in Fig. 7.2, it is shown how the Rogowski coil can be connected across a load to measure voltage by measuring the current flowing through a parallel-connected water resistor having a known resistance.

*Voltage Dividers*

Voltage dividers have been developed that can measure megavolt signals with risetimes of nanoseconds, and in some cases even hundreds of picoseconds. The literature available describes resistive, capacitive, or hybrid devices, most of which are unsuitable for making measurements in MCGs. In fact, measuring voltages at remote sites, where explosive devices are normally tested, presents an extremely complex challenge owing to the fact that normal laboratory conditions, i.e., using short cables and having good ground points available, do not exist.

### Photonic Techniques

While electromagnetic techniques still provide the most common, inexpensive, and convenient methods for measuring pulsed magnetic fields and currents in flux compressors, the last decade has witnessed the emergence of reliable and accurate photonic methods. All the drawbacks associated with electromagnetic techniques are eliminated, and the advantages of using photonic methods may be summarized as

- immunity to electromagnetic interference,
- no interaction with the pulsed magnetic fields, i.e., no forces act on the instrumentation circuits and no heating because of diffusion,
- electrical isolation,
- no ground loops,
- miniaturization,
- high bandwidth, which is limited mainly by the optoelectronic converter,
- high sensitivity,
- the ability to make measurements that cannot be made using electromagnetic techniques.

The only disadvantage of photonic techniques is the high cost of the sensors that are used. This disadvantage becomes even more important when it is considered that the equipment may be destroyed during the experiment.

#### *Faraday Rotation Effect Measurements*

This measurement technique is based on the fact that plane polarized light rotates as it passes through an electrooptical sensitive element in a sensor placed parallel to a magnetic field. The angle of rotation $\theta$ is proportional to the length of the light path $l$ in the optically active medium and the magnetic field intensity $H$; i.e., $\theta = VlH$, where $V$ is the Verdet constant of the optically active material used in the sensor. The value of the Verdet constant depends upon the particular measurement to be made, and the length $l$ is adjusted to give the required signal amplitude. A fiber optic cable is used to transmit information to and from the Faraday sensor. (See Fig. 7.3.)

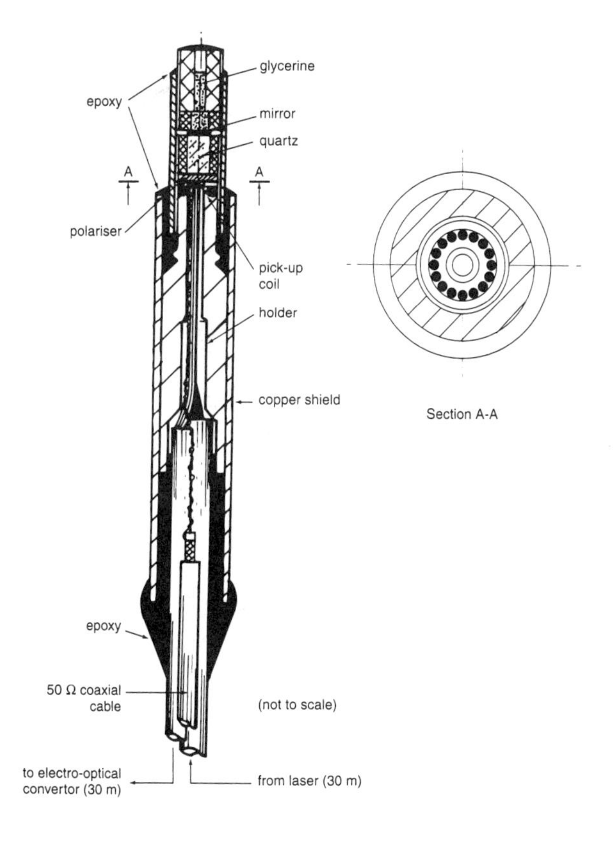


FIGURE 7.3. Miniature Faraday rotation transducer.

Faraday-rotation effect measurements are probably the only technique that can be used to measure ultrahigh magnetic fields, i.e., above 1000 T (10 MG). The highest magnetic field measured to date is 1700 T, using a sensor with flint as the active medium.

Faraday-rotation sensors that are used to measure fast rising, very-high-current pulses are constructed with special twist single-mode fibers having a low Verdet constant. Like the Rogowski coil, several fibers of the Faraday-rotation device surround the current-carrying conductor. Great care must be taken to avoid any mechanical contact with the moving conductors when using this sensor in MCGs, which may induce additional birefringes in the fiber and alter the measured signal.

*Pockels Effect*

The Pockels electrooptic effect can be used to measure the electric field strength. When an electric field is applied to a crystal in the $x$-direction, the index of refraction $n_0$ increases in that direction, while the index of refraction in the $y$-direction remains the same. As a result, a light beam

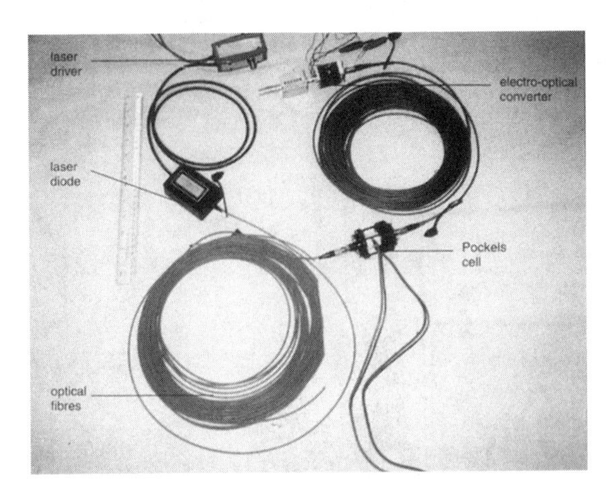


FIGURE 7.4. Pockels cell.

polarized along the $x$-axis propagates at a slower speed in the $z$-direction than a beam polarized in the $y$-direction. Therefore, a phase shift will appear between the two components that is proportional to the field strength $E$ and the crystal length $L$. This phase shift is referred to as the *induced electrical birefringence* $\Gamma(E)$, which is defined by the following equation:

$$\Gamma(E) = \frac{2\pi n_0^3 r L E}{\lambda}, \tag{7.1}$$

where $\Gamma$ is the electrooptical coefficient and $\lambda$ is the wavelength of the light. To measure the voltage, two electrodes having potential difference of $V$ are deposited on the crystal surface. The voltage difference sets up an electric field in the crystal, where $E = V/d$ and $d$ is the distance between the electrodes. The resulting phase shift can be detected by optical intensity measurements.

The Pockels cell consists of the crystal, the electrodes, and specially coated optical windows. The crystal is sometimes immersed in a liquid to reduce the piezoelectric effect and to stabilize its temperature. The input characteristics of the cell, which is a few picofarads of capacitance and more than 10 gigaohms of resistance, are better than those of most of the other recording equipment. As shown in Fig. 7.4, long optical fibers couple one end of the Pockels cell to a laser and the other end to an optoelectronic converter. Voltages greater than about 10 kV need to be reduced by using a special-purpose capacitive divider [7.1] having a risetime of less than a nanosecond. The Pockels-cell-based sensor provides all the data required from the measuring equipment, and also fully protects the recording instruments by remotely connecting them to the cell through optical fibers.

### 7.1.2 Detonic Techniques

In this section, several techniques used to monitor and control the explosives in an MCG are described [7.2]. While the design of diagnostics for laboratory experiments is rather straightforward, considerably more work must be done in developing diagnostics for large-scale experiments to measure the particular waveforms of high load currents or voltages and to measure the efficiency of converting chemical energy from high explosives into electromagnetic energy. To make these measurements, it is necessary that all the dynamic parameters of the generator are measured and controlled. Some of these parameters are related to the high explosive, i.e., the detonation velocity, while others, which are much more difficult to measure, are related to the acceleration of the conductors and insulators of the generator. Not only do their bulk velocities need to be known, but also details of how the geometry changes in time, the surface conditions, and the temperature at various times during the compression process. To make these measurements, techniques such as high-speed photography, interferometry, x-ray photography are required, along with various other methods to measure velocity, acceleration, temperature, and pressure.

*High-Speed Photography*

High-speed photography is the most common method used to obtain experimental data on the dynamics of the armature in MCGs. From these data, it is possible to determine the detonation velocity, the amount of chemical energy released by the explosive, and the uniformity with which the surfaces move.

Both rotating cameras and electronic image converters are used. They are housed in specially built laboratories, i.e., bunkers, with optical portholes so that the process being measured can be observed through the thick walls necessary to protect the measuring devices from the explosion of the MCG. Telescopic lenses are normally used to help focus the cameras on the MCG, which may be several tens of meters from the cameras. The core of the rotating camera is its rotating mirror, which can rotate at hundreds of thousands of rotations per minute when helium cooled. These cameras can be operated in the frame mode, at frame rates of millions of frames per second, or in the streak mode, with writing speeds of between 10 to 20 km/s.

Although much higher speeds can be achieved by using electronic image converters, this is usually not necessary in most detonic measurements. The advantages of using these cameras, as shown in Fig. 7.5, are the greater amount of detailed information that can be captured, because of a large image plane, and the much greater light sensitivity that permits magnification of the object being photographed. A disadvantage is the reduced number of frames that can be captured. (See Fig. 7.6.) Light flashes are needed with the cameras, and these can be produced by detonating a small

FIGURE 7.5. High-speed camera and vacuum rig.

charge of plastic explosive inside a cardboard cylinder with an inner aluminum wrapping and filled with argon gas. The shock wave produced by the explosive excites the argon atoms, producing a high intensity-flash of light with a pulse duration that depends on the distance the shock wave travels through the gas. Synchronizing the firing of the MCG, the flash of the light source, and the camera can be achieved by using high-precision detonators or detonation cords.

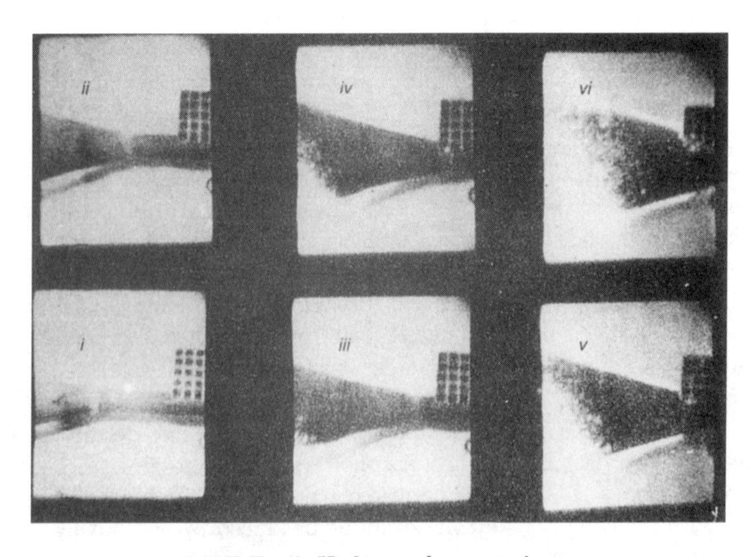


FIGURE 7.6. High-speed camera images.

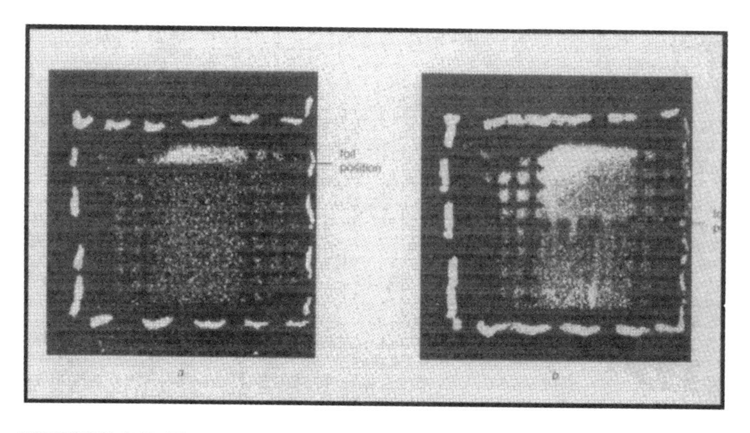

FIGURE 7.7. Tangential X-rays of a thin copper imploding cylinder.

*Interferometry*

Interferometry is a complex and costly experimental procedure that provides complete data on the hydrodynamic processes of the accelerating armature. There are several interferometric systems available including the laser-based Fabry–Perot interferometer, a system based on Michaelson interferometry, and diffuse surface interferometry, i.e., the VISAR system. The latter device has a time resolution of 1 ns and the capability to determine surface motions of any kind of surface including those that are highly tilted.

*X-Ray Photography*

Hydrodynamic data obtained from armature experiments conducted in the absence of magnetic fields are inadequate for use in designing and/or optimizing either high-energy or high-current generators or cascaded ultrahigh magnetic field generators, since the very high magnetic fields that are generated dramatically alter the shape of the accelerated conductors. Although 2D MHD codes can provide an estimate of what is occurring, the best information is obtained directly from X-ray photograph. Figure 7.7a shows a tangential X-ray still photograph of a copper cylinder. As the cylinder is imploded by electromagnetic forces, in this case, developing instabilities can be seen in Fig. 7.7b.

## Methods for Measuring Velocity and Acceleration

Almost all laboratories doing research on magnetic flux compression have developed their own slightly different approaches for measuring the detonation velocity and the armature motion. In this section, some of the more common approaches are reviewed. Other methods using devices such as phase-transition probes, displacement capacitors, and electromagnetic velocity gauges are more difficult to use and are restricted to special applications.

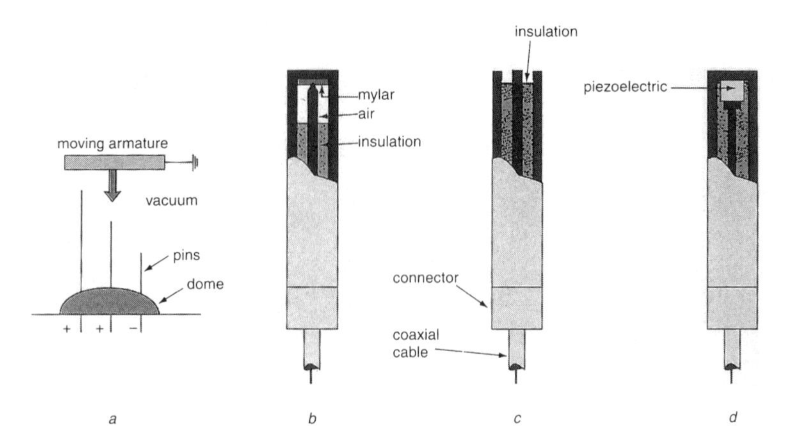


FIGURE 7.8. Four basic types of contact pins.

*Pin Contact Method*

The pin contact is one of the oldest, but still the simplest and most frequently used method for precisely measuring the armature movement and the detonation velocity. The method is straightforward, in that, when a shock wave arrives at a contact pin, the pressure closes the contact pin circuit causing a capacitor to discharge and generate a small voltage pulse.

The four basic types of contact pins are shown in Fig. 7.8. The simplest of these is shown in Fig. 7.8a. This consists of a number of thin wires, each of which is part of a circuit containing a small capacitor charged to some tens of volts. When the circuit is closed by contact with the expanding armature, the capacitor discharges. The wires can be either positively or negatively charged, and they can be identified, for example, by the shape of the discharge pulse recorded on a raster oscilloscope. A fully instrumented spherical implosion may have as many as 500 wires arranged in the form of a dome and connected to 50 raster oscilloscopes. The position measurements are accurate to within a few microns.

The second type of pin shown in Fig. 7.8b is similar to, but more complex than, the one in Fig. 7.8a. Since not all armature measurements can be conducted in evacuated chambers, account has to taken of the effects of gases on making measurements. The expanding armature generates a strong ionized shock wave that proceeds along the armature, and this wave causes the pins to be shorted prematurely. To solve the problem, special gas chambers were introduced. Other techniques may employ anodized pins or a self-shorting arrangement in which the pins have a coaxial electrode structure, with a cap insulated from the central cathode by materials like mylar tape. Such pins are useful for monitoring nonmetallic components such as the insulators in the generator.

The third pin in Fig. 7.8c is similar to that of Fig. 7.8b, but without the cap and insulator and with a fully ionized detonation wave having sufficient time to short the electrodes. The final type of pin in Fig. 7.8d is made from piezoelectric or ferroelectric materials or plastic tape, all of which produce a small current at shock wave pressures. This device is also useful in a number of special situations, such as monitoring isolators or small amplitude waves used for triggering purposes.

As an alternative to measuring the arrival time of the pulse with a contact pin connected to a raster oscilloscope, electronic counters or specially built, but costly, multichannel time measurement systems, can be used.

The most recent type of pin is the optical pin. The optical pin consists of small diameter optical fibers which are illuminated by a laser. The output light is monitored by either a photon multiplier or a fast PIN photodiode. When the light signal is lost, the time and position at which each fiber is broken by the expanding armature can accurately be determined. This technique is particularly useful in difficult situations, such as in very-high-current generators, where electric pins cannot be used. A version of the optical pin that uses the light from microspheres filled with ionized argon gas attached to the end of the fiber can be used to measure the detonation front arrival times.

### *Detotachograph Method*

The contact pin method only measures arrival times, from which the medium velocity between pins can be calculated. Continuous monitoring of the acceleration can be achieved using the detotachograph method.

There are two types of such transducers, both of which use a technique in which a high resistance wire carrying a constant current is shorted by a shock wave, which, in turn, causes the voltage to decrease proportionally to the velocity. (See Fig. 7.9a.) For relative small changes in velocity, the differential signal of the resistance-measuring probe attached to a saw-tooth generator can be recorded at the output of a differential amplifier. Velocity variation measurements can be enhanced by a factor of several tens.

The first type of detotachograph probe, which is a wire implanted into an explosive charge contained within a grounded metallic cylinder, can be seen in Fig. 7.9b. This probe uses the highly ionized detonation front to short circuit the probe that generates the output voltage. The second type of probe shown in Fig. 7.9c, consists of a small-diameter thin-wall metal tube containing an insulated conductor. The tube is deformed by the shock wave, with the point of contact between the tube and the conductor moving at the same speed as the shock wave. This type of probe can measure the velocity of a shock wave travelling through any kind of material in contact with the probe (see Fig. 7.9c), even when it changes direction.

### Pressure and Temperature Measurements

The most common methods used to measure shock wave pressures are based on using either quartz at low pressures (below several gigapascals)

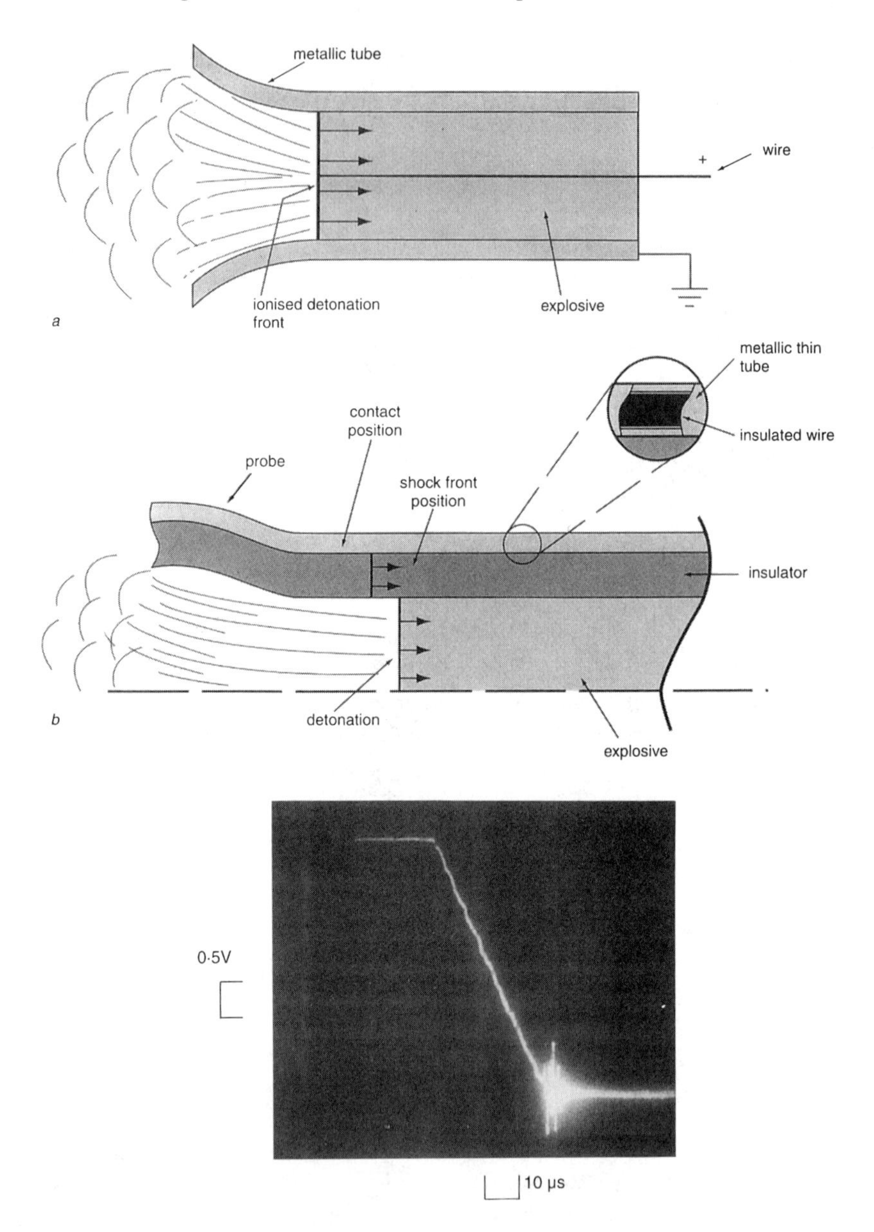


FIGURE 7.9. Detotachograph probes: (a) and (b) are different probe arrangements and (c) is a typical output signal from a type (b) probe.

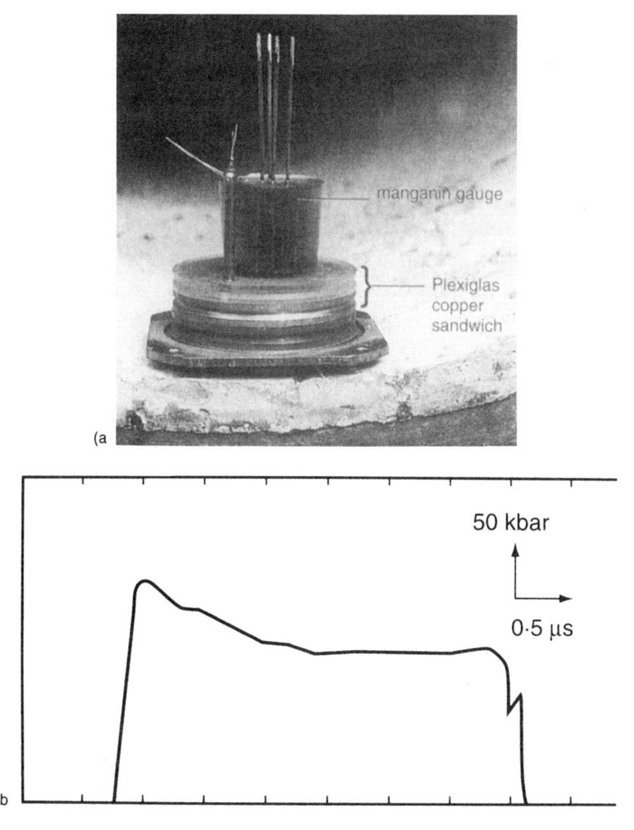


FIGURE 7.10. Simple Manganin pressure gauge (a) and a typical recorded signal (b).

or a Manganin wire at higher pressures. The pressure gauges based on a Manganin wire require a special power supply. (See Fig. 7.10.)

Temperature is a difficult parameter to measure. Normally the radiated spectrum is recorded using a spectroscope and an optical multichannel analyzer with an intensified target detector. For the case of nontransparent probes, such as those made from metal, the temperature can be measured by placing it in close contact with a material such as $Al_2O_3$. This retains the transparency of the probe under shock conditions, provides a near match relative to shock impedance, and enables the interface temperature to be measured.

## Detonation Wave Shaping

As discussed in earlier chapters, MCGs can have a number of different geometries, e.g., linear, planar, cylindrical, and spherical, and, for each of these, a special explosive device is needed to match the shape of the detonation wave to the geometry of the generator. A detonation wave with the proper shape can be formed in several ways. One of these is to use a distribution of precise bridgewire detonators, but this is both expensive and difficult to use in the presence of high currents, voltages, and electromagnetic noise generated by the MCG. A more practical approach is to use a single detonator together with a number of explosive wave generators that produce linear, planar, cylindrical, or spherical wave shapes. To generate the necessary wave shapes, the construction of the generator involves either the use of accelerating metal plates to initiate the main explosive charge, the insertion of interrupters, e.g., air holes, in the path of the detonation wave, or the use of explosive lenses, which are usually made from a combination of slow and fast explosives. A variety of methods used to shape waves are presented in Fig. 7.11. More recently, arrays of exploding foil mesh initiated by surface detonators have been used to generate the desired wave shape. Such an array is shown in Fig. 7.12.

## 7.2 Explosive Pulsed Power Laboratory

In this section, a brief description is provided of an explosive pulsed power laboratory . Such laboratories currently exist in several countries including the US, England, Russia, Japan, France, Sweden, and China. Since it is not possible to discuss how each laboratory is configured, it was decided to focus on the laboratory used by Loughborough University in England.

Experiments involving MCGs require both a laboratory in which explosive charges can be detonated and a means for providing the initial priming current, as well as to accommodate the load and its diagnostics. This laboratory should offer a means for protecting the diagnostics and, in some cases, the load from being destroyed by the explosion of the MCG. In practice, the laboratory needs three main elements:

- a firing site,
- the prime energy source, and
- a control room (bunker) as shown in Fig. 7.13.

For explosive charges that are equivalent to several tens of kilograms of TNT, the firing site, which is normally outside, has a firing table in close proximity to the laboratory buildings. If the explosive charge is equivalent to several hundreds of kilograms of TNT, the firing site needs to located further away from the laboratory buildings. However, it should be pointed out that not all shots are made outside. Some laboratories have blast containment chambers in which the MCG is detonated. The energy generated is then delivered from the containment vessel via cables to the load, which is now protected from the blast. For example, a sphere up to 15 m in diameter can be used for explosive charges up to 80 kg of TNT. Explosive

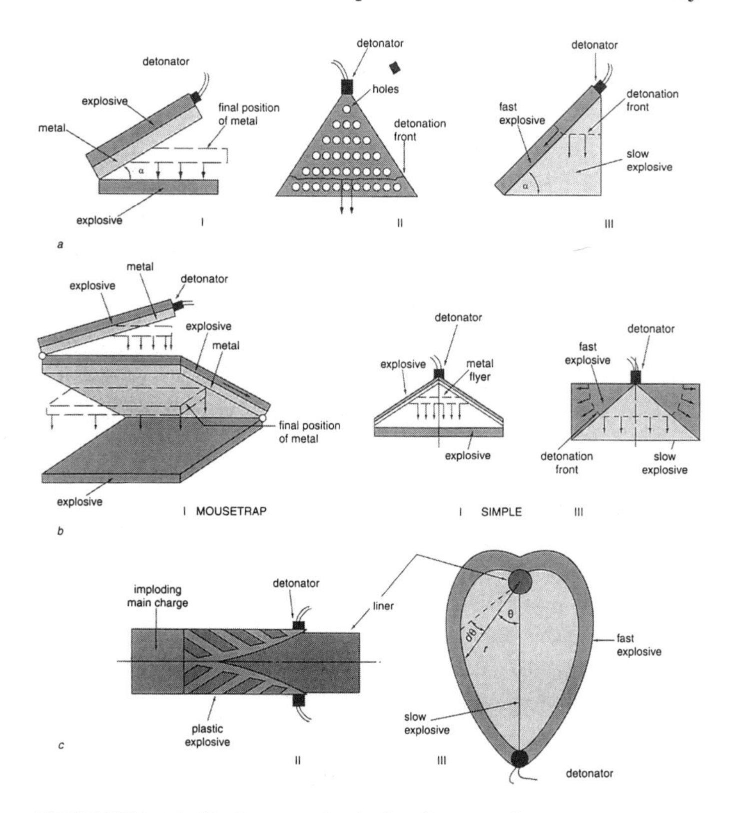


FIGURE 7.11. Various methods for shaping detonation waves.

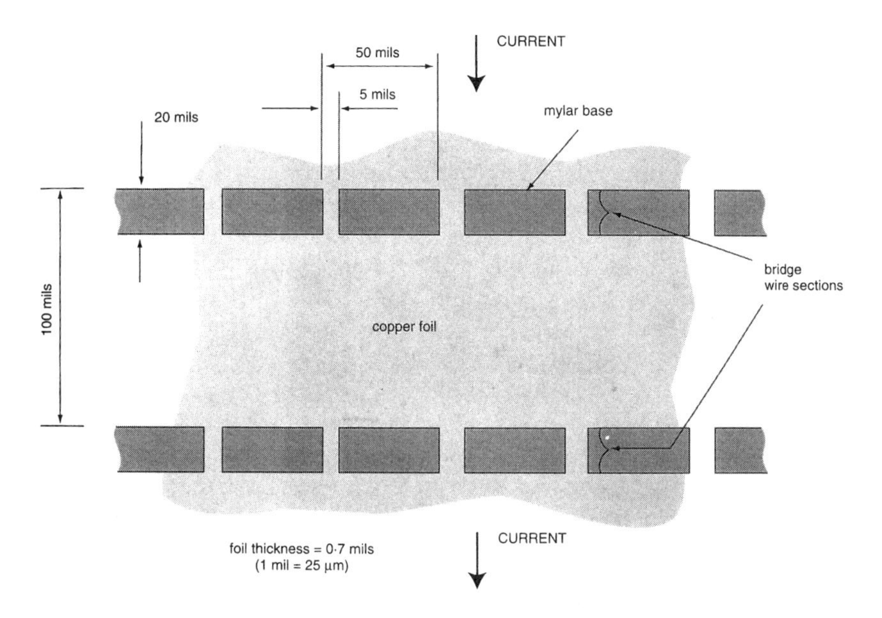


FIGURE 7.12. Array of exploding foil mesh initiated by surface detonators.

FIGURE 7.13. Photographs of main elements of an MCG laboratory at Loughborough University.

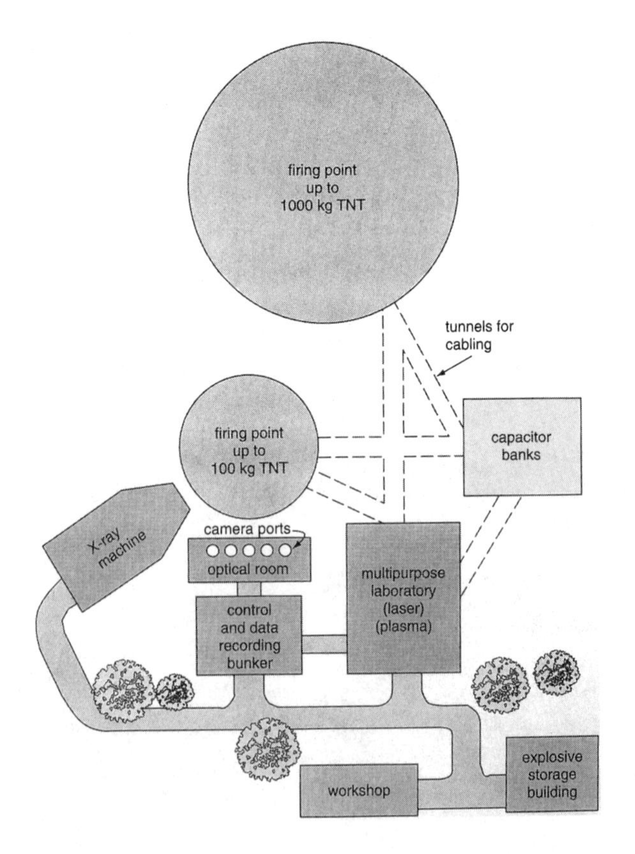


FIGURE 7.14. Layout of an ideal MCG laboratory.

containment vessels are currently in use at the Institute of Hydrodynamics in Novosibirsk, the Institute of Chemical Physics in Chernogolovka, and the Institute of High Temperatures in Moscow.

The initial energy source, which is usually a capacitor bank, may need to store up to 1 MJ of energy. The bunker contains the control room with the data recording and diagnostic equipment, and is located as near as possible to the firing point. All the major elements that make up the laboratory are connected by fiber optic cables or pneumatic systems. The laboratory also has a workshop and a place for the storage and assembly of explosive charges. The laboratory layout used by Loughborough University, where the FLEXY I was tested, is depicted in Fig. 7.13.

An ideal laboratory (see Fig. 7.14) would also have buildings to accommodate an X-ray machine and the load. This means that either hundreds of coaxial cables or a large parallel plate transmission line is required to couple the various components of the laboratory together.

A complex firing experiment can be divided into a number of phases, where the duration of a phase may vary from months to nanoseconds. For example, the design and manufacture of the components of the experiment may take up to one year, followed by a few days to assemble the MCG and its output circuits. After positioning the MCG on the firing table, the diagnostic and recording instruments must be connected, as well as the load and the prime energy source. All the instrumentation, oscilloscope settings, and connections must be checked and rechecked, in a series of tests that do not involve the explosives, in accordance with an established protocol. These preliminary tests may even include discharging the main capacitor bank. Once these tests are completed, the initial conditions, such as vacuum, gas pressure, and voltages, for the explosive test are set and the experiment can proceed. Once the firing button is pressed, all control is lost, except for some independent electronic fast delay units. The flux compressor generates an output pulse in tens of microseconds, and the output circuits condition the pulse for the load in nanoseconds. All that remains after a long and costly experimental program, apart from the debris, is a collection of photographs and oscilloscope records.

## 7.3 Testing Fast Switches and Conditioning Circuits

In this section, a description is presented of both exploding-foil opening and closing switches that are used in the output conditioning circuits of capacitor banks and MCGs. One of the major functions of these switches is to sharpen the risetime of the output of the MCG. In addition to presenting an empirical model of these switches, the results of an experiment, carried out using a 1 MJ flux compressor with an exploding-foil opening switch, are

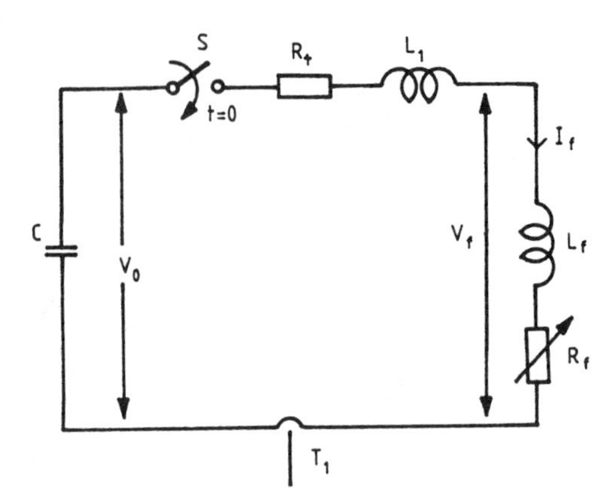


FIGURE 7.15. Equivalent circuit diagram for capacitor-powered EF experiments.

also presented. Since the material in this chapter is based on the same case study, i.e., the construction and testing of FLEXY I presented in Chapter 6, this section deals with the work done in developing and testing the switch for this particular generator.

As has been noted earlier, MCGs are powerful and inexpensive sources of high energy and high current. If they are to be used to provide fast-rising current pulses to high-impedance loads, then their sources will require conditioning. For example, if a 2 MJ MCG with a run time of 160 $\mu$s is to provide a nanosecond risetime pulse, several stages of fast exploding-foil switches followed by a single stage plasma switch are required.

To facilitate the development of a fast foil switch, a simple empirically derived plot of the dynamic resistance ratio versus input energy was obtained for a copper foil from experiments using a fast-discharge capacitor bank. These data were then used in developing a computer model, which were, in turn, used to develop foil switches for the FLEXY I MCG. A new type of closing switch was developed, based on exploding foils. A combination of exploding-foil switches was used to transfer sharp reproducible current pulses from a capacitor bank. The calculated and experimental results are in fairly good agreement.

### 7.3.1 *Exploding Foil Empirical Model*

#### Experimental Assembly

In order to obtain data for constructing a foil model, experiments were conducted using the circuit of Fig. 7.15. The energy source was a very-low-inductance 238 $\mu$F, 35 kV, 145 kJ capacitor bank ($C$). The switch ($S$) consists of two electrodes separated by two sheets of Mylar insulation and triggered by a Blumlein-driven aluminum strip positioned between the Mylar sheet. The charged capacitor bank is connected to the foil of the switch, which has an inductance of $L_f$ and a resistance of $R_f$ (which undergoes significant changes during the experiment), through a parallel plate transmission line. The inductance $L_t$ and resistance $R_t$ include the inductances and resistances of the capacitor bank and the transmission line, as well as the ballast inductance used to restrict the short circuit current to about 800 kA. The probe $T_1$, which was placed in a tunnel in the copper conductors of the transmission line, was used to measure both the current and the rate of change of current within time.

When the switch $S$ is closed at time $t = 0$, the foil voltage $V_f$ can be found by solving the equation:

$$V_f = R_f I_f + L_f \frac{dI_f}{dt}, \qquad (7.2)$$

where $I_f$ is the current in the foil. The voltage was measured with a self-integrating Pearson-type Rogowski coil, by measuring the current in a high-resistance copper sulfate resistor connected to the foil. Calibration data were obtained by measuring the current discharged from the capacitor through a foil that is much thicker, but of the same length and width as the foil used in the switch. The foil inductance $L_f$ was then found by solving Eq. 7.2, assuming that the resistance $R_f$ remains constant. To ensure good electrical contact and precise geometry, the foils were firmly secured between copper mounts and encapsulated in a plastic cassette, which was surrounded under pressure by 100 $\mu$m beads at the center of a glass fiber/glass bead assembly. In Fig. 7.16a, a foil resting on the lower section of the glass bead packing prior to an experiment can be seen, and in Fig. 7.16b, the characteristic two-layer split produced in the foil after it has vaporized and reformed, where the upper layer was slightly displaced, can be seen. In both figures, the upper section of the glass bead packing is absent.

### Experimental Results

The broken curves in Fig. 7.17a represent the experimental results for the time variation of the current discharged from the capacitor bank, when the current is discharged into three 17-$\mu$m-thick copper fuses having different widths and lengths. Table 7.1 summarizes the main parameters of the test circuits, as well as those used in a similar experiment with a 25.4 $\mu$m thick foil.

The current profile for the three foils are quite different, as are the results for the variations in the rate-of-change of foil current and the foil voltage shown in Figs. 7.17b and 7.17c, respectively. The corresponding variations in the specific action are shown in Fig. 7.17d. The values of the specific action are in good agreement with those obtained in other experiments, i.e., $1.27 \times 10^{17}$ $A^2 m^{-4} s$ [7.3]. The liquid and gaseous phases of the foil

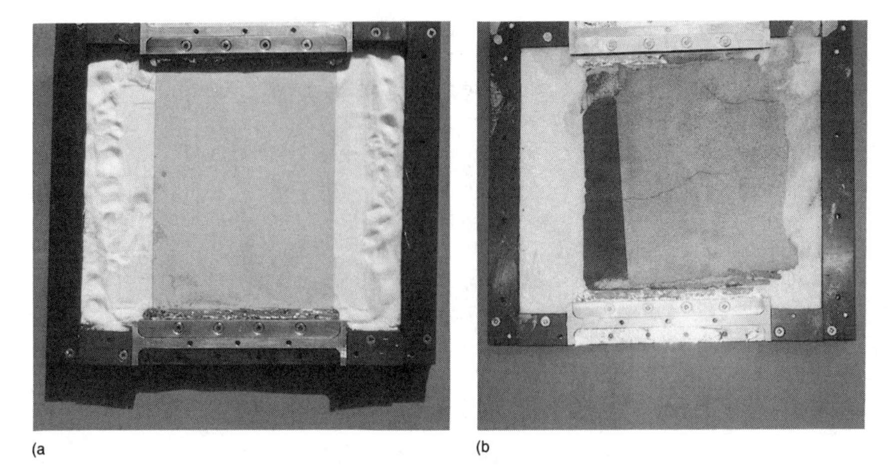


FIGURE 7.16. Foil on assembly base: (a) before experiment and (b) after experiment.

can clearly be seen in the curves in Fig. 7.17e, derived for the increase in dynamic resistance ratio with time for the foil I experiment. The dynamic ratio increases as the specific deposited energy characteristics for the three 17 $\mu$m copper foils coalesce into the single curve shown in Fig. 7.18, which is a phenomena that has been observed in other experiments using aluminum [7.4,7.5].

## Foil Model

It is well known that the behavior of an exploding foil depends on a number of parameters including the dimensions of the foil [7.5–7.8], the time profile

| | Foil | I | II | III | IV |
|---|---|---|---|---|---|
| Copper Foil | Thickness ($\mu m$) | 17.0 | 17.0 | 17.0 | 25.4 |
| | Width (cm) | 28 | 18 | 9 | 18 |
| | Length (cm) | 12 | 11 | 11 | 12 |
| | $L_t$ (nH) | 86 | 353 | 353 | 86 |
| | $R_t$ (m$\Omega$) | 3.58 | 3.20 | 3.20 | 3.50 |
| | $L_f$ (nH) | 20 | 35 | 35 | 20 |
| | $R_f(0)$ (m$\Omega$) | 0.43 | 0.62 | 1.23 | 0.45 |
| | $V_0$ | 20.0 | 15.0 | 14.5 | 20.0 |

TABLE 7.1. Parameters of exploding-foil experiments.

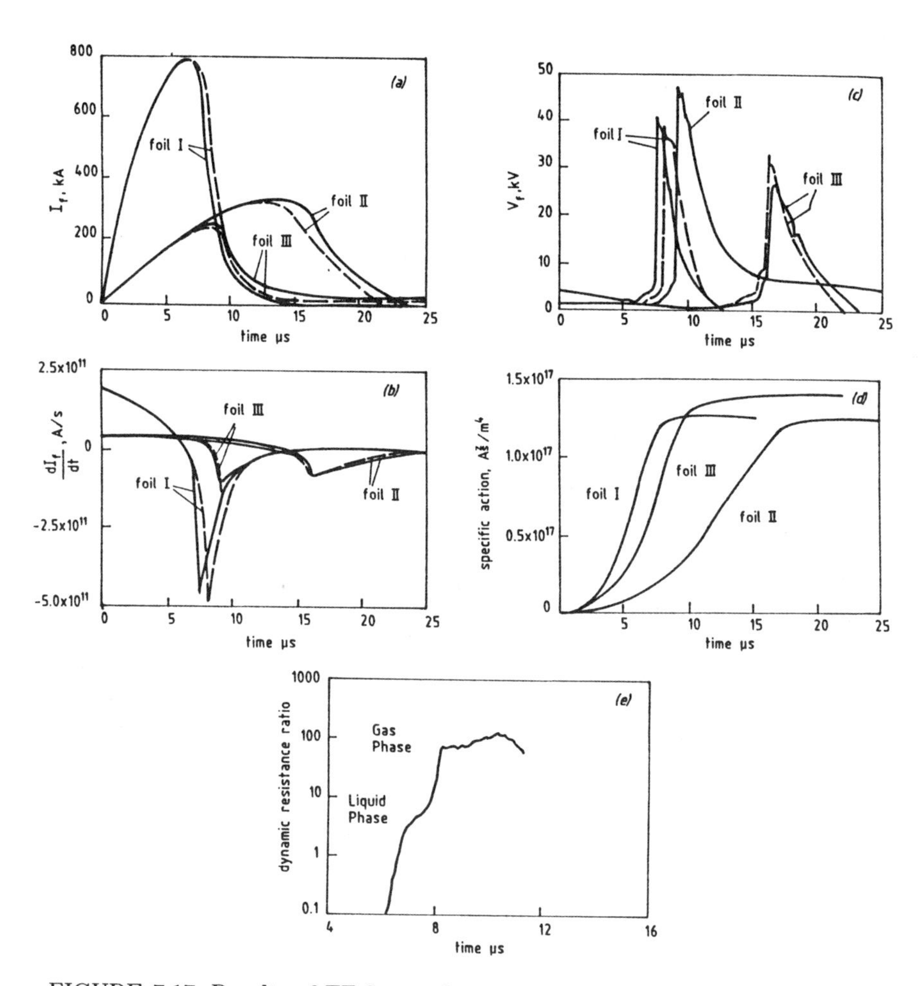


FIGURE 7.17. Results of EF-3 experiments corresponding to foils described in Table 7.1: (a) current$I_f$, (b) rate of change of current $dI_f//dt$, (c) foil voltage, (d) specific action, where + indicates burst time values, and (e) dynamic resistance ratio expressed as log(foil resistance/initial foil resistance) for foil 1. The lines (- - -) are experimental and (—) are theoretical.

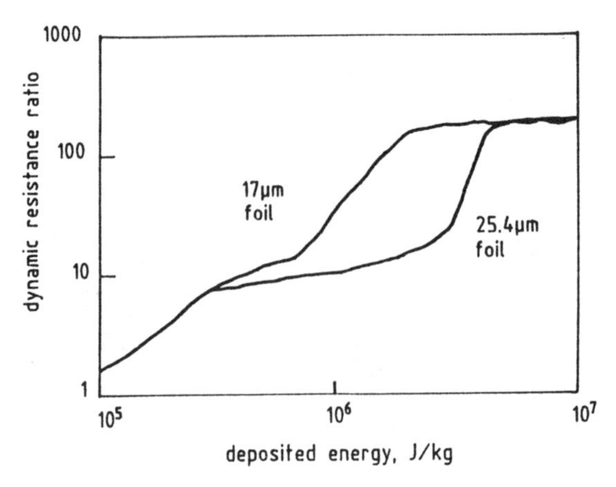


FIGURE 7.18. Variation of dynamic resistance ratio with specific deposited energy for 17 and 25 micron foils.

of the current [7.8], and the confining medium. In particular, Fig. 7.18 shows that although the characteristics of the 17 $\mu$m and 25.4 $\mu$m foils have the same general shape, that of the former reaches the same resistance ratio at a lower deposited energy. A similar phenomenon has been observed in aluminum foils [7.7].

The results presented in Fig. 7.18 for the two foil thicknesses were incorporated into a computer model to predict the experimental behavior of the foils. This model is valid for the range of foil widths and lengths used in the development of fast opening switches required by the FLEXY I. For the purposes of comparison, the calculated results for the 17 $\mu$m foil are added as the solid line curves in Figs. 7.17a–7.17e. In addition, the computed and measured current and voltage profiles of the 25.4 $\mu$m foil, described in Table 7.1, are presented in Fig. 7.19. In Figs. 7.17 and 7.19, it can be seen that the computer model is accurate in predicting the measured characteristics of the foils.

### 7.3.2 *Magnetic Flux Compressor/Opening Switch Experiments*

Having established the accuracy of the computer model, it was used to predict the experimental behavior for a 17 $\mu$m exploding foil driven by the helical eight section coil flux compressor, i.e., FLEXY I describe previously. However, as a result several improvements made in the construction of the generator, the output has been raised from 1 to 2 MJ.

Figure 7.20 shows the equivalent circuit of the generator used in the experiments. Capacitor $C$ provided the prime energy to the MCG when switch $S_1$ was closed. The generator was subsequently self-crowbarred by switch $S_2$ at $t = 0$, so that the energy was delivered to a single-turn load coil

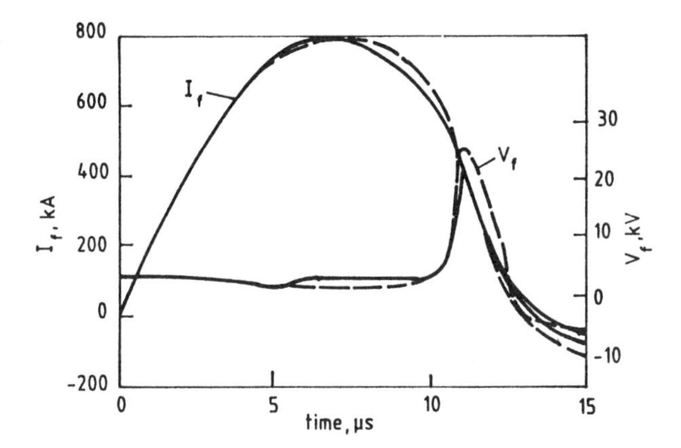


FIGURE 7.19. Comparison of experimental and predicted current and voltage for 25 $\mu$m foils. The lines (- - -) are experimental and (—) theoretical.

| | Parameter | F3 | F4 |
|---|---|---|---|
| | $I_0$ (kA) | 50.46 | 50.03 |
| | $L_l$ (nH) | 40 | 42 |
| | $L_t$ (nH) | 6 | 11 |
| Foils | No. in Parallel | 4 | 2 |
| | Thickness ($\mu$m) | 17 | 17 |
| | Width (cm) | 60 | 52 |
| | Length (cm) | 25 | 42 |

TABLE 7.2. Parameters for MCG/exploding-foil experiments.

with inductance $L_l$ through a parallel-plate transmission line with inductance $L_t$. The circuit also includes the laterally pressurized explosive-foil switch, in which parallel connected copper foils are separated by three layers of polythene faced sheets of Mylar insulation having a total thickness of 760 $\mu$m. The currents and their rate of change with time were measured by means of probes. The main circuit and foil data for the two experiments conducted are summarized in Table 7.2, and the experimental test arrangement at the firing site is presented in Fig. 7.21.

The results of experiments with FLEXY F3 and FLEXY F4 generators are presented in Figs. 7.22a and 7.22, together with the results of an earlier test with the lower performance FLEXY F1 generator with no load in the circuit, i.e., $L_f = 0$ (presented in Chapter 6). During the first 100 $\mu$s, the load parameters, including the foil resistance, do not significantly affect the operation of the generator, which means that a valid comparison can be made with results obtained in different experiments. The improved

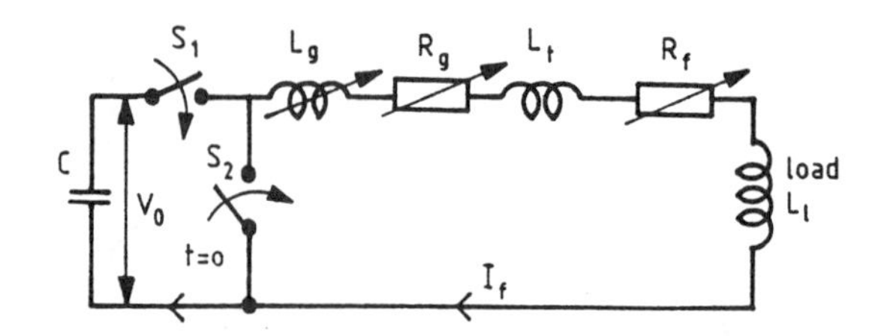


FIGURE 7.20. Equivalent circuit diagram for MCG-exploding foil experiment: $R_g$ and $L_g$ are the resistance and inductance of the MCG, respectively, $L_t$ and $L_l$ are the inductance of the transmission line and load, respectively, where $L_t$ includes the foil inductance, and $R_f$ is the fuse resistance.

FIGURE 7.21. FLEXY F3 and foil on the firing pad.

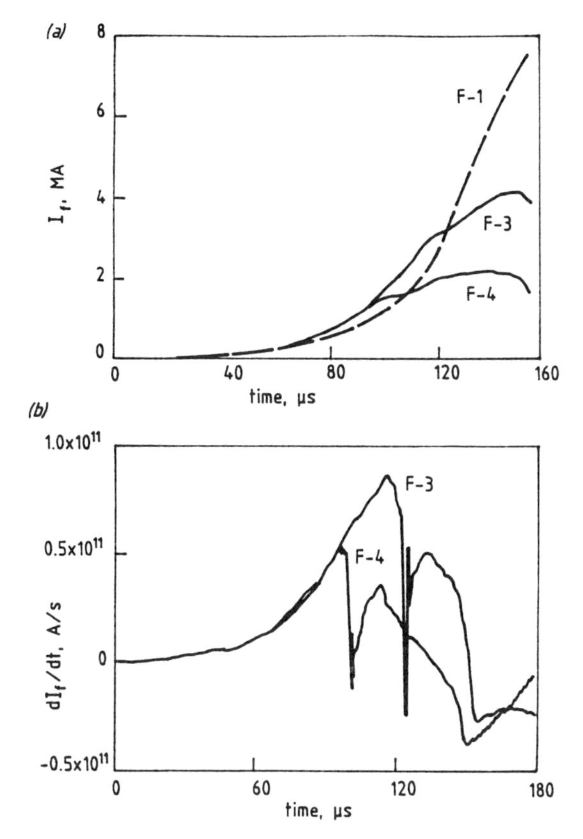


FIGURE 7.22. FLEXY-foil test results: (a) current $I_f$ and (b) rate-of-change of current with time $dI_f/dt$. Result F1 is from a FLEXY test without a foil (presented in Chapter 6) and is given for comparison.

performance of the FLEXY F3 and F4 generators and the excellent reproducibility of their output before the increasing foil resistance becomes significant can be seen in Fig. 7.22b.

In order to fit the experimental data to the theoretical predictions, it is necessary that the mathematical function that describes the change in the resistance of the generator with respect to time be modified [7.9, 7.10]. This modified function and the corresponding calculated changes in inductance and the time rate of change of inductance and their variations, are presented in Fig. 7.23 for a period of 160 $\mu$s. The results of the two experiments with the FLEXY F3 and F4 were analyzed using the data presented in Fig. 7.18. In Fig. 7.24, the experimentally measured foil voltages were compared with their calculated values. Although there is good agreement at the burst times, the foils "restrike" early in the vapor phase and at much lower voltages than expected. The maximum restrike electric field in both experiments was about 1.2 kV/cm, which is much lower than the expected

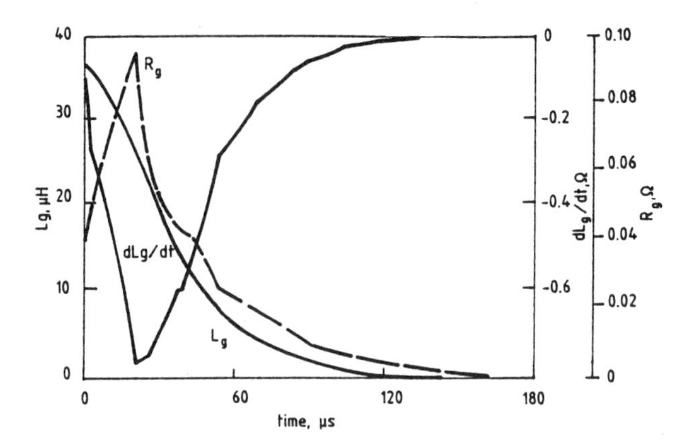


FIGURE 7.23. Time varition of MCG parameters, where $R_g$ is the resistance (fitted to the experimental data) and $L_g$ and $dL_g/dt$ are the inductance and rate of change of inductance with time (calculated), respectively.

3 to 4 kV/cm that could be sustained by the glass bead/glass fiber surrounded foils used in the capacitor powered experiments. Further capacitor powered experiments with a scaled down version of the foil package used in the MCG experiments with the same polyethylene-faced Mylar insulation confirmed that electric fields exceeding 3 kV/cm could be sustained.

A major difference in the experiments using the capacitor bank and the MCGs is that the rate at which specific energy is deposited in the foil between the melting and vaporizing phases is between 2 and 5 MJ/kg $\mu$s in the capacitor-powered experiments and only 0.4 and 0.8 MJ/kg $\mu$s in the MCG-powered experiments. The lower restrike electric fields observed in the latter case are thought to be due to differences in the thermal and hydrodynamic behavior of the foils in the vapor phase. Similar low restrike fields, i.e., 1.16 kV/cm, were observed in the early Laguna foil experiments [7.11] using the Mark IX flux compressor [7.10]. As is shown in Fig. 7.25, the predicted results using the code in [7.9] and an adjusted generator resistance [7.10] were in excellent agreement with the measured results until the foil was vaporized. After the foil vaporized, as in the experiments conducted with the FLEXY generators, the calculated maximum voltage was much higher in value than that which was measured. It has been suggested in [7.12] that to accurately model the post-burst phase, a complex code based on atomic data must be used.

### *7.3.3 Opening and Closing Exploding Foil Switches*

#### Opening Switches

Simple foils are possibly the most useful devices for the pulse sharpening required during the initial opening stages of a conditioning circuit, when the

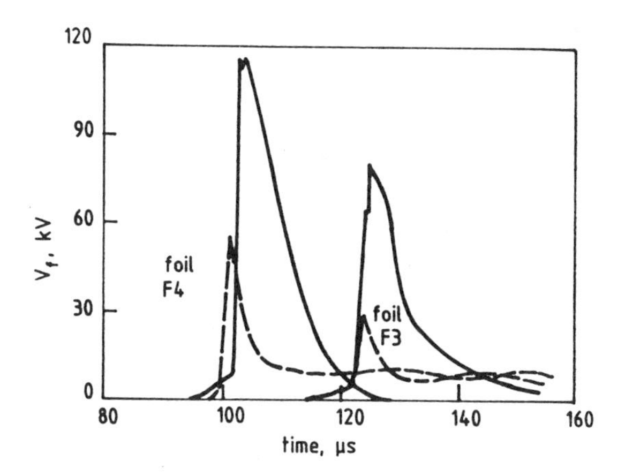


FIGURE 7.24. Foil voltage during MCG and exploding foil experiments: (- - -) experimental and (—) calculated.

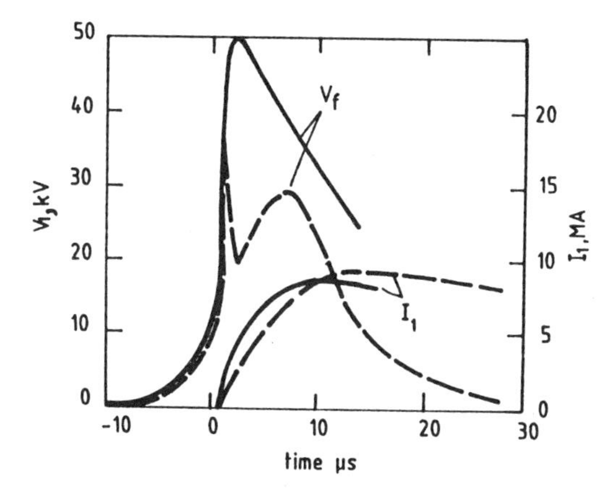


FIGURE 7.25. Foil voltage and load current for the LAGUNA foil experiment [7.11]. The lines (- - - ) are experimental and (—) theoretical predictions using the LU theory with data from [7.10].

specific energy input rates are high. They are less effective when the input rates are low. Explosively formed foils (EFFs) can be used to overcome this difficulty by forcing a thick current-carrying conductor into insulator grooves to form thin conductors that act in a similar manner as conventional foils. This enables the EFF to control the instant at which the circuit is opened. However, a complex triggering circuit must be used to initiate the explosion, which further complicates the use of the MCG [7.13].

Exploding foils are simple, operate automatically, and give reproducible results. The inherent deterioration in performance at low rates of specific energy input can be minimized by using multiple stages of thin opening fuses. Automatic conditioning circuits using opening and closing exploding foil switches can be designed for a wide range of input current profiles.

### Closing Switches

The most difficult problem associated with using conditioning circuits that employ closing switches is synchronization. In capacitor powered experiments, the optimal time for closure can be established by performing preliminary tests, but this is not practical in single-shot MCG experiments. One possible solution is to use dielectric switches that have a preset breakdown voltage [7.14], but significant difficulties arise in high-current and low-inductance systems. Surface tracking switches are another possible solution, although at present there are a number of unresolved difficulties that arise because of their dependence on the polarity profile of the switch voltage [7.15]. The most practical solution is to use an explosive foil switch, which does not require synchronization, since it operates automatically once its correct dimensions have been established.

In Fig. 7.26, a circuit diagram is presented in which the explosive foil is used as both an opening and a closing switch, the latter consisting of an assembly of parallelly connected, small dimension aluminum bridges, as shown in Fig. 7.27a, connected in series to a main copper foil. The geometry of the small bridges enhances their explosive action, by an effect that is probably similar to that used in detonics for shaped charge devices. The dimensions of the bridges are such that they burst when the current flowing through them is near its maximum. They subsequently penetrate the insulation between them, with the following conditioning stage introduced into the circuit just before the main foil completely vaporizes.

The bridges were made by removing sections from a 100 $\mu$m aluminum foil, thus leaving a number of strips that are 9 mm long and 5 mm wide to carry the current. The action of the bridges is enhanced, by bending them into a 2 mm wide by 2 mm deep slot milled into a 15 mm wide insulator as shown in Fig. 7.27b. Brass electrodes are attached to each side of the insulator to complete the connections between the foil and the main stripline. Mylar sheets are used to insulate initially the fuse assembly from an anvil connected to the stripline of the circuit, to which contact is

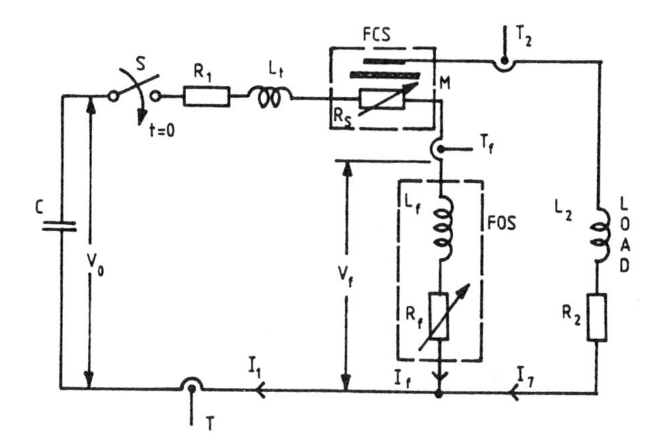


FIGURE 7.26. Circuit using exploding foil opening (FOS) and closing switches (FCS): $R_s$ is the resistance of the aluminum bridges and $M$ is the insulator.

subsequently made. Explosive debris is released into a groove in the anvil, while a 100 $\mu$m aluminum foil laid across it helps to maintain a low switch resistance. In Fig. 7.27c, the results of the foil explosion on the switch insulation, which is made from six sheets of Mylar having a total thickness of 762 $\mu$m, can be seen.

The resistance of the closing switch (FCS) in Fig. 7.26 which introduces the load with resistance $R_2$ and inductance $L_2$, into the circuit is insignificant relative to the other resistive factors in the circuit. After the action of the closing switch, the system can be described by the following system of equations:

$$\frac{dI_1}{dt} = \frac{dI_f}{dt} + \frac{dI_2}{dt}, \tag{7.3}$$

$$\frac{dI_f}{dt} = \frac{L_2\left(V_0 - \frac{Q}{C}\right) - I_f R_f (L_t + L_2) - I_1 R_t L_2 + I_2 R_2 L_t}{L_t L_2 + L_f (L_t + L_t)}, \tag{7.4}$$

$$\frac{dI_2}{dt} = \frac{R_f I_f}{L_2} - \frac{R_2 I_2}{L_2}\frac{L_f}{L_2}\frac{dI_f}{dt}, \tag{7.5}$$

$$\frac{dQ}{dt} = I_1, \tag{7.6}$$

$$\frac{dW}{dt} = \frac{I_f^2 R_f}{M_f}, \tag{7.7}$$

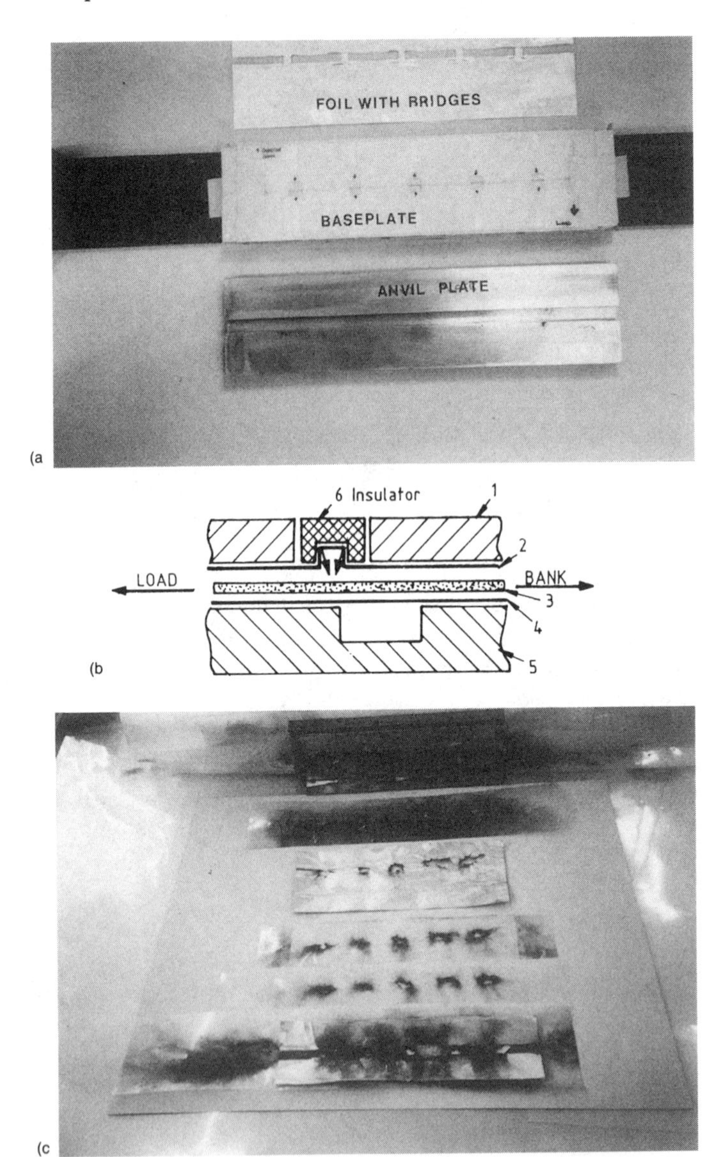


FIGURE 7.27. Exploding-foil closing switch: (a) components, (b) diagram of switch, and (c) components after explosion. The switch consists of 1 – Brass foil-forming base plate, 2 – aluminum foil with five bridges, 3 – Mylar insulator, 4 – aluminum foil, 5 – brass anvil plate with groove, and 6 – insulator.

| Parameter | Unit | Value |
|---|---|---|
| Thickness | ($\mu$m) | 17 |
| Width | (cm) | 12 |
| Length | (cm) | 10 |
| $L_t$ | (nH) | 353 |
| $R_t$ | (m$\Omega$) | 3.2 |
| $L_f$ | (nH) | 35 |
| $R_f(0)$ | (m$\Omega$) | 0.84 |
| $L_2$ | (nH) | 60 |
| $R_2$ | (m$\Omega$) | 1 |
| $V_0$ | (kV) | 15 |

TABLE 7.3. Parameters of exploding copper foil switch experiments.

$$\frac{dA}{dt} = \frac{I_f^2}{S_f^2}, \tag{7.8}$$

where $Q$ is the charge released by the capacitor, $M_f$ and $S_f$ are the mass and cross-sectional area of the fuse, respectively, $A$ is the specific action, and $W$ is the specific energy deposited in the fuse.

Solving the above set of first-order differential equations using a FORTRAN subroutine [7.16], the amount of specific energy deposited in the foil was calculated at each time step of the solution. The data in Fig. 7.18 are used to determine the corresponding resistance.

Experimental Results

The results of three identical experiments using the circuit in Fig. 7.26, together with the results of theoretical predictions, are presented in Fig. 7.28. The main parameters for the experiments are given in Table 7.3.

In Fig. 7.28a, it can be seen that the reproducibility of the operating time is good, where in two of the cases the fast closing switch (FCS) functioned about 11.25 $\mu$s after the main switch $S$ in Fig. 7.26 functioned and in the third case the delay was about 11 $\mu$s. This small difference enables accuracy with which the predictions made by numerical models describe the actual switch behavior to be considered. The small jump in the rate-of-change of current with respect to time, denoted by the arrow in Fig. 7.28b, which was correctly predicted when the switch operated after 11.25 $\mu$s, was absent from both the theoretical and measured results when the switching occurred after 11 $\mu$s. In addition, Fig. 7.28c shows that the maximum recorded foil voltages can be closely predicted for both operational times.

It has been observed that when the current is interrupted before it reaches its first maximum, both the computed and measured values of

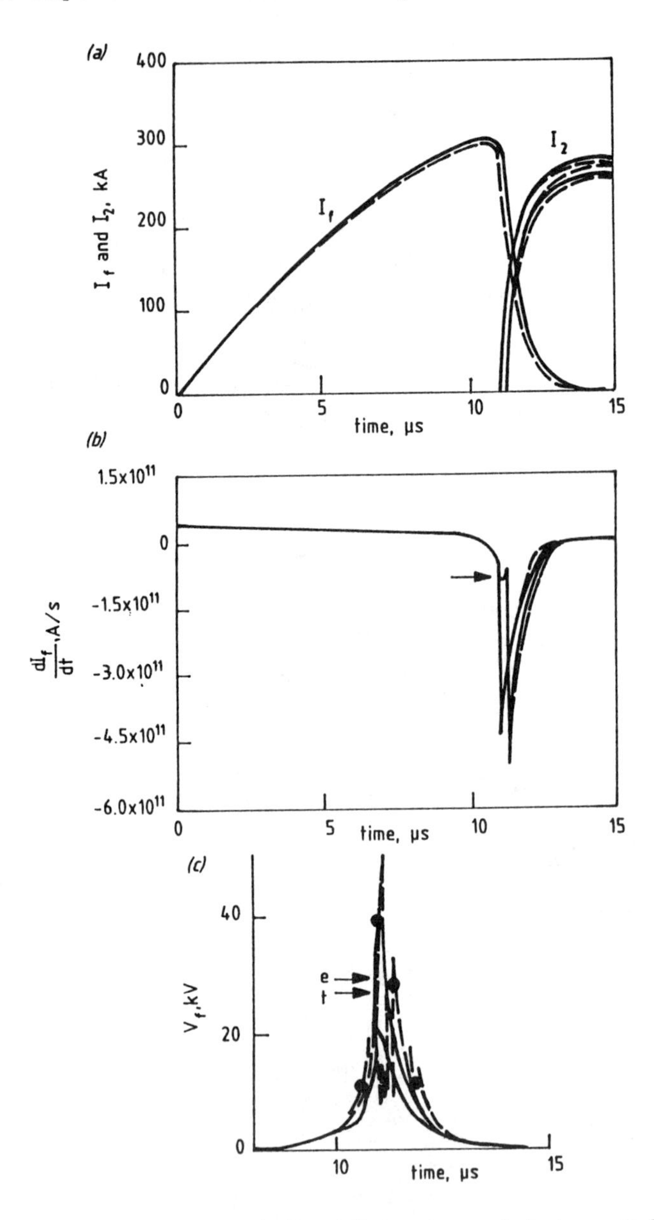


FIGURE 7.28. Results from exploding foil opening and closing switch experiments: (a) foil current, $I_f$, and transferred currents, $I_2$; (b) rate of change of foil current, $dI_f/dt$, and (c) fuse voltage, $V_f$, where "e" is the maximum experimental value for the 11 $\mu$s experiment and "t" is the maximum theoretical value for the same experiment. The lines are (- - -) experimental and (—) theoretical.

the maximum foil voltage exceed those predicted by earlier models [7.3], whereas there is good agreement if the interruption occurs after the current maximum. This is due to the influence of the nonlinear interaction between the foil and the capacitor discharge on the behavior of the circuit, which is significant when the interruption is early in the discharge cycle.

### 7.3.4 Faster Switching Techniques

Transformers

When an MCG drives a high-impedance load, an intermediate transformer is required to match the impedance of the load to that of the generator. It has been found that it is best to connect the load to the transformer through a switch, which shapes the waveform of the power source and applies a fast-rising high-voltage pulse to the load [7.17].

A transformer was developed on the basis of previous work described in [7.17]. Both the primary and secondary windings of the transformer were made from two 16.5-cm-wide copper strips and were laid side-by-side. The interwinding insulation consisted of six layers of Mylar having a thickness of 125 $\mu$m. The strip used in the 25-cm-diameter single-turn primary had a thickness of 500 $\mu$m, while that used in the six-turn secondary had a thickness of 51 $\mu$m. The secondary turns were thermally bonded between Mylar foil having a thickness of 50 $\mu$m and coated with polyethylene having a thickness of 100 $\mu$m. Additional layers of the same material were used to separate the turns by about 600 $\mu$m.

The equivalent circuit for an experiment in which the above transformer was connected in series with a foil is presented in Fig. 7.29. If $I_1$ and $I_2$ are the currents in the primary and secondary turns, respectively, then the behavior of the overall system after closing the solid dielectric switch (SDS) can be predicted by using Eqs. 7.6–7.8 and the following equations:

$$\frac{dI_1}{dt} = \frac{1}{M}\left((L_2 + L_s)\frac{dI_2}{dt} + R_2 I_2\right) \tag{7.9}$$

and

$$\frac{dI_2}{dt} = \frac{M(R_t + R_f)I_1 - L^* R_2 I_2 - M\left(V_0 - \frac{Q}{C}\right)}{L^*(L_2 + L_s) - M^2}, \tag{7.10}$$

where $I_f$ in Eqs. 7.7 and 7.8 is replaced by $I_1$ and $L^* = L_t + L_f + L_p$.

The primary, $L_p$, and secondary, $L_s$, inductances of the transformer and the primary/secondary mutual inductance, $M$, can be calculated in the following manner. A typical view of a helical transformer is presented in Fig. 7.30. If initially a transformer that has a 2:1 turn ratio is considered,

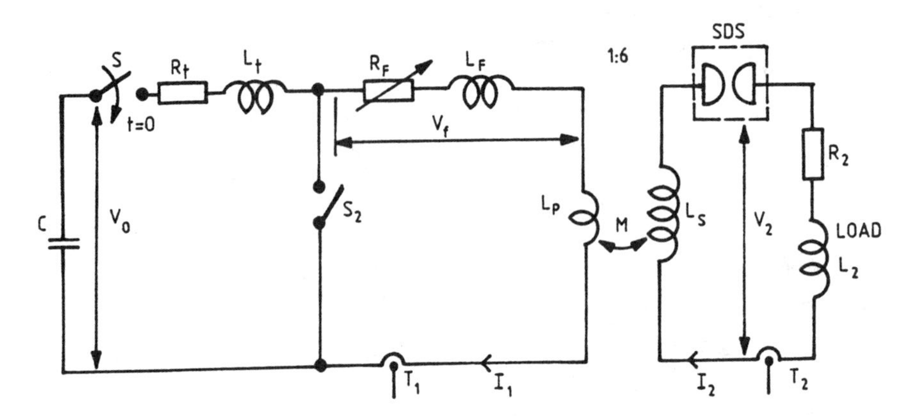


FIGURE 7.29. Equivalent circuit diagram for transformer-fuse experiments. SDS is the solid dielectric switch.

then, to calculate the inductance, it is assumed that the single-turn primary winding is a cylinder and that the secondary winding is two cylinders having radii equal to the minimum radii of the two winding turns in the actual transformer. The field distribution, $H_z(r, z)$, in the $z$-direction inside the cylinder, which has a length $l$, radius $a$, and carries a current $I$, is [7.18]

$$\begin{aligned} H_z(r,z) &= \frac{I}{4\pi l}\int_0^{\pi}\left(\frac{1-K\cos\Psi}{1+K^2-2K\cos\Psi}\right) \\ &\quad\times\left(\frac{1-Z'}{\left[y^2(1+K^2-2K\cos\Psi)+\frac{(1-Z')^2}{4}\right]^{\frac{1}{2}}}\right. \\ &\quad\left.+\frac{1+Z'}{\left[y^2(1+K^2-2K\cos\Psi)+\frac{(1+Z')^2}{4}\right]^{\frac{1}{2}}}\right)d\Psi, \end{aligned} \tag{7.11}$$

where $y = a/l$, $K = r/a$, and $Z' = Zl/2$. The inductance of the primary winding can be found by integrating Eq. 7.11 over the surface of the cylinder, i.e.,

$$L_p = \frac{\mu_0}{I}\int_0^a 2\pi H_z(r,0)r\,dr. \tag{7.12}$$

Relative simple approximations for $H_z(r, z)$ enable Eq. 7.12 [7.18] to be integrated, but it is normally solved numerically. The mutual inductance, $M_{12}$, between the two concentric cylinders can be calculated from

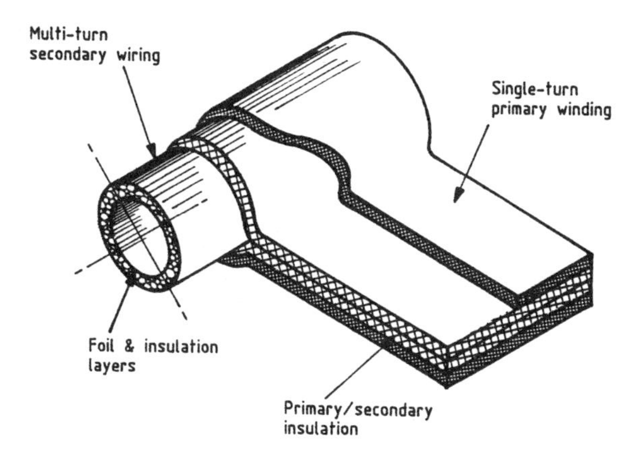


FIGURE 7.30. Diagram of a helical transformer.

$$M_{12} = \frac{\mu_0}{I_2} \int_0^{a_1} 2\pi H_z'(r,0) r dr, \tag{7.13}$$

where $a_1$ is the radius of the inner cylinder, $I_2$ is the current in the outer cylinder, and $H_z'(r,0)$ is the magnetic field that is generated by the current.

In general, the inductance of a multiturn secondary winding represented by $n$ concentric cylinders is

$$L_s = \sum_{i=1}^{n} L_i + \sum_{i,j=1}^{n} M_{ij}, \tag{7.14}$$

where $L_i$ is the inductance of the $i^{th}$ secondary cylinder and $M_{ij}$ is the mutual inductance between the $i^{th}$ and $j^{th}$ cylinders. The primary-to-secondary mutual inductance is

$$M_{ps} = \sum_{i=1}^{n} M_{pi}, \tag{7.15}$$

where $M_{pi}$ is the mutual inductance between the primary and the $i^{th}$ secondary cylinder.

The above transformer model was validated experimentally using data from [7.17]. The main parameters of the circuit in Fig. 7.29 calculated with the model are presented in Table 7.4. The theoretical mutual inductance is $M_t$ and the measured inductance is $M_e$.

| Device | Parameter | Value |
|---|---|---|
| Transformer | $L_p$ (nH) | 180 |
| | $L_s$ ($\mu$H) | 5.9 |
| | $M_t$ ($\mu$H) | 0.98 |
| | $M_e$ ($\mu$H) | 0.88 |
| Copper Foil | Thickness ($\mu$m) | 25.4 |
| | Width (cm) | 7 |
| Circuit Parameters | Length (cm) | 18.3 |
| | $L_t$ (nH) | 141 |
| | $R_t$ (m$\Omega$) | 2.7 |
| | $R_f$ (m$\Omega$) | 1.77 |
| | $L_f$ (nH) | 40 |
| | $L_2$ (nH) | 100 |
| | $R_2$ ($\Omega$) | 30 |
| | $V_0$ (kV) | 12 |

TABLE 7.4. Parameters for transformer experiment.

The solid dielectric switch (SDS) in the secondary circuit of the transformer contains a 305 mm$^2$ polythene film with a thickness of 1.59 mm. There are a number of 0.75 mm indentations in one side of the film. The switch breakdown voltage between the circular electrodes is 70 kV.

The calculated and measured results obtained from an experiment in which the crowbar switch $S_2$ remained open are presented in Fig. 7.31. From Figs. 7.31a and Fig. 7.31b, it can be seen that the model proposed above determines most of the circuit parameters accurately, although in Fig. 7.31c, the rate-of-change of the secondary current is predicted accurately only for the very short time period prior to electrical breakdown.

### *7.3.5 Optimizing Exploding Foils*

A high voltage must be developed by the exploding foil if it is to drive a transformer-based power conditioning system. Although the parameters of exploding foils that can generate voltages up to six times that of the initial capacitor voltage can be determined experimentally, this is not practical because it is time consuming, costly, and may potentially damage the experimental circuits. A preferred approach is to use theoretical models of the exploding foils and the experimental results obtained for the 17 $\mu$m foils discussed earlier. The parameters of these foils are summarized in Table 7.5, when $C = 238$ $\mu$F, $L_t = 86$ nH, $R_t = 3.0$ m$\Omega$, $L_f = 20$ nH, and $V_0 = 20$ kV for the circuit in Fig. 7.26. The objective of this study is to optimize the exploding foils provided the maximum current is not less than 600 kA. The calculations in Table 7.5 show that the for each width there is an optimal length which gives the maximum voltage.

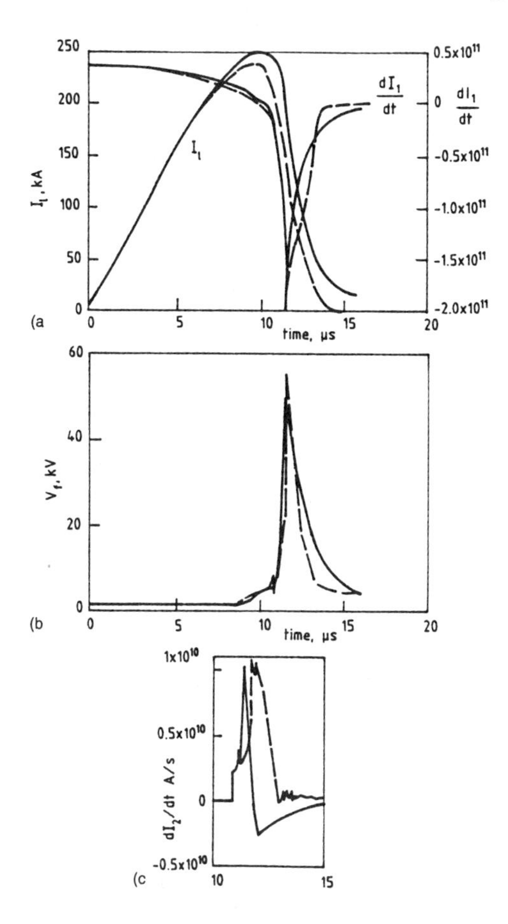


FIGURE 7.31. Results of transformer experiments: (a) primary current, $I_1$, and rate of change of current, $dI_1/dt$, (b) foil voltage, $V_f$, and (c) rate-of-change of transferred current, $dI_2/dt$. The lines are (- - -) experimental and (—) theoretical.

| Width (cm) | Optimal Length (cm) | $V$ (kV) | $I$ (kA) | $dI/dt$ (TA/s) |
|---|---|---|---|---|
| 15 | 22 | 100 | 600 | 1.00 |
| 16 | 30 | 116 | 600 | 1.15 |
| 17 | 40 | 122 | 606 | 1.20 |
| 18 | 39 | 119 | 624 | 1.16 |
| 19 | 40 | 116 | 640 | 1.12 |
| 22 | 38 | 98 | 691 | 0.96 |

TABLE 7.5. Maximum circuit voltages and currents for 17 micron foils.

| $V_m$ (kV) | $t$ at $V_m$ ($\mu$s) | $dI_f/dt$ (A/s) |
|---|---|---|
| 222 (1.48 kV/cm) | 125 | $-9 \times 10^{11}$ |

TABLE 7.6. Foil performance data.

In order to check the model predictions, while protecting the system, an experiment was performed using foils that were 22 cm long, 17 cm wide, and 17 $\mu$m in thickness. The foils were connected to the capacitor bank presented earlier with an initial voltage of 17 kV. The theoretically predicted maximum voltage of 78 kV agreed well with the corresponding measured voltage of 82 kV, and the predicted maximum value of the rate-of-change in current with respect to time of $-0.78$ TA/s was also in good agreement with the measured value of $-0.75$ TA/s. It is important to note that when optimized exploding foils are used in switching experiments, the foils do not generate voltages greater than 80 kV, and thus, do not damage the capacitor bank circuits. Experiments using transformers need to be conducted by using a crowbar technique to protect the power system.

### Crowbar Switching Techniques

The insight gained by investigating flux compressor/exploding-foil circuits led to a novel method for increasing significantly the value of the negative voltage derivative in the output of the power conditioning circuit of Fig. 7.29, when powered by an MCG rather than by a capacitor bank. If the foil and transformer circuit are crowbarred by switch $S_2$, when the foil voltage has reached its maximum value, then the final stage of the circuit is effectively disconnected from the MCG and the early stages of the conditioning circuit operation. The current in the primary circuit will remain approximately the same, but its rate of change with time will become more negative due to the reductions in the total circuit inductance. The crowbarring can be provided either by the armature cone of the MCG or the exploding foil closing switch.

To illustrate the crowbar technique, the dimensions of a foil that yields a maximum voltage of $V_m$ at 2 MA of current were calculated on the basis of the FLEXY I MCG given in Chapter 6 and in [7.9] and the 17 $\mu$m foil data in Fig. 7.18. The foil cross section is the same as that used in the F3 test in Table 7.2, but the length was increased by a factor of six. The foil performance results are presented in Table 7.6, and were somewhat surprising, since the 17 kJ/cm$^3$, i.e., a total energy of 1 MJ, energy density dissipated at the maximum voltage is only about one-third of that expected in capacitor bank experiments to fully vaporize the foil. This also demonstrates the ability of the MCG to deposit large quantities of energy into a resistive load.

Figures 7.32a and 7.32b present results predicted for the transfer of current and the foil voltage in the circuit of Fig. 7.29, when crowbarred but using an MCG as the power source. Figure 7.32c shows that the effect is

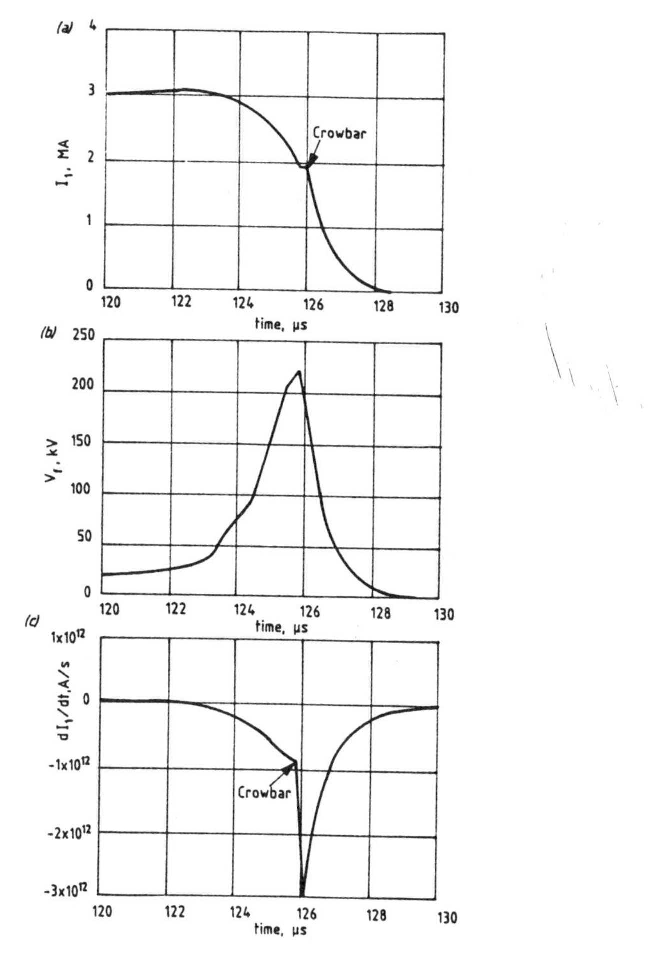


FIGURE 7.32. Predicted waveforms for FLEXY-foil conditioning: (a) FLEXY current, $I_1$, (b) foil voltage, $V_f$, and (c) rate-of-change of FLEXY current, $dI_1/dt$.

to increase the negative rate of change of current to a maximum value of −3 TA/s, which is ample for operating a plasma erosion opening switch (PEOS) connected in the secondary circuit. A foil closing switch (FCS) can serve to connect and close the secondary before the foil voltage has achieved its maximum value.

In summary, both theoretical and experimental data on the performance of foils used in opening and closing switches over a wide range of conditions have been examined. Both capacitor banks and MCGs were used to drive the power conditioning circuits in which these switches were used. In particular, attention was focused on the increase in the resistance ratio as energy is deposited into the foil. It was found that thinner foils required less energy for the same dynamic resistance ratio. It was also found that the reduced rate of energy input by an MCG, as opposed to a capacitor bank, resulted in a reduced voltage generated in the foil. In addition, it was found that both opening and closing switches can be used in simple power conditioning circuits of both capacitor banks and MCGs. These conditioning circuits operate automatically and provide very sharp output current and voltage waveforms. In particular, by predicting the performance of the output conditioning circuits for an MCG using the techniques described in this section, followed by experimental tests with capacitor banks, the probability of a first time firing success can be maximized.

## 7.4 Magnetic Coupling between MCGs

When using MCGs at remote sites, where capacitor banks are unavailable to provide their seed fields, energy multiplication is required, to enable the experiments to be energized by low-power sources such as batteries [7.19, 7.20] or permanent magnets [7.21, 7.22]. The maximum energy multiplication that can be achieved from a single generator is thus of considerable importance. It is shown later in Section 7.5, that if the maximum energy multiplication factor, $k$, is related to the global energy efficiency, $\eta$, by

$$\Lambda = \eta k, \tag{7.16}$$

then $\Lambda$, which is termed the *activity*, is $\leq 100$ for the state-of-the-art generator designs. One of the highest values of $k$ reported in the literature is about 1000 [7.23], which was obtained with only about 4% of the energy stored in the explosive charge being converted into electrical energy. Equation 7.16 confirms that for values of $\Lambda$ limited by present technology, very high values of $k$ are only possible if the generator efficiency is very low. In practice, this would require an unacceptably large generator, with a correspondingly large quantity of high explosive.

To attain the increased energy multiplication required for many applications ($10^4$–$10^6$ or higher), it is necessary that inductive coupling between cascaded generators be introduced. One common method for doing this is to use air-cored transformers [7.24], while another is to use a *dynamic transformer* (also called the *flux-trapping* method) [7.22, 7.25, 7.26]. From a practical standpoint, air-cored transformers are preferable at high energies and dynamic transformers at lower energies [7.27–7.30]. Initially, it was believed that opening switches had to be used in the input circuits of transformers [7.30], and only recently has it been shown that the presence of these switches is not so important [7.22, 7.25, 7.26]. While most experimental work has been done at low energy levels (kJ) [7.19, 7.21, 7.22, 7.25, 7.26, 7.28, 7.30–7.35], some high energy (MJ) experiments have also been conducted [7.25, 7.36].

Dynamic transformers have also been used in the production of high current pulses with short rise times [7.37, 7.38], and, related to this, the

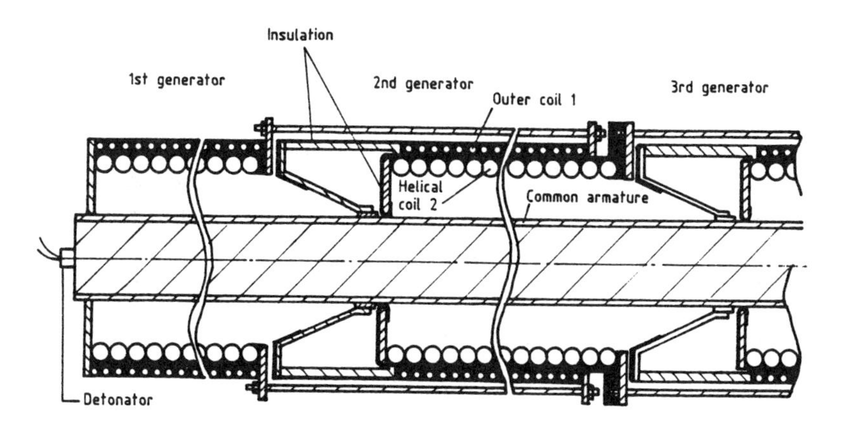


FIGURE 7.33. FLUXAR generating system.

generation of very high voltage pulses for special applications [7.39]. Efficient use of opening switches, such as the exploding foil switch, requires that the current pulse risetime should be less than about 20 $\mu$s [7.40], which is not achievable with most booster generators. When two generators are cascaded through a dynamic transformer, the load current begins to rise only after crowbarring of the second faster generator, so that the required compression time of the output pulse is achieved.

The complexity of dynamic transformers has prevented the development of simple numerical codes. In this section, a complete numerical model for a helical dynamic transformer, called FLUXAR, will be presented. In addition, examples of its use to feed energy into an MCG by either AC or DC excited internal coils, with and without opening switches in its circuit, or by permanent magnets are examined.

## *7.4.1 The FLUXAR System*

The FLUXAR system in Fig. 7.33 was developed at Loughborough University. The first generator is fed by a low-energy seed source and the generators in the cascade are coupled to each other through dynamic transformers. Alternating or pulsed (AC) or direct (DC) current seed sources can be used, and they can either be coupled directly or inductively to the first generator of the FLUXAR chain. The advantages and disadvantages of both methods are described later.

In a single FLUXAR stage, the primary of the transformer (which is the load of the previous stage) forms an outer coil into which the generator is introduced, with the secondary winding being the helical coil of the generator. The operation of a stage can be divided into two distinct phases, which are discussed below.

Phase A: Injection and Capture of the Initial Field

If the seed source of a FLUXAR stage primary is either an alternating or a pulsed source, then an EMF is generated between the crowbar and the armature of the generator. If no steps are taken to limit this by design or to insulate the crowbar electrode (or even omitting the crowbar), then this may cause a premature closure of the secondary circuit, causing the generator to fail.

The diffusion of an outer magnetic field into a hollow conductor and the decay of the field within a hollow cylinder are modeled by the equation

$$H = H_0 \exp\left(-\frac{t}{\tau}\right), \tag{7.17}$$

where $H_0$ is the initial magnetic field. For the majority of the FLUXAR systems, the characteristic diffusion time is approximately [7.25]

$$\tau = \frac{1}{4}(r_a - r'_a)(r_a + r'_a)\mu_0\sigma_0, \tag{7.18}$$

where $r'_a$ and $r_a$ are the internal and external radii of the armature and $\mu_0$ and $\sigma_0$ are the magnetic permeability and electrical conductivity, respectively.

For most systems with a copper armature, the diffusion time is of the order of milliseconds, which is why, with short pulses (tens of microseconds), a circular $\theta$-current distribution is produced on the surface of the armature to mirror that in the outer (primary) coil. The geometry and equations that govern the action during this phase are similar to those used by Miura [7.18] in a $\theta$-type flux compression experiment. When the armature makes normal contact with the crowbar, which is usually synchronized with the peak current in the outer coil, Phase B of the FLUXAR system begins and the trapped (or captured) seed magnetic flux remains in the generator for tens of microseconds.

Phase B: Creation of Field in the Next Stage

Because a larger number of turns are used in the secondary circuit of the transformer (the helical coil of the next generator) than in the primary, the magnetic flux trapped in the secondary circuit is much greater than that in the primary circuit. This leads to the somewhat misleading term *flux multiplication*, since two different circuits are involved. It should be noted however that the magnetic energy captured is less than that stored in the outer coil, so that the flux multiplication is always accompanied by a loss of energy.

Coupling between the helical coil of the generator and the outer coil is maintained if an opening switch is not included in the first circuit. However, no coupling exists between the outer coil and the armature, since it can be viewed as being inside a metal cage formed by the helical coil. Owing to the millisecond time decay of the trapped field, it does not decay significantly during the tens of microseconds of Phase B. Because the load current is zero at the beginning of Phase B, the only important source of EMF is the armature movement through the magnetic field. When the source field is permanent magnets, this is the only source of EMF. This source of EMF continues to play a major role if a crowbar is present, by delaying the shorting of the helical coil. On the basis of the 2D model developed in [7.41] and presented in Chapter 6, the normal generator EMF term in Phase B coexists with three additional EMF terms. One of these is due to the coupling of the armature with the captured magnetic field described above, and the other two to the transformer coupling between the outer and the helical coils of the generator. It should be noted that two $\theta$-distributions of current are now superimposed on the armature, one that mirrors the currents in the outer coil and the other that in the helical coil. The sense in which the superposition occurs is clearly important during the initial stages of operation of the generator. If the generator has a coaxial part, both components disappear during the final coaxial stage of operation of the generator.

If the load of the generator is the outer coil of the next generator stage, a magnetic field and a corresponding $\theta$-current distribution are set up in the armature in the next generator. Operation of the FLUXAR system continues by crowbarring and then capture of the field in this stage, and so on along the chain of generators. As is evident from Fig. 7.33, synchronization of the system is automatic and the common explosive charge is initiated only once. This is a characteristic of the $Z$-type system geometry [7.22, 7.26, 7.28, 7.30, 7.35]. The $\theta$-type geometry, in which the outer coil is simply a single turn coil fed by the previous generator through a parallel plate transmission line, has also been used, but initiation of each generator must be synchronized with the others in the chain [7.19, 7.32–7.34,7.36]

### *7.4.2 FLUXAR Working Equations*

The analysis below uses the same 2D model and notation used in Chapter 6, where the superimposed magnetic flux densities are produced, respectively, by $Z$- and $\theta$-directed currents. According to this model, the armature and both coils of the transformer are decomposed into the same number ($N$) rings for convenience. Such filamentary models have successfully been used to model flux compression [7.18] and electromagnetic launchers [7.42].

The complete equivalent circuit diagrams for a FLUXAR stage in both Phases A and B are presented in Fig. 7.34a and 7.34b. The working equations for each phase are now presented.

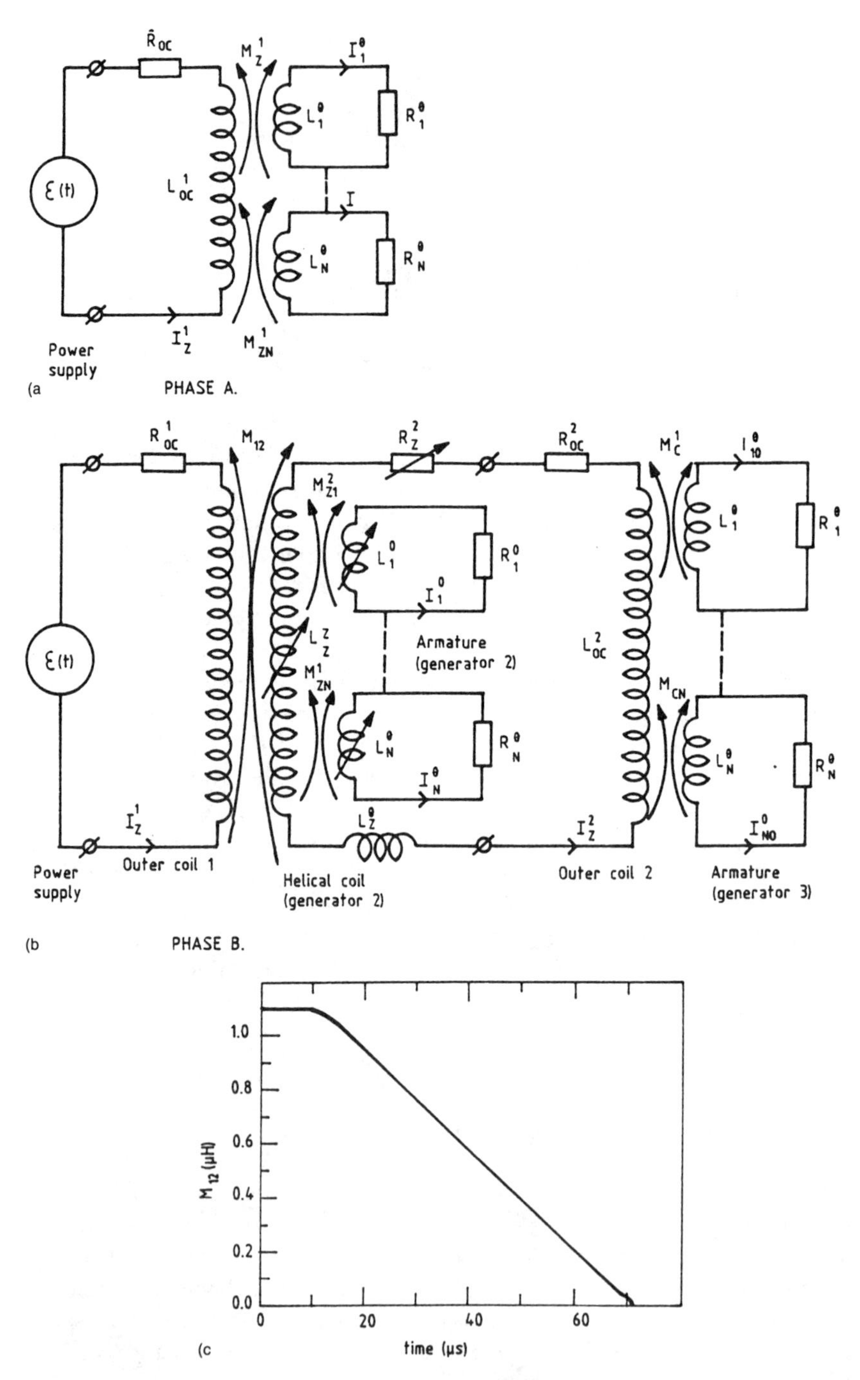


FIGURE 7.34. Equivalent circuit diagram for FLUXAR: (a) Phase A, (b) Phase B, and (c) variation of mutual inductance between transformer elements. Time measured from second generator crowbar.

Phase A: N + 1 Equations

The equations that describe the $Z$- (for the outer primary coil) and $\theta$- (for the armature rings) circuits in Phase A of the dynamic transformer are, respectively

$$\varepsilon(t) = L_{0c}^{1} \frac{dI_z^1}{dt} + \sum_{i=1}^{N} M_{zi}^{1} \frac{dI_i^\theta}{dt} + R_{0c}^{1} I_z^1 \tag{7.19}$$

and

$$0 = L_i^\theta \frac{dI_i^\theta}{dt} + M_{iz}^{1} \frac{dI_z^1}{dt} + \sum_{\substack{j=1 \\ j \neq i}}^{N} \left( M_{ij}^{\theta} \frac{dI_j^\theta}{dt} \right) + R_i^\theta I_i^\theta, \tag{7.20}$$

where $i = 1, 2, ..., N$, $L_{0c}^1$ and $R_{0c}^1$ are the inductance and resistance of the outer coil, respectively, $M_{zi}^1$ is the mutual inductance between an armature ring and the outer coil, $L_i^\theta$ and $R_i^\theta$ are the inductance and resistance of an armature ring, $M_{ij}^\theta$ is the mutual inductance between two armature rings, and

$$\varepsilon(t) = -\left[ \frac{Q(t)}{C} - V_0 + L_i \frac{dI_z^1}{dt} + R_t I_z^1 \right], \tag{7.21}$$

if a capacitor bank with capacitance $C$ and initial voltage $V_0$ provides power to the outer coil of the first generator of the FLUXAR chain. The charge released to the circuit is $Q(t)$, where $dQ/dt = I_z^1$ and the subscript $t$ denotes the transmission line. Since there is no armature movement, all the inductances in Eqs. 7.19 and 7.20 are constant.

When one generator feeds the outer coil of the next generator in the chain, then

$$\varepsilon(t) = -\left[ L_z^1 \frac{dI_z^1}{dt} + \frac{dL_z^1}{dt} I_z^1 + R_z^1 I_z^1 \right], \tag{7.22}$$

where $L_z^1$ includes both the armature $Z$-current inductance and the helical coil inductance. Solving Eqs. 7.19 and 7.20 for the initial currents $I_z^1(0) = I_i^\theta(0) = 0$, one obtains the initial values of the current for Phase B of the transformer: i.e., $I_z^1(1)$ and $I_i^\theta(1)$, where $i = 1, 2, ..., N$.

Phase B: $N$+2 Equations

By analogy with the equations for a simple helical generator, the equations that govern the transformer (neglecting interaction with the armature of the next stage) during Phase B are

$$\varepsilon(t) = L_{0c}^1 \frac{dI_z^1}{dt} + M_{12} \frac{dI_z^2}{dt} + \frac{dM_{12}}{dt} I_z^2 + R_{0c}^1 I_z^1, \tag{7.23}$$

$$L_z^2 \frac{dI_z^2}{dt} + \sum_{i=1}^{N} \left( M_{zi}^2 \frac{dI_i^\theta}{dt} + \frac{dM_{zi}^2}{dt} I_i^\theta \right) + \left( R_z^2 + \frac{dL_z^2}{dt} \right) I_z^2 \tag{7.24}$$

$$+M_{12} \frac{dI_z^1}{dt} + \frac{dM_{12}}{dt} I_z^1 = -\left( L_{0c}^2 \frac{dI_z^2}{dt} + R_{0c}^2 I_z^2 \right),$$

and

$$0 = L_i^\theta \frac{dI_i^\theta}{dt} + \left( \frac{dL_i^\theta}{dt} + R_i^\theta \right) I_i^\theta + \sum_{\substack{j=1 \\ j \neq i}}^{N} \left[ M_{ij}^\theta \frac{dI_j^\theta}{dt} + \frac{dM_{ij}^\theta}{dt} I_j^\theta \right] \tag{7.25}$$

$$+M_{iz}^2 \frac{dI_z^2}{dt} + \frac{dM_{iz}^2}{dt} I_z^2 + \frac{dM_{iz}^1}{dt} I_z^1(1),$$

where the superscript 2 is for the secondary $Z$-circuit [helical generator coil and the second outer coil (load) circuit], $L_z^2 = L_z^z + L_z^\theta$, $L_z^z$ and $L_z^\theta$ are the inductances of the helical coil and the ($Z$-current) armature, respectively, the mutual inductance between the primary and secondary circuits is

$$M_{12} = \sum_{i,j=1}^{N} M_{ij} \left( r_i^c, r_j^{0c}, d_{ij} \right), \tag{7.26}$$

$r_j^{0c}$ and $r_i^c$ are the radii of the $j$th ring of the outer coil and the $i$th ring of the armature of the second generator helical coil, respectively, $d_{ij}$ is the axial separation between the two rings, and $M_{zi}^2$ is the mutual inductance between the $i$th armature ring and the helical coil of the second generator. In Eq. 7.24, it can be seen that in addition to the normal generator terms, there are two further terms related to the coupling ($M_{12}$) with the outer coil. In Eq. 7.25, the supplementary term $dM_{iz}^1/dt I_z^1(1)$ accounts for the coupling of the $i$th armature ring with the trapped magnetic field, rather than $B_i(1) ds_i(t)/dt$ or $2\pi B_i(1) r_i^a dr_i^a/dt$, where $B_i(1)$ is the magnitude of the trapped magnetic field in the region of the $i$th armature ring and the ring is expanding with a rate-of-change of surface $ds_i/dt$ or velocity $v_i = dr_i^a/dt$. In many cases, the extremely small radial variation of the external coil induced magnetic field intensity in the armature region can be neglected, so that $B_i$ is constant. Very accurate calculations of $B_i$ can be

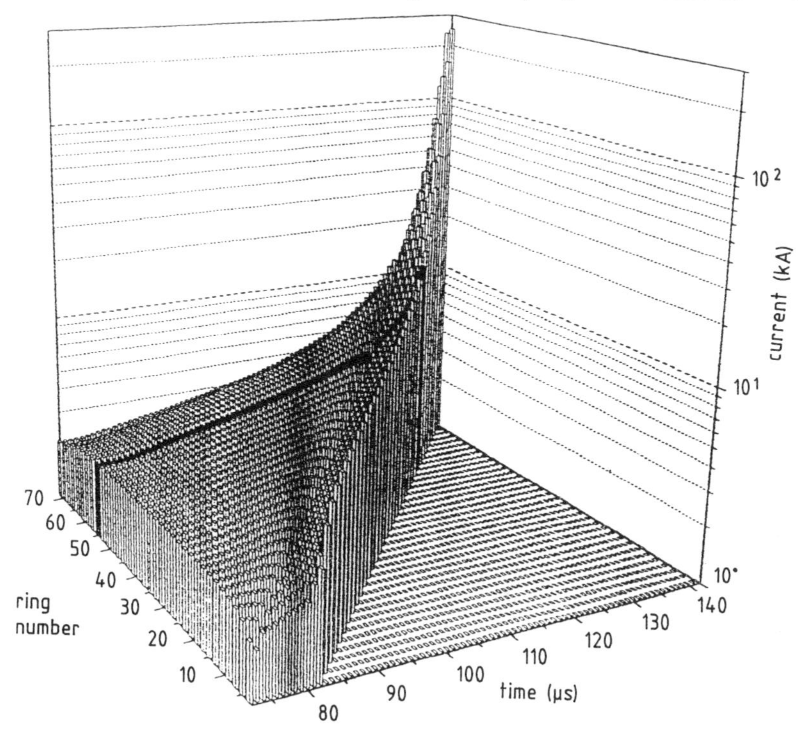


FIGURE 7.35. $\theta$-current history (Phase B only). Time measured from first generator crowbar.

made by adding the contributions of all the rings (both coil and armature) and using the technique proposed by Miura [7.18] and Novac [7.41]. Time variations in the self and mutual inductances of a typical generator were addressed by Novac [7.41], with the exception of $M_{12}$ presented in Fig. 7.34c.

### *7.4.3 FLUXAR Techniques and Performance*

By way of example, the results of the analysis presented above will be applied to a FLUXAR stage using the MCG described in [7.41] and presented in Chapter 6. This will enable one to study the relative gain that can be achieved in using different techniques to inject the same initial magnetic flux into a generator. The outer coils of a stage of the FLUXAR system presented in Fig. 7.33 has a length of 0.5 m, an inside diameter of 74 mm, and a pitch of 160 mm, which gives a total inductance of about 100 nH. The time history of all the currents in this stage are presented in Figs. 7.35 and 7.36 with the initial $\theta$-current at the beginning of Phase B on the left-hand-side of Fig. 7.35.

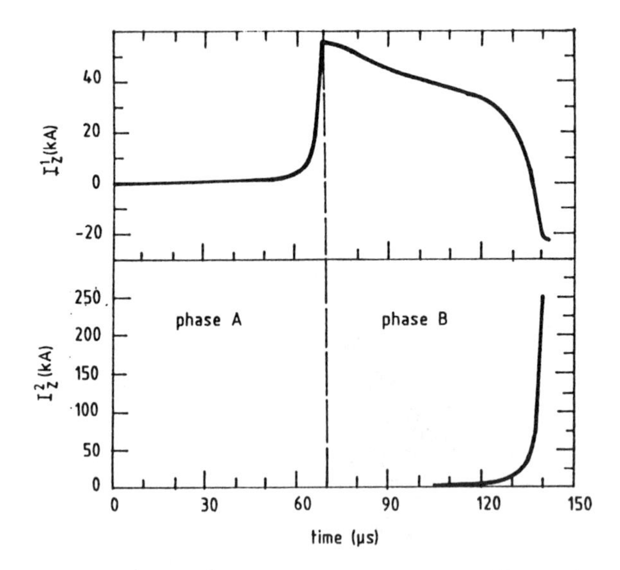


FIGURE 7.36. $Z$-circuit currents for dynamic transformer (Phase A and B). Time measured from first generator crowbar.

### Opening Switch

The influence on the relative gain of the opening time of a switch interrupting the current in the outer coil is shown in Fig. 7.37, and the time variation of the currents when the opening time is chosen to give the maximum gain are presented in Fig. 7.38. To model this effect once the switch has been opened at time $t = t_s$, Eq. 7.23 and the term $M_{12}dI_z^1/dt$ in Eq. 7.24 are discarded. Since the total flux in the generator has to be conserved, the term $I_z^1(t_s)dM_{12}/dt$ remains, with $I_z^1(t_s)$ being the value of the current at the instant the switch begins to open. This term can by replaced by $\sum_{i=1}^{N} 2\pi B_i(t_s)r_i^a dr_i^a/dt$, which represents further compression of the externally produced (trapped and slowly decaying) magnetic field.

In Fig. 7.37, it can be seen that, even with the optimal switch opening time, the relative improvement in the final magnetic flux for this particular system design in only about 12%. Taking into account the technical complications that are involved in adding opening switches, they are only useful with high efficiency systems in order to restrict the number of stages needed and to minimize the amount of explosive required and the overall mass of the system.

### Premature Crowbarring

The influence of electrical breakdown between the crowbar and armature, which prematurely closes the secondary (generator) circuit, on the final performance of the system must be taken into account. Once breakdown

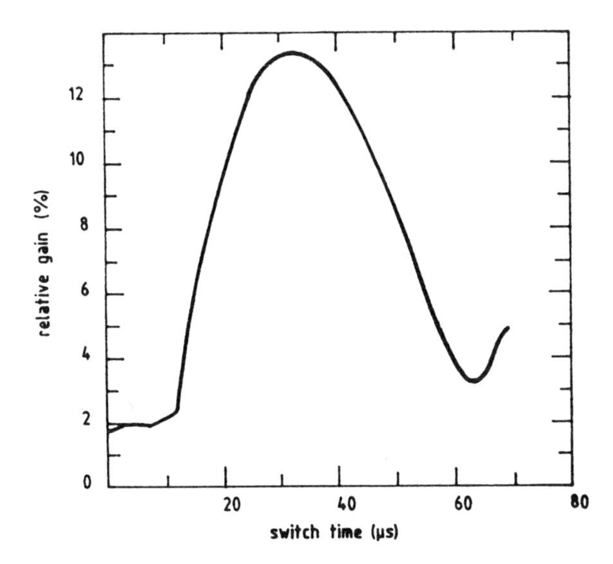


FIGURE 7.37. Gain in final load current (relative to a capacitor bank direct feed case) provided by an opening switch in first circuit. Time scale from second generator crowbar. Peak value corresponds to figure given in Table 7.7.

has occurred at time $t = t_B$, Eqs. 7.19 and 7.20 of Phase A are interchanged with Eqs. 7.23–7.25 of Phase B, where, because the second generator is still unaffected by the explosive, all self and mutual inductances are constant and where $I_z^1(1) = I_z^1(t_B)$ from Eq. 7.25 now represents the current in the outer coil at the moment that breakdown occurs.

The results for the case of breakdown are presented in Fig. 7.39. Most generators achieve the maximum value for $dI_z^1/dt$ at the moment the current $I_z^1$ is about 70% of its peak value. Maximum voltage (given by $M_{12}dI_z^1/dt$) is then generated in the secondary circuit, which most likely occurs at the moment breakdown occurs. The time history of the generator currents resulting from breakdown are shown in Fig. 7.40. The effect this has on the primary current, $I_z^1$, shown in Fig. 7.40, can be explained by considering that, after the breakdown has closed the secondary circuit, the coupling factor between the outer coil and the secondary generator ($k_{12}$) changes rapidly from a low value (due to poor coupling with armature alone in Phase A) to the characteristic high value of Phase B (due to the strong coupling with the helical coil). The *reflected* value of the load inductance seen by the first generator, $L_{load}^1 = L_{0c}^1(1 - k_{12}^2)$, is thus lowered by the breakdown and an increase in the current $I_z$ occurs.

Another interesting feature is related to the generation of a premature small reversal in the current $I_z^2$, which begins to be cancelled once the armature cone makes contact with the helical coil and begins to short out the turns (Fig. 7.40).

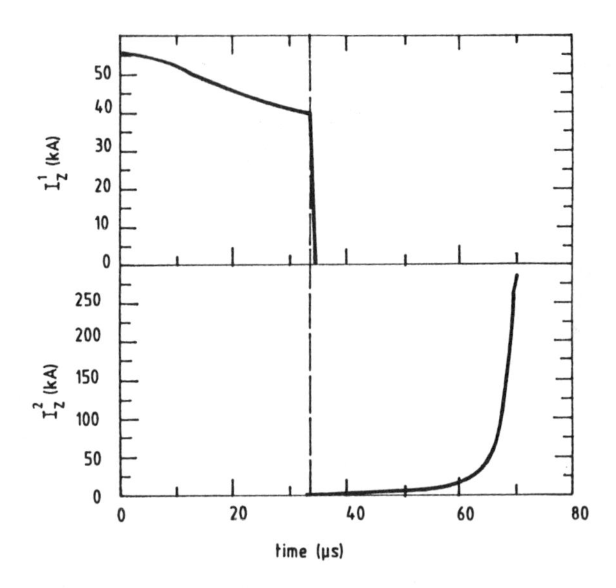


FIGURE 7.38. $Z$-circuit currents for dynamic transformer with optimized opening switch in the first circuit.

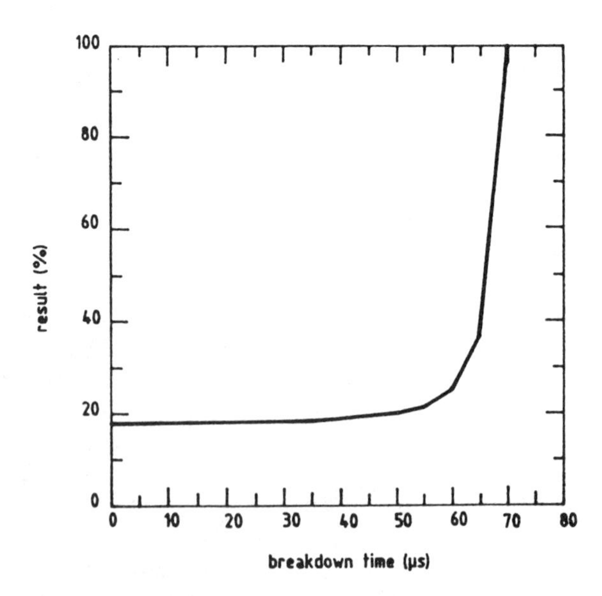


FIGURE 7.39. Relative final load currents obtained with electrical breakdown causing premature closure of secondary circuit.

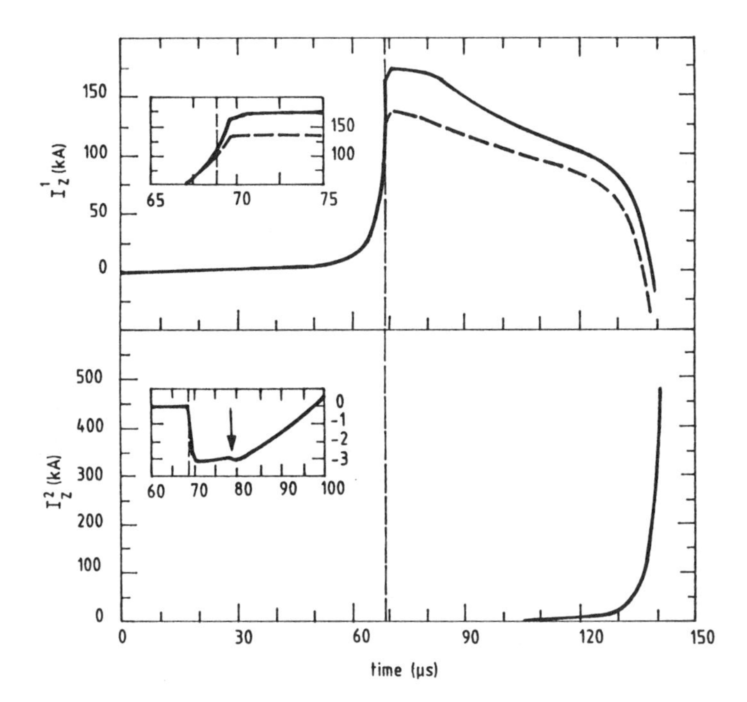


FIGURE 7.40. $Z$-circuit currents for dynamic transformer with electrical breakdown closing the secondary circuit at maximum $dI_{z1}/dt$. Upper window: $I_{z1}$ current with (—) and without (- - -) breakdown. Lower window: effects of breakdown on $I_{z2}$ current: arrow shows initial armature/coil contact.

## Seed Source and Coupling Circuit

The influence of the seed source and/or coupling circuit on the performance of the generator also needs to be taken into account. Although these effects are related to the first generator of the FLUXAR chain, the results are applicable to any helical generator design.

The following seed sources are considered: capacitor banks (AC), batteries (DC), permanent magnets (M), and direct (D) and inductive (I) coupling through an outer coil. In general, a DC source is one that does not induce circular currents in the armature, so that a source providing a very long pulse (with a risetime of many milliseconds) is regarded as a DC source for both direct and inductive coupling. Inductive coupling was also investigated with an opening switch (O) in the outer circuit synchronized for optimal performance and also as a superconducting outer coil (S). It is assumed in all cases that the same amount of magnetic flux is injected initially and captured by the generator, where the direct feed (AC and DC) of the same generator with the same load was investigated.

The results of this investigation are summarized in Table 7.7, which shows the improvements attained in the relative gain of the load current

| Source | Coupling | Improve–ments | $I_z^1$ (kA) | $I_z^2$ (kA) | $I^{\theta(*)}$ (kA) | Relative Gain in Load Current (%) |
|---|---|---|---|---|---|---|
| AC | D | – | – | 1.50 | –1/35 | 0.0 |
| | I | – | – | – | – | 2.0 |
| | | O | 55.6 | 0.0 | –2.52 | 1.3 |
| | | S | – | – | – | 2.6 |
| DC | D | – | – | 1.57 | 0.0 | 35.1 |
| | | – | – | – | – | 2.6 |
| | I | O | 37.0 | 0.0 | 0.0 | 18.6 |
| | | S | – | – | – | 3.0 |
| M | | – | – | 0.0 | 0.0 | 4.4 |

TABLE 7.7. Comparison of the increase in load current obtained using various generator excitation schemes.

over that when the direct feed is from a capacitor bank (cases AC–D). In Table 7.7, the currents $I_z^1$, $I_z^2$, and $I^\theta$ are the initial current in kA, and the * indicates the mean of the $\theta$-currents. For permanent magnets, the equations used are the same as for a DC–I–O system, with the opening switch time coinciding with the crowbar action. The axial distribution of the magnetic field, $B_i(1) = B_i(t_s)$, at this instant was used to represent the permanent magnetic field distribution.

The most efficient method for providing the seed magnetic flux is to use a battery coupled directly to the generator (DC–D), and this also happens to be the simplest method for a very low energy input. Using superconducting coils does not offer any significant advantages, when the high costs of using superconducting materials and coolants is taken into account, except in outer space because of its ability to store very high energy densities.

The results obtained with permanent magnets are in good agreement with those of Bojko [7.21], where the total conservation of magnetic flux of 71.11% is remarkably close to the 71.22% calculated using the 2D model.

### 7.4.4 A Case Study

In this section, the model developed above will be used to model a complex FLUXAR system designed by Lyudaev [7.30]. Most of the required input data are available, as well as sufficient experimental results to enable one to validate the model used.

Phase A

In this phase, $N = 112$ equations are used to describe the armature $\theta$-currents and one to describe the $Z$-current through the first generator and its load. The inductance of a helical coil, made from two different sections, is

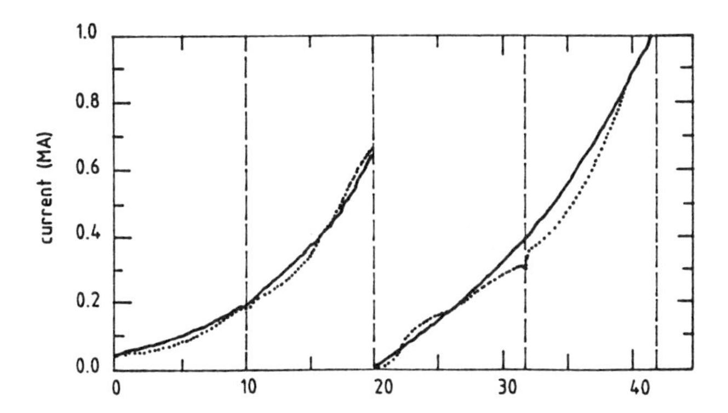


FIGURE 7.41. Calculated (- - -) and experimental (—) $Z$-circuit currents for a high-efficiency FLUXAR system [7.30]. Broken vertical lines show end times for generator sections.

$$L_z = \sum_{i,j=1}^{N} M_{ij}\left(r_i^c, r_j^c, d_{ij}\right) \frac{1}{prl_i^c} \cdot \frac{1}{prl_j^c} \tag{7.27}$$

and the mutual inductance between the outer coil and the second generator is

$$M_{12} = \sum_{i=1}^{\frac{N}{2}} \sum_{j=1}^{N} M_{ij}(r_i^{0c}, r_j^c, d_{ij}) \frac{1}{prl_i^{0c}} \cdot \frac{1}{prl_j^c}, \tag{7.28}$$

where

$$\begin{aligned} prl_i^{0c} &= \frac{N}{2}, \quad i = 1, ..., \frac{N}{2}, \\ prl_j^c &= \begin{cases} \frac{N}{6}, \; j = 1, ..., \frac{N}{2} \text{ (first section)} \\ \frac{N}{4}, \; j = \frac{N}{2}, ..., N \text{ (second section)}, \end{cases} \end{aligned} \tag{7.29}$$

represent the number of parallel wires in the section and $M_{ij}$ is the mutual inductance between two coaxial rings [7.41]. Similar formulas can be derived for $M_{iz}$ and other inductance terms. The total calculated initial inductance is extremely close to the 0.9 $\mu$H reported by Lyudaev [7.30].

Examination of the generator design shows that it does not experience either electrical breakdown or $2\pi$-clocking, so that only 1D magnetic field diffusion needs to be taken into account when calculating the ohmic resistance. Since the coils are made from the same type of conductors, the proximity effect is approximated by a factor of 6.9 [7.43]. From Fig. 2 in [7.30], the initial current is estimated to be 37 kA, and it is assumed that the seed source is a capacitor bank. Examining the diagram of Lyudaev's FLUXAR system, it can be seen that the first generator does not have a

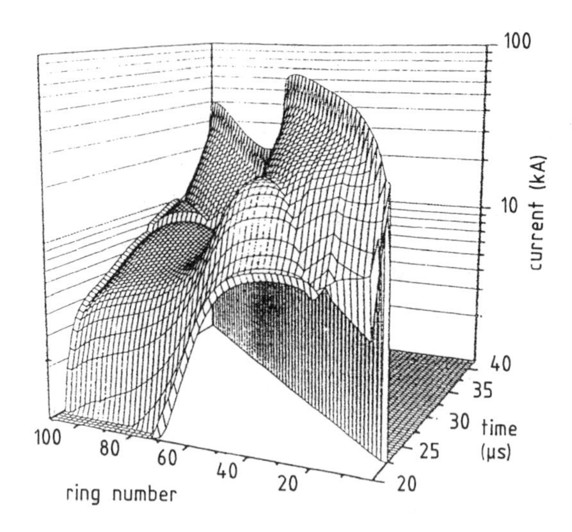


FIGURE 7.42. Second generator armature $\theta$-current history (Phase B only).

crowbar, and therefore, the inductance of the generator at the moment the armature touches the coil ($t = 0$) is only about 0.6 $\mu$H. Using the same reasoning, the $\theta$-currents induced in the armature prior to $t = 0$, e.g., during capacitor bank discharge, is calculated for the moving rings by using time dependent inductances in Eqs. 7.19 and 7.20. In Fig. 7.41, it can be seen that there is excellent agreement between the calculated and experimental $Z$-currents for the first generator ($t = 0$ up to $t = 20$ $\mu$s), which confirms that there are no unusual losses.

Based upon the above calculations, the currents induced in the armature of the second generator at the end of Phase A were calculated. The position of the crowbar of the second generator allows some movement of the armature before the beginning of Phase B ($t = 20$ $\mu$s). Therefore, higher currents are induced in the moving rings. The outer feeding coil covers only one-half of the armature of the second generator and interacts strongly with only one-half of the armature rings. Both effects contribute to the interesting axial current distribution at the beginning of Phase B presented in Fig. 7.42.

Phase B

In this phase, 112 equations are also required to describe the armature $\theta$-currents (Eq. 7.25) and one equation to describe the $Z$-currents (load currents)(Eq. 7.24) together with a supplementary equation to describe the outer coil circuit [Eq. 7.23 with $\varepsilon(t) = 0$]. Calculations cease when the armature/coil contact point arrives at the end of the helical coil ($t = 42$ $\mu$s). In Fig. 7.41, it can be seen that there is excellent overall agreement with the measured data, although the very short negative current predicted at the beginning of Phase B is not evident in the experimental data. The Lyudaev design has very strong magnetic coupling between the concentric coils, and predictions show that an opening switch in the outer coil circuit, if properly activated, would almost double the final current.

## 7.5 Limitations of Helical MCGs

The ratio of the maximum final magnetic energy to the initial magnetic energy, $k = W_m/W_0$, can be written in terms of inductance, $k = L_o/L_l$, where $L_0$ and $L_l$ are the corresponding initial and final inductances, respectively [7.9, 7.22]. The maximum current that can be supported by the generator is $I_m = 2\pi r_c i_m$, where $r_c$ is the radius of the coil and $i_m$ is the maximum linear current density. The global efficiency of the generator is $\eta = W_m/Q$, where $W_m = L_i I_m^2/2$ and $Q = \pi r_{ex}^2 t_m Y$ is the chemical energy stored in an explosive charge having a radius of $r_{ex}$. The quantity $Y$ is called the *intensity of the explosive* and includes the initial mass density, $\rho_0$, the detonation velocity, $D$, and the characteristic heat of detonation, $\Delta H_{ex}$, ($Y = \rho_0 D \Delta H_{ex}$). If the operating time of the generator is $t_m$, then

$$\eta = \frac{2\pi L_l i_m^2}{t_m Y}\left[\frac{r_c}{r_{ex}}\right]^2. \tag{7.30}$$

The basic limitation of energy multiplication by MCGs is the maximum induced voltage that can be sustained without breakdown occurring. If a linear rate-of-change in current is assumed throughout time $t_m$, then $dI/dt = I_m/t_m$ and the initial and final voltages inside the generator are $V_0 = L_0 I_m/t_m$ and $V_m = L_l I_m/t_m$, respectively. The corresponding internal electric fields are $E_0 = V_0/(r_c - r_a)$ and $E_m = V_m/(r_c - r_a)$, where $r_a$ is the outer radius of the armature. If $r_a$ is considered to be close in value to $r_{ex}$, then $k = E_0/E_m$ and the activity of the generator is $\Lambda = \eta k$ is

$$\Lambda = \frac{E_0 i_m R(r-1)}{Y}, \tag{7.31}$$

where $R = r_c/r_a$ is the maximum expansion ratio of the armature of the generator. A high value of the activity clearly denotes a good initial generator design.

The presence of the explosive intensity term in the expression for $\Lambda$ indicates that a better explosive will produce a lower value of the activity, and that to obtain the same activity requires an improved design in which higher electric fields can be sustained. The following range of basic parameters have been reported for a variety of practical MCGs:

- 20 kV/cm $< E_0 <$ 150 kV/cm [7.9, 7.22, 7.23]

- 0.2 MA $< i_m <$ 1 MA/cm [7.9, 7.44, 7.45]
- $Y = \begin{cases} 37 \text{ nitromethane} \\ 88 \text{ composition B [7.46]} \\ 106 \text{ PBX–9404} \end{cases}$
- 2$< R <$ 2.5 [7.9, 7.23, 7.44]

These indicate that the limits for the activity of a helical generator are

$$1 < \Lambda < 100. \tag{7.32}$$

It should be noted that for flux compressors with a constant rate-of-change in inductance with time ($dL/dt$ = const), such as in a single-pitch helical generator, the maximum theoretical value of $\Lambda$ is one [7.47].

As the definition of the activity shows ($\Lambda = \eta k$), a choice can be made between a high-efficiency, low-energy multiplication generator (a very-high-current design) or a low-efficiency, high-energy multiplication generator (a booster). As an example, the best results reported for the first case are $\Lambda = 13$ with $\eta = 30\%$ and $k = 43$ [7.44] and for the second case $\Lambda = 40$ with efficiency $\eta = 4\%$ and $k = 1000$ [7.23].

## 7.6 Summary

The purpose of Chapters 6 and 7 was to relate the theoretical concepts of the previous chapters to the 'real' world. To this end, it was decided to focus on the design and construction of one class of generator, namely those built at Loughborough University. In the next chapter, several specific cases in which MCGs have been use to drive actual loads will be examined. Both the nature of the load and how it influences the properties of the explosive power source will be considered.

**Acknowledgements:** Figures 7.1-7.14 are reproduced from *Engineering Science and Education Journal*, **5**, pp.212-219 (1996), with permission of IEE Publishing Department, Michael Faraday House, Herts., U.K. Tables 7.1-7.6 and Figs. 7.15-7.32 are reproduced from *J Phys D*, **28**, pp. 2619-2630 (1995), with permission of IOP Publishing Ltd., Bristol, U.K. Figures 7.33-7.42 and Table 7.1 are reproduced from *Laser and Particle Beams*, **15**, pp. 397-412 (1997), with permission of Cambridge University Press.

# References


[7.1] B.M. Novac, M. Ganclu, M.C. Enache, I.R. Smith, H.R, Stewardson, and V.V. Vadher. "A Fast Electro-optic High-Voltage Senso," Meas. Sci. Technol., **6**, pp. 241–242 (1995).

[7.2] B.M. Novac, I.R. Smith, H.R. Stewardson, and P. Senior, "Experimental Methods with Flux-compression Generators," Engineering Science & Technology Journal, October, pp. 211–222 (1996).

[7.3] R.E. Reinovsky, I.R. Lindermuth, and J.E. Vorthman, "High Voltage Power Condition Systems Powered by Flux Compression Generators," *Seventh IEEE Int. Pulsed Power Conference*, (eds. B.H. Bernstein and J.P., Shannon, pp. 971–974 (1989).

[7.4] N.F. Roderick, B.J. Kohn, W.F. McCullough, C.W. Beason, J.A. Lupo, and J.D. Letterio, "Theoretical Modelling of Electromagnetically Imploded Plasma Liners," Laser and Particle Beams, **1**, pp. 181–206 (1983).

[7.5] R.E. Reinovsky, "Fuse Opening Switches for Pulsed Power Applications," *Opening Switches* (ed. A. Guenther), Plenum, New York, pp. 209–232 (1987).

[7.6] G.W. Wilkinson and A.R. Miller, "Generation of Sub-microsecond Current Rise Time into Inductive Loads, using Fuses as Switching Elements," *Fifth IEEE Int. Pulsed Power Conf.* (eds. P.J. Turchi and M.F. Rose), pp. 280–282 (1985).

[7.7] J.C. Bueck and R.E. Reinovsky, "High Performance Electrically Exploded Foil Switches," *Fifth IEEE Int. Pulsed Power Conf.* (eds. P.J. Turchi and M.F. Rose), pp. 287–290 (1985).

[7.8] J.V. Parker and W.M. Parsons, "Foil Fuses as Opening Switches for Slow Discharge Circuits," *Fifth IEEE Int. Pulsed Power Conf.*, (eds. P.J. Turchi and M.F. Rose), pp. 283–286 (1985).

[7.9] B.M. Novac, I.R. Smith, H.R. Stewardson, P. Senior, V.V. Vadher, and M.C. Enache, "Design, Construction, and Testing of Explosive-Driven Helical Generators," J. Phys. D: Applied Phys. **28**, pp. 807–823.

[7.10] C.M. Fowler and R.S. Caird, "The Mark IX Generator," *Seventh Institute of Electronic and Electrical Engineers Pulsed Power Conference*, New York, pp. 475–488 (1989).

[7.11] R.E. Reinovsky, I.R. Lindermuth, J.H. Goforth, R.S. Caird, and C.M. Fowler, "High-performance, High-current Fuses for Flux Compression Generator Driven Inductive Store Power Conditioning Applications," *Megagauss Field and Pulsed Power Systems* (eds. V.M. Titov and G.A. Shvetsov), Nova Science, New York, pp. 453–464 (1990).

[7.12] I.R. Lindermuth, J. Brownell, A.E. Greene, G. Nickel, T. Oliphant, and D. Weiss, "A Computational Model of Exploding Metallic Fuses for Multimegajoule Switching," J. Appl. Phys., **57**, pp. 4447–4460 (1985).

[7.13] J.H. Goforth, et al., "Review of the Procyon Explosive Pulsed-power System," *Ninth IEEE Int. Pulsed Power Conf.* (eds. A. Prestwick and R. White), pp. 36–42 (1993).

[7.14] J. Vedel, J. Bernard, J. Boussinesq, J. Morin, C. Nazet, and C. Patou, "Communication Ultra-rapide de Courants Intenses Superieurs au Mega-amperes," Rev. Gen. Electr., **80**, pp. 873–877 (1971).

[7.15] R.E. Reinovsky, J.H. Goforth, A.E. Greene, and J. Graham, "Characteristics of Surface Discharge Switches and High Performance Applications," *Sixth Int. Pulsed Power Conf.* (eds. P.J. Turchi and B.M. Bernstein), pp. 544–547 (1987).

[7.16] D. Kahaner, C. Moler, and S. Nash, *Numerical Methods and Software*, Prentice Hall, New York (1989).

[7.17] R.E. Reinovsky, R.G. Colchester, J.M. Welby, and E.A. Lopez, "Energy Storage Transformer Power Conditioning Systems for Megajoule Class Flux Compression Generators," *Megagauss Technology and Pulsed Power Applications* (eds. C.M. Fowler, R.S. Caird, and D.J. Erickson), Plenum, New York, pp. 575–582 (1987).

[7.18] N. Miura and S. Chikazumi, "Computer Simulation of Megagauss Field Generation," *Megagauss Technology and Pulsed Power Applications* (eds. C.M. Fowler, R.S. Caird, and D.J. Erickson), Nova Science Publ., New York, pp. 137–147 (1987).

[7.19] J.E. Vorthman, C.M. Fowler, R.F. Hoeberling, and M.F. Fazio, "Battery-Powered Flux Compression Generator System," *Megagauss Fields and Pulsed Power Systems* (eds. V.M. Titov and G.A. Shvetsov), Nova Science Publ., New York, pp. 437–440 (1990).

[7.20] A.S. Kravchenko, S.T. Nasarenko, A.M. Shuvalov, L.T. Burenko, V.V. Chernyshev, and D.N. Nevzorov, "Explosive Magnetic Generator-Based Self-Contained Energy Source," *Megagauss Magnetic Field Generation and Pulsed Power Applications* (eds. M. Cowan and R.B. Spielman), Nova Science Publ., New York, pp. 583–589 (1994).

[7.21] B.A. Bojko, V.E. Gurin, R.Z. Lyudaev, and A.I. Pavlovskii, "Autonomous Cascade MC-System with Constant Magnets," *Megagauss Magnetic Field Generation and Pulsed Power Applications* (eds. M. Cowan and R.B. Spielman), Nova Science Publ., New York, pp. 467–473 (1994).

[7.22] A.I. Pavlovskii, R.Z. Lyudaev, V.A. Zolotov, A.S. Seryoghin, A.S. Yuryzhev, M.M. Kharlamov, A.M. Shuvalov, V. Yegurin, G.M. Spirov, and B.S. Makaev, "Magnetic Cumulation Generator Parameters and Means To Improve Them," *Megagauss Physics and Technology* (ed. P.J. Turchi), Plenum, New York, pp. 557–583 (1980).

[7.23] V.K. Chernyshev, E.J. Zharionov, V.A. Demidov, and S.A. Kazakov, "High Inductance Explosive Magnetic Generators with High Energy Multiplication," *Megagauss Physics and Technology* (ed. P.J. Turchi), Plenum, New York, pp. 641–649 (1980).

[7.24] R.Z. Pavlovskii, Lyudaev, L.N. Pljashkevich, A.M. Shuvalov, A.S. Kravchenko, Yu. I. Plyuschchev, D.I. Zenkov, V.F. Bukharov, V. Yegurin, and V.A. Vasyukov, "Transformer Energy Output Magnetic Cumulation Generators," *Megagauss Physics and Technology* (ed. P.J. Turchi), Plenum Press, New York, pp. 611–626 (1980).

[7.25] V.K. Chernyshev and V.A. Davydov, "Generation of the Magnetic Flux by Multicascade Capture," *Megagauss Physics and Technology* (ed. P.J. Turchi), Plenum Press, New York, pp. 651–655 (1980).

[7.26] A.I. Pavlovskii, R.Z. Lyudaev, L.I. Sel'chenkov, A.S. Seryoghin, V.A. Zolotov, A.S. Yuryzhev, D.I. Zenkov, V. Ye Gurin, A.S. Borkiskin, and V.F. Basmanov, "A Multiwire Helical Magnetic Cumulation Generator," *Megagauss Physics and Technology* (ed. P.J. Turchi), Plenum Press, New York, pp. 585–593 (1980).

[7.27] V.K. Chernyshev, V.A. Davydov, and V.E. Vaneev, "The Investigation of Magnetic Cumulation Process in the Magnetic Flux Interception System," *Ultrahigh Magnetic Fields Physics Techniques and Applications* (eds. V.M. Titov and G.A. Shvetsov), Nauka, Moscow, pp. 278–280 (1984).

[7.28] E.I. Bichenkov, S.D. Gilev, S. Prokopiev, V.I. Telenkov, and A.M. Trubachev, "Cascade MC-Generation with Flux Trapping," *Megagauss Technology and Pulsed Power Applications* (eds. C.M. Fowler, R.S. Caird, and D.J. Erickson), Plenum Press, New York, pp. 377–388 (1987).

[7.29] V.K. Chernyshev, E.I. Zharinov, V.E. Vaneev, A.I. Ionov, V.N. Buzin, and Y.G. Bazanov, "Effectiveness Comparison of Explosive Magnetic Cascade Systems," *Megagauss Fields and Pulsed Power Systems* (eds. V.M. Titov and G.A. Shvetsov), Nova Science Publ., New York, pp. 355–365 (1990).

[7.30] R.Z. Lyudaev, A.I. Pavlovskii, A.S. Yuryzhev, and V.A. Zolotov, "MC-Generators with Magnetic Flux Trappers (Dynamic Transformers)," *Megagauss Magnetic Field Generation and Pulsed Power Applications* (eds. M. Cowan and R.B. Spielman), Nova Science Publ., New York, pp. 607–618 (1994).

[7.31] I. Ursu, M. Ivascu, B. Novac, V. Zoita, V. Zambreanu, D. Preotescu, M. Butuman, and A. Radu, "Pulsed Power from Helical Generators," *Megagauss Fields and Pulsed Power Systems* (eds. V.M. Titov and G.A. Shvetsov), Nova, New York, pp. 403–410 (1990).

[7.32] A.I. Pavlovskii, R.Z. Lyudaev, A.S. Yuryzhev, S.S. Pavlov, G.M. Spirov, and N.P. Biyushkin, "MC-2 Multisectional Generator," *Ultrahigh Magnetic Fields Physics Techniques and Applications* (eds. V.M. Titov and G.A. Shvetsov), Nauka, Moscow, pp. 312–320 (1984).

[7.33] R.E. Reinovsky, P.S. Levi, and J.M. Welby, "An Economical 2-Stage Flux Compression Generator System," *5th IEEE Pulsed Power Conference*, pp. 216–219 (1985).

[7.34] I. Ursu, M. Ivascu, A. Ludu, B.M. Novac, I. Panaitescu, D. Preotescu, R. Radu, N. Verbuta, V. Zambreanu, and V. Zoita, "Helical Magnetic Flux Compression Generators," *Megagauss Technology and Pulsed*

*Power Applications* (eds. C.M. Fowler, R.S. Caird, and D.J. Erickson), Plenum Press, New York, pp. 389–396 (1987).

[7.35] V.E. Fortov, Y.V. Karpushin, A.A. Leontyev, V.B. Mintsev, and A.E. Ushnurtsev, "Testing of Compact Magnetocumulative Generators with Flux Trapping," *Megagauss Magnetic Fields Generation and Pulsed Power Applications* (eds. M. Cowan and R.B. Spielman), Nova Science Publ. New York, pp. 947–954 (1994).

[7.36] M. Cnare, R.J. Kaye, and M. Cowan, "An Explosive Generator of Cascaded Helical Systems," *Ultrahigh Magnetic Fields Physics Techniques and Applications* (eds. V.M. Titov and G.A. Shvetsov), Nauka, Moscow, pp. 50–56 (1984).

[7.37] G.A. Shvetsov, G.A. and A.D. Matrosov, "Explosive Magnetocumulative Generator with Outer Excitation," *Ultrahigh Magnetic Fields Physics Techniques and Applications* (eds. V.M.. Titov and G.A. Shvetsov), Nauka, Moscow, pp. 263–264 (1984).

[7.38] R.S. Caird and C.M. Fowler, "Conceptual Design for a Short-Pulse Explosive-Driven Generator," *Megagauss Technology and Pulsed Power Applications* (eds. C.M. Fowler, R.S. Caird, and D.J. Erickson), Plenum Press, New York, pp. 425–431 (1987).

[7.39] A.G. Zherlitsin, V.P. Isakov, M. Vlopatin, G.V. Melenikov, V.B. Mintsev, S.A. Timchenko, V.E. Fortov, and B.T. Tsvetkov, "High Voltage Pulse Generation Using an Explosive Magnetic Generator with Axis Initiation," *Megagauss Fields and Pulsed Power Systems* (eds. V.M. Titov and G.A. Shvetsov), Nova Science Publ., New York, pp. 607–613 (1990).

[7.40] B.M. Novac, H.R. Stewardson, I.R. Smith, and P. Senior, "Analysis of Helical Generator Driven Exploding Foil Opening Switch Experiments," *10$^{th}$ IEEE Pulsed Power Conference* (eds. W. Baker and G. Cooperstein), Vol. 2, pp. 1182–187.

[7.41] B.M. Novac, I.R. Smith, M.C. Enache, and H.R. Stewardson, "Simple 2-Dimensional Model for Helical Flux–Compression Generators," Laser and Particle Beams (1997).

[7.42] R.J. Kaye, E.L. Brawley, B.W. Duggin, E.C. Cnare, D.C. Rovang, and M.M. Widner, "Design and Performance of a Multi-Stage Cylindrical Reconnection Launcher," IEEE Trans. on Magnetics, **27**(1), pp. 596–600 (1991).

[7.43] A.H.M. Arnold, "The Resistance of Round-Wire Single-Layer Inductance Coils," Proc. IEE, **98**, pp. 94–100 (1951).

[7.44] J. Morin and J. Vedel, "Generators de Courants Intenses par Conversion d'energie Explosive en Energie Electrique," C.R. Acad. Sci., Tome 272, Serie GB, pp. 1232–1235 (1971).

[7.45] B.L. Freeman, M.G. Sheppard, and C.M. Fowler, "A Numerically Designed, Experimentally Tested, High-Current Coaxial Generator," *Megagauss Magnetic Field Generation and Pulsed Power Applications* (eds. M. Cowan and R.B. Speilman) Nova Science Publ., New York, pp. 565–572 (1994).

[7.46] B.M. Dobratz, "Properties of Chemical Explosive and Explosive Simulants," Lawrence Livermore National Laboratory Report, UCRL-51319 (Rev 1) (1974).

[7.47] D.B. Cummings and M.J. Morley, "Electrical Pulses from Helical and Coaxial Explosive Generators," *Megagauss Magnetic Field Generation by Explosives and Related Experiments* (eds. H Knoepfel and F. Herlach), pp. 451–471 (1966).

# 8
# Applications: Lasers and Microwaves

In this chapter, two specific applications in which MCGs have been used as the power source are examined. This chapter is not intended to be an extensive literature review, but rather a discussion based only on a limited number of published papers that focus on the physics of the MCG used in these particular applications.

The use of MCGs for a particular application depends, of course, on the nature of the load. Many different loads have been considered in the literature, including electromagnetic launchers (railguns), high-power lasers, particle accelerators, high-power microwave sources, lightening simulators, capacitor banks, shock wave sources, neutron sources, and X-ray sources. For example, most of these loads have been discussed in papers published in the proceedings of the Megagauss conferences [8.1–8.6] that have taken place about every four years since the 1960s. However, only two types of loads—high-power lasers and high-power microwave sources—are discussed in this chapter. The two laser systems for which MCGs have been used as the power source, and which have been built and tested, are the neodymium glass laser and the photodissociation iodine laser. Recently, it has been reported that there are two ways for generating high power microwaves:

- conventional methods based on the use of high-power microwave tubes such as the virtual cathode oscillator, multiwave Cerenkov generator, magnetically insulated linear oscillator, and relativistic backward wave oscillator.

- unconventional methods such as the transition radiation generator and direct-drive devices,

In the case of tube devices, the virtual cathode oscillator and the magnetically insulated linear oscillator are the most suitable for use with the MCG. In the case of the direct-drive devices, the energy is delivered from the MCG to an antenna and then radiated. This particular scheme is interesting, since these microwave sources can be very compact and lightweight.

## 8.1 Lasers

In this section, two laser systems, in which MCGs have been used as the prime power sources, are discussed. These two laser systems are the solid state neodymium glass laser and the photodissociation iodine laser, both of which were developed, built, and tested at the All-Russia Scientific Research Institute of Experimental Physics (Arzamas-16). These are not the only laser systems that can be powered by MCGs, but they are the only ones that are considered.

### *8.1.1 Neodymium Solid-State Lasers*

The primary difficulty in using an MCG to drive a solid-state laser system is that a prolonged pulse of input energy (milliseconds) is required, that is substantially longer than the operating time of the MCG (microseconds). To circumvent this problem, the MCG is used to charge an intermediate inductive energy store, which, in turn, delivers electrical pulses with the proper waveform to the pumping lamps of the laser. The required pumping time depends on the attenuation constant of the "inductive store-lamp" circuit [8.7].

Arzamas-16 built its first MCG powered neodymium glass laser system in 1968. This consisted of 25 modules and generated 36 kJ of laser energy in a pulse with a length of 600 $\mu$s. Arzamas-16 built a second multichannel laser system that consisted of neodymium glass laser rods with a diameter of 45 mm and a length of 920 mm, a coaxial pumping lamp, and a parallel mirror cavity. The coaxial construction of the pulsed lamps was chosen so that the modules could be densely packed into the laser system and it consisted of 48 modules mounted in eight metallic cassettes. This construction also meant that the lamp surfaces could be placed very close to the active laser rods, so that their energy could be more efficiently coupled into the rods. The energy output of each module was 2 to 3 kJ in a pulse having a length of 1 ms.

In [8.7], the authors briefly describe the three experimental setups shown in Fig. 8.1. The first of these, Fig. 8.1a, used a 6 mF, 1.6 MJ capacitor bank to power the laser system, the second, Fig. 8.1b, a C-160 model MCG, and the third, Fig. 8.1c, a H-320 model MCG. For the case of the C-160 setup, the MCG delivered 3 MJ of electric energy to the 3 $\mu$H load. The energy

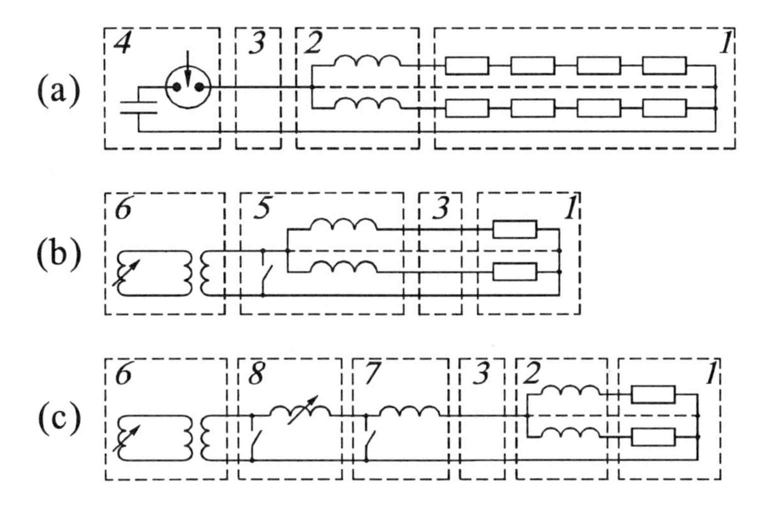


FIGURE 8.1. Equivalent circuit diagram for three glass laser facilities: (a) capacitor bank, (b) C-160 MCG, and (c) H-320 MCG. The major components are: 1 – laser lamps, 2 – inductors, 3 – energy transmission line, 4 – capacitor bank, 5 – switch and inductors, 6 – C-160 MCG, 7 – switch and storage inductors, and 8 – H-320 MCG.

from the MCG was initially stored in an inductive store that consisted of individual solenoids connected in series with a laser lamp. The solenoid–lamp chains were connected in parallel. After the MCG completed its operation, it was shunted with an explosive switch. In experiments conducted with a 25 module laser system, the total inductance of the system was 4.2 $\mu$H. A one megaamp current pulse with a risetime from zero to the peak of the current pulse at 200 $\mu$s and a decay time of 600 $\mu$s was delivered to the laser system. The laser generated 33 kJ of light energy with a full-width, half-maximum pulse length of 600 $\mu$s.

For the case of the H-320 setup, the major portion of the energy store inductance was in a copper plate loop, having a width of 500 mm and a thickness of 2 mm. A small portion of the inductance was in the 10 $\mu$H solenoids that were coupled to individual lamps to separate the lamps inductively. The total inductance of the solenoids, lamps, and transmission line was 0.4 $\mu$H. The inductance of the loop could be changed in order to change the input time of the energy into the lamps. The MCG was connected to the laser system with cables, having a resistance of 0.7 mOhm and a working voltage of 30 kV at a frequency of 5 kHz. In one experiment, the H-320 generator delivered 1.27 MA to a 2.7 $\mu$H storage loop inductance. The average current delivered to each lamp was about 27 kA. The total energy delivered to the lamps was 1.75 MJ, 4.1% of which, i.e., 72 kJ, was released as light energy in a pulse having a full-width, half-maximum pulse length of 640 $\mu$s. In a second experiment, a peak current of 1.45 MA was delivered to a 1.6 $\mu$H loop inductance. Each lamp received 31 kA. The total energy delivered to the lamps was 1.65 MJ, 3.8% of which, i.e., 62 kJ, was released as light energy in a pulse having a full-width, half-maximum pulse length of 540 $\mu$s.

The above experiments demonstrated that it is feasible to use MCGs to power solid-state laser systems. It was pointed out in [8.7] that the conditions under which these tests were done were not optimal. The resistance of the transmission line needed to be reduced, and the load inductance of 2–3 $\mu$H did not match that of the H-320 generator. If the impedance was matched, then the dimensions of the MCG could be reduced. It was found that the length of the electrical pulse delivered to the laser could be increased to 2–3 ms, provided that the Q-factor of the inductive store was increased to its optimal value.

### *8.1.2 Photodissociation Iodine Laser*

A pulsed light source (PLS) was first demonstrated when a C-160 MCG was used to provide megaamps of current to a plasma load to create a quasistationary self-compressed discharge with time-varying parameters [8.8]. Since those initial experiments, multichannel PLS systems powered by cascaded MCGs have been built and tested.

In the block diagram presented in Fig. 8.2, the MCG system was a three-stage cascade consisting of the C-80, C-180, and H-320 MCGs, interconnected using transformers. The load, an electrodischarge generator (EDG), was coupled to the final stage of the MCG cascade, i.e., the H-320, with a cable transmission line designed to tolerate voltages up to 150 kV and current pulses with peak values up to 50 MA. Their inductance was 30 nH, and their resistance at 10 kHz was 0.2 mohm.

A multichannel pulsed discharge was created in the EDG by a strong electrical discharge. The EDG was filled with a mixture of $CF_3$, $C_3F_7I$, $SF_6$, $CO_2$, and noble gases. The discharge chamber of the EDG consisted of electrically independent sections. The electrical energy from the MCG was converted into light energy in each autonomous section of the EDG. The active portion of each section had a length of 1.5 m and a diameter of 0.9 m.

In order to optimize the energy transfer to the load, a technique was developed to close electrically the MCG–EDG circuit at the moment the cascade began to operate. Thin metallic conductors located in the intra-electrode region of the discharge chamber were used for this purpose. A high-voltage source and a low-power capacitor helped to initiate the cascaded MCG system.

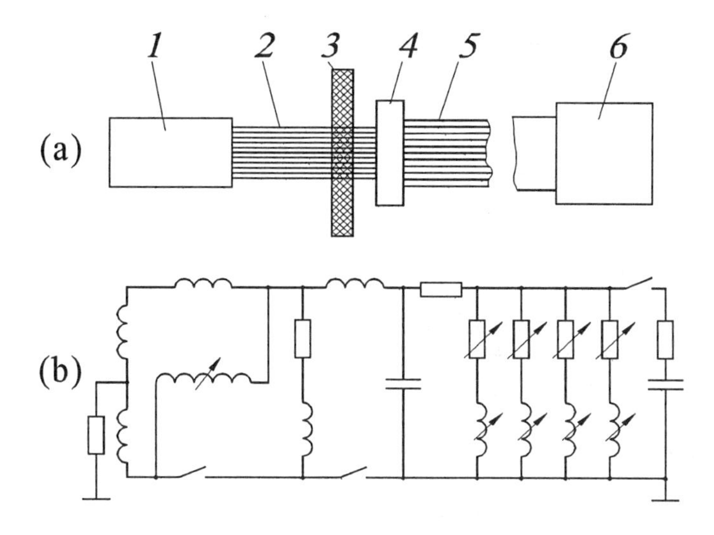


FIGURE 8.2. (a) Block diagram and (b) equivalent circuit diagram for MCG–PLS system. Block diagram: 1 – MCG, 2,5 – transmission lines, 3 –shielding, 4 – collector, and 6 – load. Equivalent circuit diagram: $L_3$ – H-320; $C_1$, $R_3$, and $L_5$ – transmission line parameters; $K_1$ and $K_2$ – switches; and $R_2$ and $L_4$ – pulse sharpening device parameters.

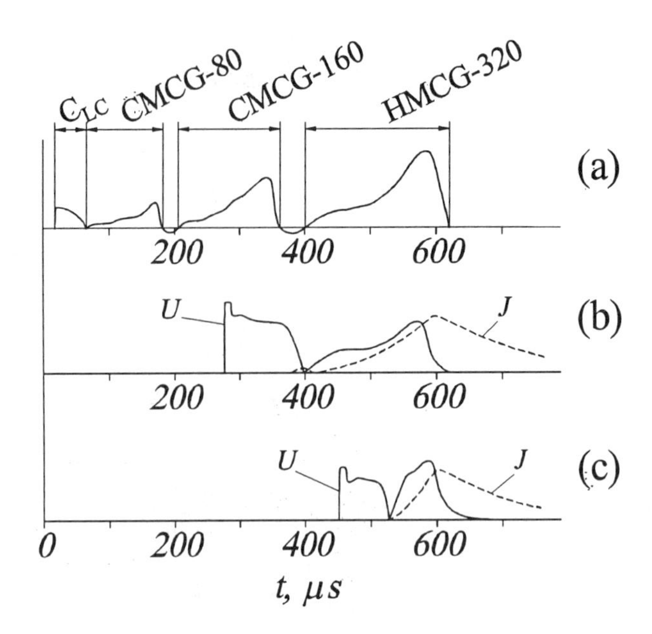


FIGURE 8.3. Time history of cascaded MCG–PLS system: (a) time history of the cascade, (b) time history of switching H-320 MCG into load, and (c) time history of capacitor bank.

The time lines for operation of the MCG cascade are presented in Fig. 8.3. These plots show the turn-on times of the three stages of the cascade, the behavior of the final stage during its operation, and the shape of the output pulse.

It was confirmed in one experiment that 14 MJ of electrical energy was transferred through 35 m of low loss transmission line to the load during the operating time of the final stage of the cascade, i.e., 200 μs. Approximately 0.4% of the chemical energy from the high explosive was converted into light energy. This efficiency was not optimized and could have been significantly increased by optimizing the operating parameters and the impedance matching between the MCG, transmission line, and load.

## 8.2 High-Power Microwave Sources

In this section, several high-power microwave (HPM) sources that have been powered by MCGs are discussed. These sources include the *virtual cathode oscillator (vircator)*, *relativistic Cherenkov generator*, and *magnetically insulated linear oscillator (MILO)*. In addition, a new electromagnetic energy source that could potentially be driven by an MCG is the *transition*

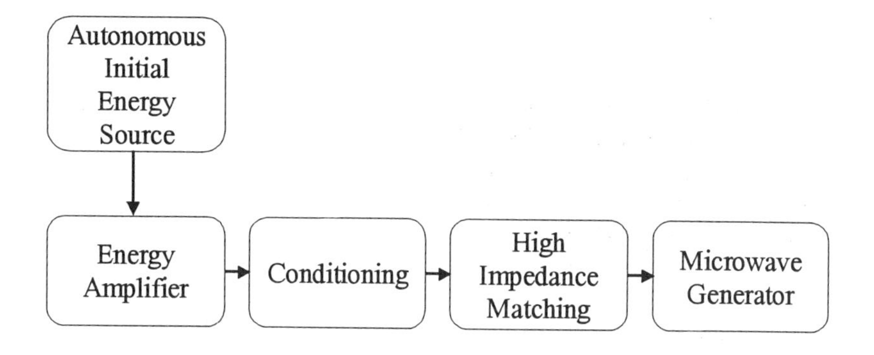


FIGURE 8.4. Block diagram of an autonomous power source for a high-power microwave generator.

*radiation generator (TRG),* and this is also discussed. As in the previous section on lasers, only a brief description of the load is presented, since the goal is to focus on the MCG and the role that it plays. However, prior to discussing these various generators, the requirements for a lightweight, compact autonomous single-shot power supply for microwave sources will be discussed.

### 8.2.1 Autonomous Power Supplies for Microwave Sources

In designing lightweight, compact autonomous single-shot power supplies for microwave sources [8.9], special attention must be paid to the conditioning and matching elements of the overall system due to the fast-rising, high-voltage output pulse that is required. There are fundamentally two situations in which explosive-driven flux compression generators are used as energy sources: proof-of-principle experiments (nuclear fusion experiments and X-ray simulators) that require tens of megajoules of energy and single-shot rocket borne experiments that require lightweight, compact systems. This section addresses the technical issues associated with the second situation.

Taking, as an example, the MILO [8.10], an autonomous high-power source is required to provide a fast-rising, 500 kV pulse to a high resistance (Ohms) load and to maintain the voltage as constant as possible for at least hundreds of nanoseconds. The basic scheme for achieving this is shown in Fig. 8.4. The technologies for each block in this diagram are discussed below, followed by a discussion of various system designs. All the computational results presented in this section were obtained using the previously verified computer models presented in Chapter 6.

Major Components

*Helical MCGs*

Two types of helical MCGs will be considered for use in the power source. The first is a slow HMCG (SHMCG), which has a pulse length of tens to hundreds of microseconds, with a simple multisection helical coil. Other features include initiation at only one end, no armature or coil shaping, and a crowbar located close to the armature. The input current is either fed directly, with the armature closing the circuit, or through a dynamic transformer (DT) [8.11]. The action of the SHMCG is decoupled from that of the other components of the system. Its role is simply to amplify the energy from the seed source and to compress the final flux into a static inductive load, which can be either a ballast inductor, $L_b$, when an explosive forming fuze is used to condition the output, or the external coil of a DT, when a fast HMCG (FHMCG) (see below) is used to sharpen the output pulse. It is assumed that the SHMCG has the following conservative parameters:

- energy multiplication – $k = 100$,
- flux conservation – $\lambda = 0.30$,
- global (chemical to electromagnetic) energy efficiency – $\eta = 4\%$,
- explosive charge initial density – $\rho = 1.8$ g/cm$^3$,
- explosive heat of detonation – $q = 4.5$ MJ/kg,
- maximum liner current density – $\Delta \leq 0.2$ MA/cm,
- expansion (coil/armature diameter) ratio: $(d_c/d_a) \approx 2$.

It follows from these data that the ratio of the initial inductance to the final (ballast) inductance is $L_g(t = 0)/L_b = k/\lambda^2 = 1100$, the ratio of the total explosive mass to the maximum energy delivered to the load is $M_{\text{ex}}/W_m = 1/\eta q = 5.6$ kg/MJ, and a first estimate of the armature diameter is $d_a = I_m/\pi\Delta$, where $I_m$ is the maximum current. The length of the device depends on the maximum allowable internal voltage and can be found by using the 2D numerical design code in Chapter 6.

The FHMCG is a fast pulse (less than 10 microseconds) helical MCG having the same basic design described by Caird and Fowler [8.12]. When a number of technological problems have been solved relating to the very-high-voltage stresses induced with the generator, the duration of the output pulse can be less than 1 $\mu$s. The solution to this problem appears to be to used magnetic self-insulation in a high vacuum vessel [8.13], when the resulting very compact device requires very little initial energy. It should be noted that whenever a FHMCG is used, the initial field is established by

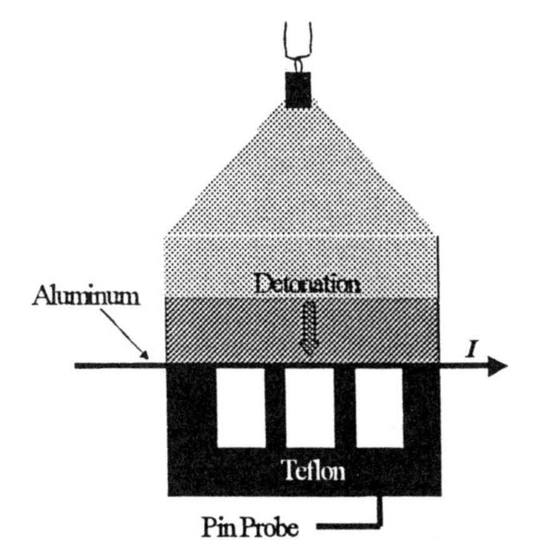


FIGURE 8.5. Schematic drawing of an explosively formed fuse (courtesy of Los Alamos National Laboratory).

an outer cylindrical coil powered through a DT from a SHMCG. Because of the low initial energy requirement, the SHMCG can be considered to be part of the seed source for the system.

A slower, but still compact, version of the FHMCG that does not require a vacuum has already been shown to provide an output of 10 MA with a risetime less than 10 $\mu$s [8.14]. Alternative designs also exist for fast HMCGs [8.15].

Although complex, the numerical simulation of these generators have been performed by two of the authors Novac and Smith [8.16]. To simplify this discussion, it is assumed that after crowbarring the time dependence of the inductance can be described by the relationship: $L_g(t) = DL(t-T) + LF$, where $DL$ is the time rate of change of inductance ($DL < 0$) and is considered to be constant, $T$ is the compression time, $t$ is the time measured from the crowbar action, and $LF$ is the final residual inductance, which usually has a very small value of about 1 nH. The generator resistance, including all losses, is assumed later to be much less than 10 m$\Omega$. It is estimated that the length of the FHMCG will be less than 0.5 m [8.12] and that the diameter will depend on the maximum current that is required.

*Explosively Formed Fuse*

The explosively formed fuse (EFF) (Fig. 8.5) is the best opening switch available for medium energy applications [8.17], since it can conduct megaamps of current for tens of microseconds with negligible loss, before opening in a few microseconds. Its main advantages over electromagnetically exploded foils are that the action is independent of the power source, allowing for a separate design, and that a small supplementary amount of energy is introduced as the result of flux compression occurring during the explosive forming stage. The EFF uses high explosives to extrude a solid conductor, which would otherwise conduct current for a very long time. When the switch is activated, the cross section of the conductor is reduced sufficiently fast that it behaves like a conventional electrically exploding fuse in which the resistance rapidly increases. The EFF has been identified as a "command" fuse in that the time at which it opens is related to the time of detonation of the high explosive.

A conservative set of design data for a standard EFF [8.17,8.18] Teflon made groove (6.5 mm wide, 1.5 mm die, and 13 mm deep) is

- voltage – 6.5 kV,
- linear current density – 0.15 MA/cm,
- deposited energy – 0.6 kJ/cm width.

For small EFFs (linear dimensions up to some tens of centimeters), it is simpler, from the point of view of explosive techniques, to use a plane geometry, while cylindrical geometry provides the best solution for large EFFs.

*Plasma Opening Switch*

The plasma opening switch (POS) is the fastest opening switch available that is capable of producing multimegavolt output. Of particular interest is the plasma opening erosion switch (PEOS), which usually needs an input current with a rise time less than 1 $\mu$s [8.19]. The main drawback of using a PEOS is that an additional power source is required to drive the plasma source, although this can be overcome if another type of POS becomes available. For simplicity, a PEOS having the following characteristics [8.20] will be considered:

- cathode radius – 2.25 cm,
- plasma length – 5 cm,
- plasma density – 1.35 $\times 10^{19}$ ions/m$^3$,
- plasma velocity – 4 cm/$\mu$s.

*High-Voltage Air Core Transformer*

The concept of air core, cylindrical high-voltage transformers (HVT) originated with Martin and Smith [8.21] in 1968. It has an $N$-turn secondary winding of internal radius $r$, inside a single-turn primary winding, both of width $\omega$. With an interwinding Mylar insulation thickness of $t$, and a primary–secondary insulation thickness of $Nt$, the transformer geometry is completely defined. However accurate the calculations, the actual coupling coefficient, $k_c$, between the primary and secondary windings will be lower than that predicted, and a value of 0.85 is assumed for the overall system predictions. If the actual voltage multiplication factor is $k_v$ and the primary inductance is $L_p$, the secondary inductance is $L_s = [k_v/k_c]^2 L_p$ and the mutual inductance is $M = k_v L_p$.

*Microwave Generator (Load)*

The parameters of the load, which is the MILO, are considered to be time independent, with a resistance of 5 Ω and inductance of 100 nH [8.10]. The voltage pulse required from the high voltage power source should have a rise time of less than 100 ns, an amplitude of about 500 kV, and a pulse duration of 1 μs.

*Autonomous Initial Energy Source (AIES)*

If permanent magnet are not considered, the autonomous initial energy source (seed source) can be realized by using one of two basic arrangements:

- battery powered small capacitor connected either directly or through a DT to a HMCG [8.22],
- explosively driven shock-wave piezoelectric generator coupled to two small HMCGs connected through a compact static transformer [8.23]. One version of AIES uses a 6 J piezoceramic generator and has an output of 0.4 MJ, a length of 0.7 m, and a requirement for 2 kg of high explosive.

The second arrangement above has a number of advantages including being lightweight, compact, and not requiring a battery. It provides energy to the amplifier section (SHMCG) or directly to a FHMCG, through either a dynamic or static transformer.

*Closing Switches*

The closing switch required to connect the load (for HVT systems), or the POS, could be a voltage activated $SF_6$ pressurized spark gap or, preferably, a simple detonator activated switch.

## System Designs

Having identified the major components required to develop a high voltage pulsed power system for high power microwave sources, the optimal overall system design must be identified. Four designs are considered.

From among the various techniques used in high-energy pulsed power for the generation of fast-rising, long-duration high-voltage pulses in high impedance loads, the most suitable appear to be either a plasma opening switch or a high voltage transformer. Both these high-impedance matching elements need a preconditioned input pulse, in order for the POS to properly function or for the HVT to provide the required voltage output. Two different kinds of power conditioning are considered, the first based on an

EFF and the second on a FHMCG. By combining these two power conditioning techniques with the two matching elements, four different high power systems become possible.

*EFF Conditioning*

In this arrangement, the output of the energy amplifier, i.e., a SHMCG, is fed to a ballast inductor, $L_b$. When the action of the SHMCG is complete and its inductance is negligible in comparison to the ballast inductance, the EFF opens the circuit by producing a high voltage across the closing switch connecting the matching element.

*FHMCG Conditioning*

In this case the AIES injects energy into the outer (primary) coil of the DT of an FHMCG. Then the FHMCG (representing the secondary of the DT) crowbars, capturing the magnetic flux, and compresses it into a ballast inductor. The matching element is connected by a closing switch at an appropriate time.

*Complete High-Power Systems*

Designs have been generated for all four possible power systems using the data presented above. The components required in these systems and their parameters are presented in Table 8.1.

| Component | Parameter | System I | System II | System III | System IV |
|---|---|---|---|---|---|
| SHMCG | | Yes | Yes | No | No |
| | $L_g(0)$ ($\mu$H) | 100 | 400 | – | – |
| | $l$ (mm) | 500 | 1000 | – | – |
| | $d_a$ (mm) | 140 | 120 | – | – |
| | $d_c$ (mm) | 300 | 250 | – | – |
| | $M_{ex}$ (kg) | 14 | 20 | – | – |
| | $I_m$ (MA) | 6 | 2.5 | – | – |
| | $W_m$ (MJ) | 2.5 | 3.8 | – | – |
| FHMCG | | No | No | Yes | Yes |
| | $DL$ ($\Omega$) | – | – | – 1.4 | – 0.6 |
| | $I_m$ (MA) | – | – | 2.5 | 0.9 |
| | $W_m$ (MJ) | – | – | 0.6 | 0.65 |
| EFF | | Yes | Yes | No | No |
| | Geometry | Plane | Cylindrical | – | – |
| | Dimensions (mm) | 400 x 280 | 80 X 960 | – | – |
| | No. of Grooves | 35 | 120 | – | – |
| | Max Voltage (kV) | 200 | 800 | – | – |
| | Energy Dissipated (MJ) | 0.84 | 1.7 | – | – |
| HVT | | Yes | No | Yes | No |
| | $k_v$ | 3 | – | 8 | – |
| | $N$ | 4 | – | 10 | – |
| | $\omega$ (mm) | 400 | – | 150 | – |
| | $t$ (mm) | 1.2 | – | 0.7 | – |
| | $r$ (mm) | 95 | – | 230 | – |
| | $L_p$ ($\mu$H) | $L_b$ | – | 0.7 | – |
| | $L_s$ ($\mu$H) | 1.3 | – | 62 | – |
| | $M$ ($\mu$H) | 0.3 | – | 5.6 | – |
| Ballast Inductor | $L_b$ ($\mu$H) | 0.1 | 0.35 | 0.7 | 1.9 |
| Transmission Line | $L_t$ ($\mu$H) | No | 1.5 | No | 1 |
| Seed Energy | (kJ) | 30 | 400 | 220 | 400 |

TABLE 8.1. Components and parameters of the four autonomous pulsed power systems being considered for driving high-power microwave generators.

The first design, which is referred to as System I, consists of a EFF and a HVT. An equivalent circuit diagram is presented in Fig. 8.6, and the

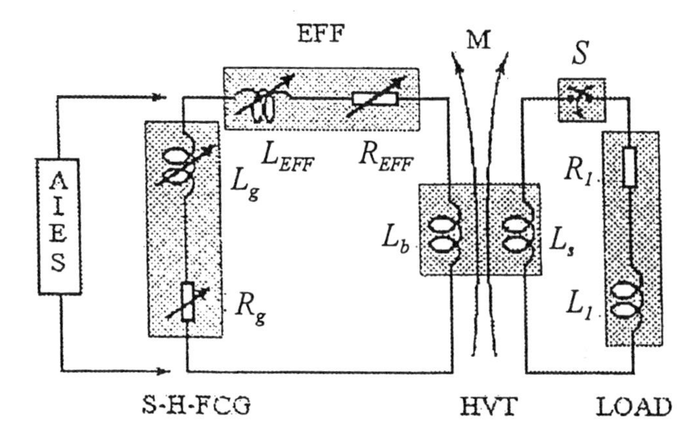


**FIGURE 8.6. Equivalent circuit diagram for System I.**

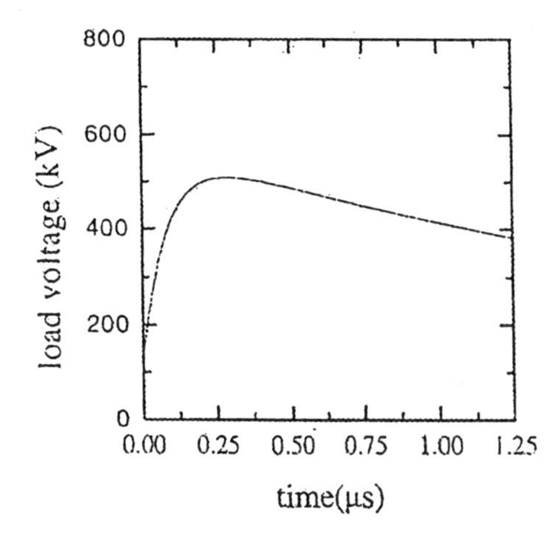


FIGURE 8.7. Output voltage for System I (time measured from the closing of switch S).

output voltage generated across the load is presented in Fig. 8.7. The ballast inductor in this case is the primary winding of the HVT, i.e., $L_p = L_b$, where only 1.7 MJ of the 2.5 MJ generated by the SHMCG is stored. The remainder of the energy is dissipated in the EFF. Either a capacitor or piezoelectric AIES can be used to provide the required seed energy of about 30 kJ.

The equivalent circuit diagram of System II, which consists of a EFF and POS, is shown in Fig. 8.8, and the output voltage that it produces across the load in Fig. 8.9. The transmission line inductance, $L_t$, serves as an energy store in maintaining the voltage following the initial step. This system requires the full 400 kJ of energy from the AIES. Therefore, a capacitor seed source cannot be used.

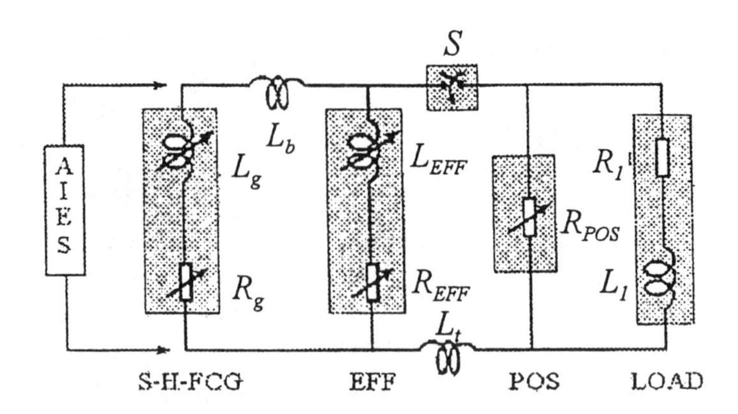


FIGURE 8.8. Equivalent circuit diagram for System II.

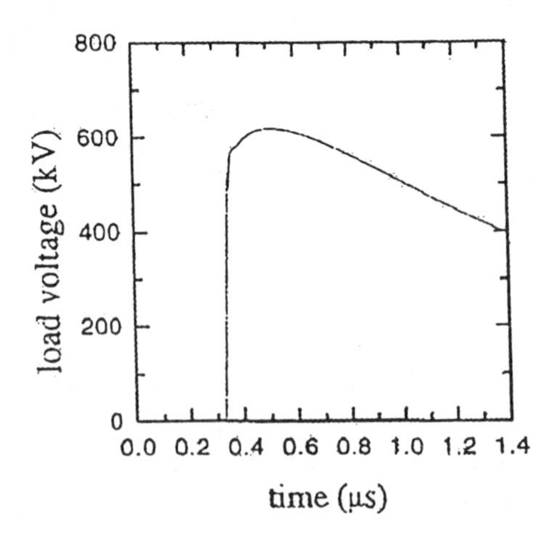


FIGURE 8.9. Output voltage of System II (time measured from the closing of switch S).

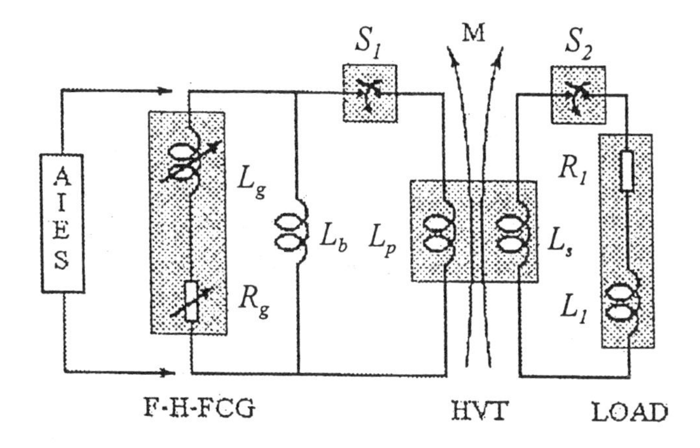


FIGURE 8.10. Equivalent circuit diagram for System III.

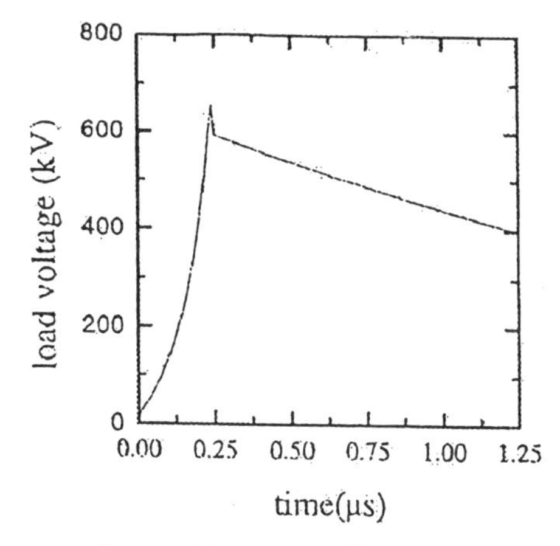


FIGURE 8.11. Output voltage of System III (time measured from the closing of switch $S_2$; switch $S_1$ is closed at all times).

The equivalent circuit diagram of System III, which consists of a FHMCG and HVT, is shown in Fig. 8.10. A similar arrangement was used to generate peak voltages up to 1 MV with long (microseconds) risetimes in high-impedance loads [8.24]. The best results were obtained from a system in which a closing switch was used in the secondary winding (rather than in the primary winding) of the transformer when the current through the generator reached 0.5 MA and are presented in Fig. 8.11. In order to produce this waveform, the FHMCG required a seed energy of 220 kJ and had to produce 2.50 MV across the ballast inductor.

The equivalent circuit diagram of System IV, which consists of a FHMCG and POS, is presented in Fig. 8.12, and the output voltage that it produces in the load in Fig. 8.13. The closing switch activates when the generator current reaches 5.0 MA. The ballast inductor controls the amplitude of the

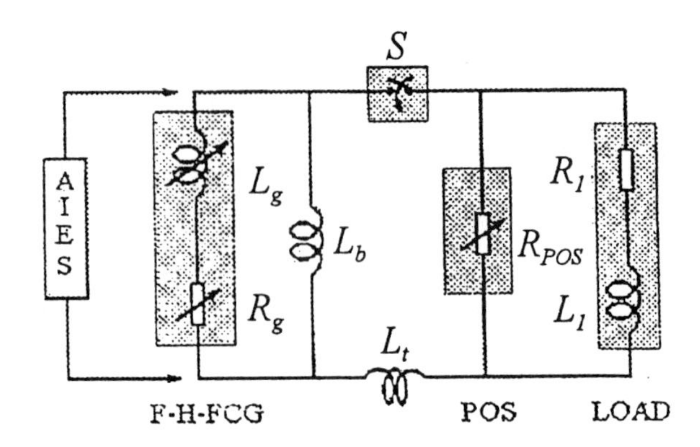


FIGURE 8.12. Equivalent circuit diagram for System IV.

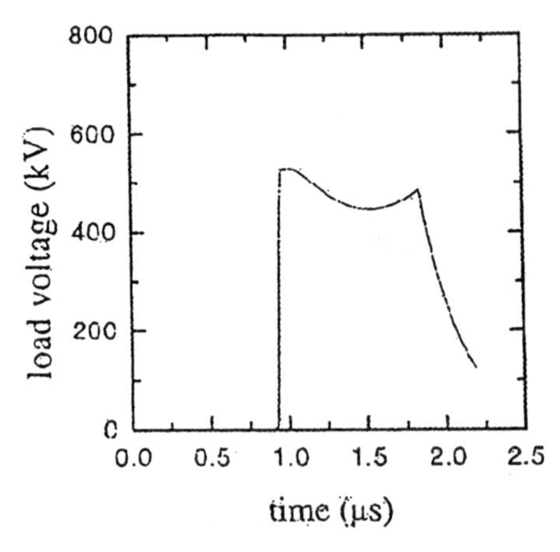


FIGURE 8.13. Output voltage of System IV (time measured from the closing of switch S).

voltage pulse and the transmission line its pulse length, as well as the time at which the POS begins to function. The maximum voltage generated by the FHMCG is about 600 kV.

*System Assessments*

In this section, four possible designs of an autonomous single-shot power source for high-power microwave generators were discussed. The question is "which design provides optimal performance within the constraints imposed by the application." Examining these four designs, it can be concluded that Systems II and III can immediately be disregarded, since the first is too bulky and complex and the second imposes unduly high-voltage requirements on the FHMCG. Within the constraints of existing technology, System I clearly provides the most robust solution, since it employs already established techniques and the overall system development is straightforward. The entire system presented in Fig. 8.4 will weigh less than 200 kg and occupy a volume of only 1.5 $m^3$. System IV is an increasingly attractive alternative as technology advances, since it can potentially generate much higher voltage pulses with an extremely short risetime, while being lighter and far more compact. To take full advantage of the outstanding characteristics it offers requires the development of a vacuum-based FHMCG capable of withstanding internal voltages of hundreds of kilovolts and a long conduction time POS. It is important to note that these two requirements are universally related and that the greater the conduction time that can be obtained from the POS the lower the internal voltage stress in the FHMCG will be.

In the rest of this section, several high-power microwave sources and their associated pulsed-power systems will be discussed. The material presented in this section is to help put into perspective what has already been done.

### *8.2.2 Virtual Cathode Oscillators*

Advances in pulsed-power technology and HPM sources have made it possible to generate intense electron beams capable of producing gigawatts of microwave power in pulses with hundreds of joules of energy [8.25]. One such HPM source capable of generating these powers is the vircator. Unlike other HPM sources, the vircator requires a current that exceeds the *space-charge-limiting current*, which is the maximum beam current that can pass through the drift space. If a beam current exceeding the space-charge-limiting current is injected through a transmitting anode into a drift space of sufficiently large radius, a strong potential barrier forms at the head of the beam. This potential barrier, from which excess space charge in the beam is rejected from the drift space, is known as a *virtual cathode* [8.25]. In the schematic drawing of the vircator presented in Fig. 8.14, the virtual cathode oscillates axially. The motion of the space charge and the potential well of the virtual cathode generates electromagnetic radiation. The bandwidth of this radiation is rather large, but if the virtual cathode is formed in a cavity with a resonant frequency that lies within the bandwidth of the emitted radiation, then an interaction between the radiation and the cavity modes can yield much narrower bandwidths.

Another type of space-charge device that is similar to the vircator is the *reflex triode*. As shown in Fig. 8.15, the anode, rather than the cathode, is attached to the center conductor of the transmission line that feeds it power. If the anode consists of a transmitting foil, electrons pass through the foil and create a virtual cathode. Follow-on electrons are trapped in the potential well formed between the cathode and the virtual cathode, where they oscillate. This *reflexing* back and forth by the electrons generates microwave radiation.

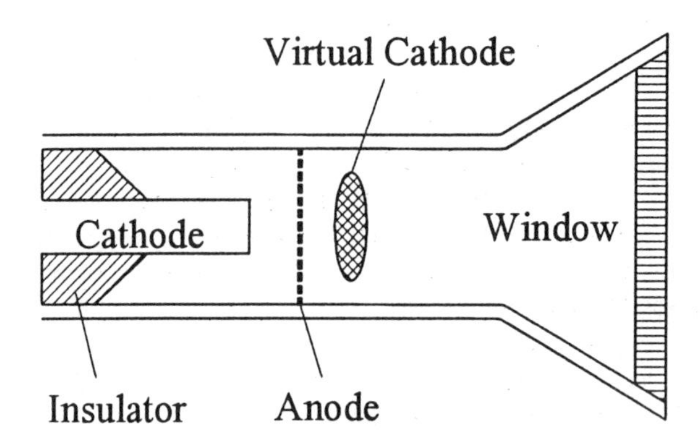


FIGURE 8.14. Schematic diagram of the vircator. When the space-charge-limiting current is exceeded, a virtual cathode is created. This cathode oscillates axially and generates electromagnetic radiation.

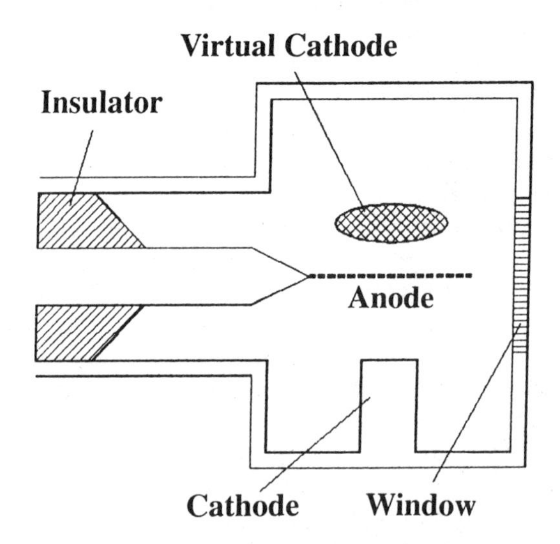


FIGURE 8.15. The reflex triode differs from the vircator in that the anode, rather than the cathode, is attached to the center conductor of the transmission line that feeds it power. If the anode is a transmitting foil, electrons that pass through it form a virtual cathode. Follow-on electrons are trapped in the potential well between the cathode and the virtual cathode and oscillate. This reflexing back and forth generates microwaves.

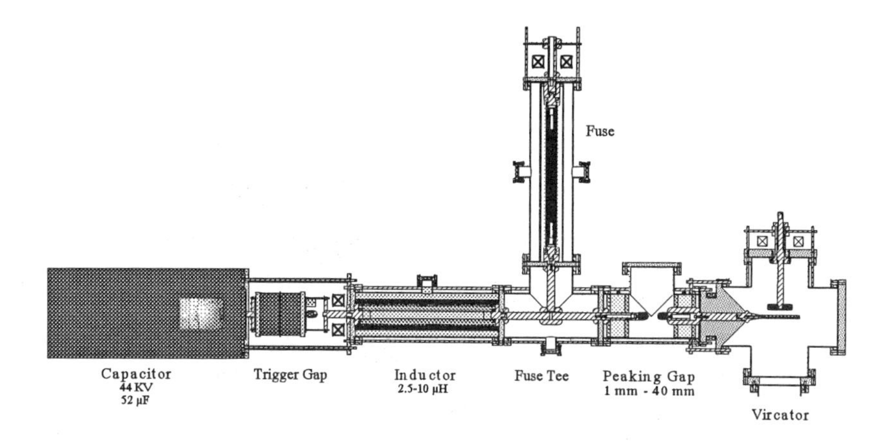


FIGURE 8.16. Drawing of the overall experimental setup at TTU of the prime power source, pulse forming system, and vircator. The capacitor will eventually be replace by an MCG. (Courtesy of Texas Tech University.)

Texas Tech University (TTU) has been awarded a Multidisciplinary University Research Initiative (MURI) to optimize the operation of MCGs and its associated power conditioning equipment for the purpose of driving high-power microwave sources, such as the vircator. A drawing of a capacitor-driven vircator system developed by TTU is shown in Fig. 8.16, and a more detailed drawing of the vircator is shown in Fig. 8.17. This system is being used to verify the feasibility of the pulse shaping system and to validate numerical simulations. The ultimate goal of this project is to replace the capacitor bank with an optimized MCG.

A consortium of institutes in Russia including the High Energy Density Research Center in Moscow, the Institute of Chemical Physics in Chernogolovka, and the Institute of High Voltages in Tomsk developed an MCG-driven vircator, i.e., a reflex triode, system. A diagram of the system is presented in Fig. 8.18, and although this had an operating frequency of approximately 3 GHz, the frequency could in fact be varied between 0.5 and 30 GHz [8.25,8.26].

The vircator demands much of its power supply, which must be capable of providing high-voltage pulses in excess of 300 kV, with risetimes of less than 100 ns and currents in excess of 10 kA. Usually, these parameters can be achieved by Marx banks with a pulse-forming line that includes some type of switch, such as the electroexplosive opening switch (EEOS), which functions basically as a power amplifier by increasing the voltage in the circuit. But for the EEOS to operate optimally, it must be supplied with energy in times $\leq 10$ $\mu$s.

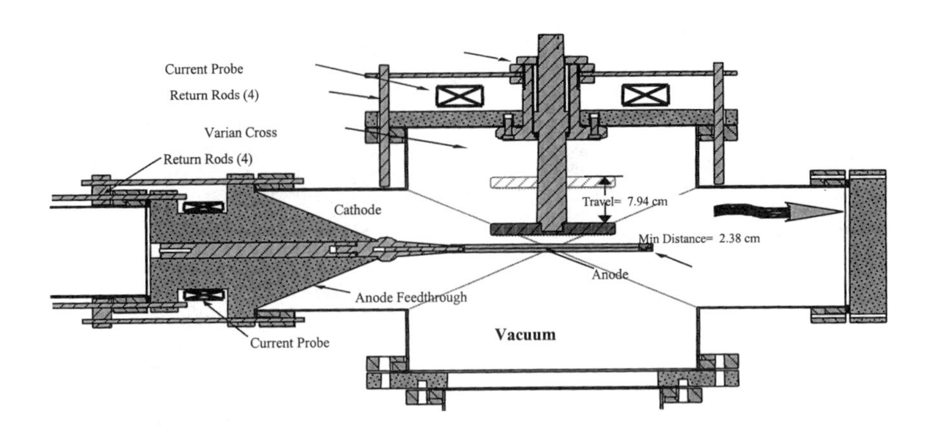


FIGURE 8.17. Drawing of the TTU vircator. (Courtesy of Texas Tech University.)

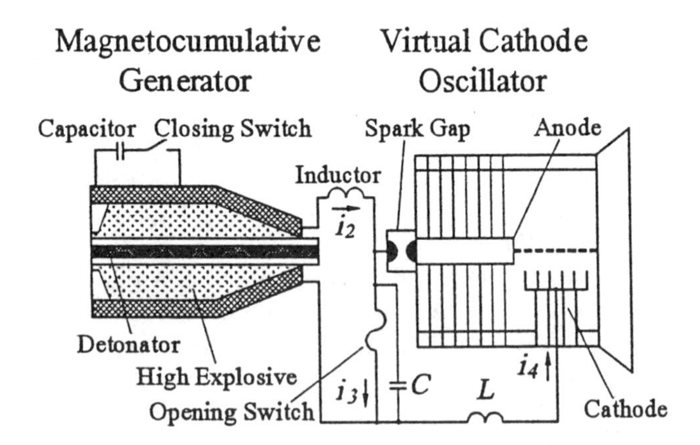


FIGURE 8.18. Schmatic diagram of MCG–reflex triode system developed by Fortov and his team [8.25,8.26].

Owing to the bulky nature of the Marx bank, it is not practical as a power source for the vircator in some applications. The only other power source that could meet the requirements in these applications is the MCG. To meet the timing requirements of the EEOS, two types of compact high-voltage helical MCGs with *flux trapping* were used in experiments conducted by the Russian consortium. The first was a simultaneous axially initiated helical generator and the second was a small conical generator with a moving contact point. The generators contained between 200 and 600 g of high explosive, and generated voltage pulses of 50–200 kV due to rapid changes, i.e., within approximately 5 to 15 $\mu$s, in their high initial inductance. When the internal coil is crowbarred, the inductance begins to change as a result of the movement of the armature and flux trapping takes place.

It has been found that to drive vacuum diodes in microwave generators with MCGs, it is best to use a transformerless scheme based on flux trapping [8.25,8.26] (see Section 4.5). The flux trapping method was first used in cascaded systems consisting of many MCGs. An MCG that uses flux trapping consists of an outer solenoid $L_1$, an inner solenoid $L_2$, and a cylindrical armature $L_3$, as shown in Fig. 4.23. An initial magnetic flux is generated within solenoid $L_2$ by an external energy source such as a capacitor or a "booster" MCG, $L_p$. At the moment the current created by the prime energy source is at its peak value, the secondary circuit is closed by initiating the explosive charge (HE). As the armature expands, the trapped magnetic flux is compressed and pushed out into the load. Diagrams of both axially and end initiated schemes for MCGs with flux trapping are presented in Fig. 4.23.

The inductance of the inner helix of MCGs that use flux trapping must be high, in order for the MCG to work effectively. However, owing to the sharp increase in the magnetic field strength within the volume of the MCG, these high inductances also result in high flux losses. In addition, since the MCG output voltage exceeds 50 kV, it is necessary that the insulation between the turns, between the armature and inner helix, and between the helices is of sufficient quality. However, the available reliable insulation that is used in small MCGs also lowers its inductance.

In Fig. 8.18, a 100 $\mu$H capacitor, $C_0$, is charged to 3 kV to provide the seed current to a cylindrical "booster" MCG with flux trapping, $L_{11}$. The "booster" MCG contains between 200 and 700 g of high explosive and delivers between 5 and 10 kJ of energy to the external coil with inductance $L_1$ of the high voltage MCG. If axially initiated MCGs are used in place of cylindrical generators, they offer the advantage that they are simple and can deliver up to 60 kJ to the high voltage generator.

At the moment the current is at its peak value in $L_1$, the secondary circuit of the high-voltage generator is crowbarred. Expansion of the liner of the high-voltage generator leads to flux trapping by the internal solenoid $L_2$, thus generating current in the EEOS circuit. The operating time of the high-voltage MCG is between 6 and 8 $\mu$s, the voltage output is between 50 and 200 kV, and the current output can be as high as 30 kA.

The switch consists of several tens of copper wire with a diameter of 40–50 $\mu$m and a length of 0.5–1.0 m. The wires are in parallel and are immersed in nitrogen at a pressure of 0.5 MPa. The wire sizes were selected such that the maximum rate of increase in their resistance occurs during the final stages of operation of the generator.

Since the high-voltage generator could only withstand voltages up to 200 kV, an inductive store is inserted between the MCG and the EEOS. When an overvoltage is applied to the spark gap $P$, a high-voltage pulse is delivered to the anode $A$ of the vircator, which in turn results in an explosive emission of electrons from the cathode $C$. The resulting electron beam is used to generate microwaves.

In experiments carried out with the MCG–vircator system, voltage pulses up to 600 kV with a pulse length of 180–250 ns and a risetime of approximately 60 ns were supplied to the input of the vircator. The peak current in the vircator of 16 kA, corresponds to a relativistic beam power of approximately 10 GW. The peak power of the microwaves radiated into the atmosphere was 100–200 MW, at a wavelength of $10 \pm 0.5$ cm, for a pulse duration of 100–200 ns.

### *8.2.3 Multiwave Cerenkov Generators*

Pavlovskii, Chernyshev, Selemir, and others at the All-Russian Scientific Research Institute of Experimental Physics in Arzamas-16 and at the Institute of Radiotechniques and Electronics in Moscow used an MCG to drive a *multiwave Cerenkov generator* (MWCG) [8.27]. The goal of these experiments was to investigate the efficiency of converting chemical energy into microwave energy using this system.

The MWCG belongs to the same family of microwave sources as does the backward wave oscillator (BWO) and the traveling wave tube (TWT), both of which are conventional tubes that have been in use for at least 40 years. All are based on the *Cerenkov effect.* Cerenkov radiation is generated when charged particles move through a dielectric medium at speeds greater than the speed of light in that medium. In the case of the MWCG, a resonator takes the place of the medium, where the speed of light is the axial phase velocity of the resonant normal mode of the resonator. A slow-wave structure is used to reduce the axial phase velocity of the normal modes below that of the speed of light in vacuum so that it can exchange energy directly with a beam of electrons drifting through the resonator. When the velocity of the electrons is equal to the phase velocity of the resonator, each electron is either accelerated or decelerated, causing electron bunching but no net energy exchange since equal numbers of electrons are accelerated or decelerated. But if the velocity of the electrons is greater than the phase velocity, more electrons will be decelerated than accelerated, and there is a net transfer of energy to a space-charge wave when the beam current is large [8.25].

The MWCG is classified as an O-type Cerenkov device, as opposed to an M-type Cerenkov device. In the former case, electrons drift along a magnetic field, while in the latter case they drift across a magnetic field. MWCGs with a slow wave structure consisting of periodic variations in the wall radius have generated powers of 15 GW in a 500 J pulse at 9.4 GHz [8.27].

Two important features of the MWCG are their:

- large cross section,
- use of two slow-wave structures separated by a drift space, as shown in Fig. 8.19.

The large diameter, which is at least several free-space wavelengths, helps to reduce the average power intensity and increase the power handling capability of the structure. The use of two sections aids the process of mode selection, i.e., the first section is a buncher and the second section is the output section in which the bunched beam of electrons radiate.

In experiments conducted in Russia, the basic experimental system consisted of the six basic components shown in the equivalent circuit diagram in Fig. 8.20 [8.27]. The components are an MCG with a transformer (1) that serves as the initial energy source, an intermediate inductive store (2) with an opening switch (3) that generates a megavolt pulse, a sharpening discharge switch (4), a high-current diode (5), and a solenoid (6) that generates an axial magnetic field. The electron beam energy was converted into microwaves by using the electrodynamic structure in Fig. 8.19.

The primary energy source was the C-160 helix MCG, with a five section multiwire helix. These sections were manufactured from insulated copper wire having a diameter of 6 mm. The winding in the first section consisted of four wires with a step size of 24 mm. The number of wires doubled in number from one section to the next, and the winding step size decreased correspondingly. The liner was a copper tube with a length of 64 mm and a wall thickness of 4 mm. The fifth section was connected to a 355-mm-long cylinder. In the final section, the generator was conically expanded at a $4^o$ half angle over the 355 mm length of the cylinder. The tube was filled with high explosive encased in an organic glass envelope with a thickness of 4 mm. The explosive had a mass of 7.50 kg and a detonation velocity of 7.6 km/s. The initial inductance of the MCG was 6.5 $\mu$H. The requisite initial magnetic field in the MCG was created by a 900 $\mu$F capacitor charged to a voltage of 35 kV. The initial current in the generator reached a peak value of 270 kA and the flux a value of 1.6 Wb. The output current of the MCG reached a peak value of 4.5 MA.

The transformer was made of high-voltage cable that could withstand pulsed voltages up to 300 kV, and it consisted of four cylindrical coils. The

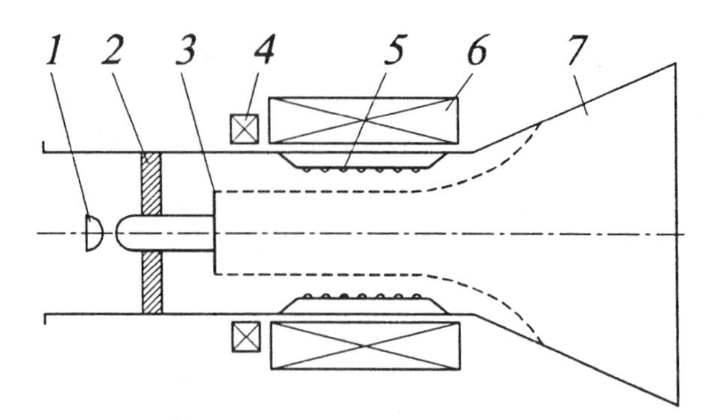


FIGURE 8.19. Multiwave Cerenkov microwave generator: 1 – spark gap switch, 2 – insulator, 3 – cathode, 4 – electromagnet, 5 – slow-wave structure, 6 – electromagnet, and 7 – antenna. The MWCG offers the advantages for having large cross sectional areas, which reduces its average power intensity and increases the power handling capability, and two slow-wave structures, which aids in mode selection.

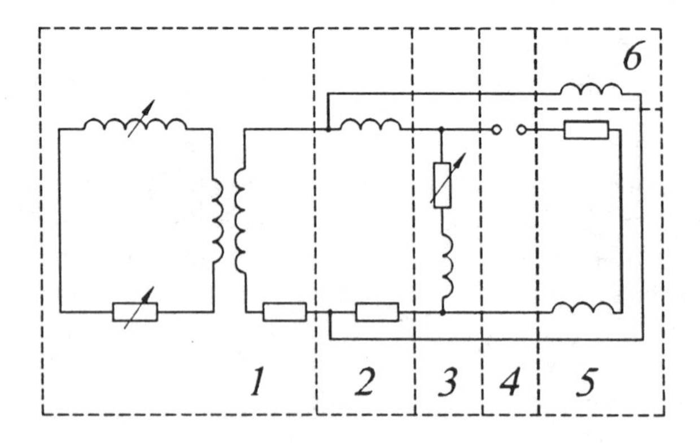


FIGURE 8.20. Equivalent circuit diagram for MCG–Cerenkov generator system: 1 – MCG and transformer, 2 – intermediate inductive store, 3 – opening switch, 4 – pulse sharpening discharge switch, 5 – high-current diode, and 6 – solenoid.

outer diameter was 320 mm and the length was 380 mm. The estimated inductance of the primary winding of the transformer was 26.5 nH, and the measured inductance of the 32-turn secondary winding was 27.5 $\mu$H. The mutual inductance between the primary and secondary windings was estimated as 0.8 $\mu$H.

A 25 $\mu$H solenoid was used as the intermediate energy storage system. It was connected to the MCG with a 20-m-long 300 kV transmission line. In order to generate a 2 T axial magnetic field, the solenoid had a diameter of 640 mm and a length of 1.6 m. In the central part of the solenoid, which was 1 m long, the nonuniformity of the magnetic field did not exceed 10%. The inductance of the energy storage system, including the transmission line, did not exceed 45 $\mu$H. The opening switch of the energy storage system consisted of five copper conductors with a diameter of 0.72 mm, were coated with polyethylene, to yield an outside diameter of 4 mm. The conductors had a length of 4.5 m. The initial active resistance of the switch was 39 mOhm. To reduce the length of the switch, as well as its inductance, each of the five conductors was made in the form of a helix with a diameter of 5 cm and a step size of 4 cm. This reduced the length of the switch to 1.1 m and the inductance to 1 $\mu$H.

An electrically exploding copper conductor insulated with polyethylene sealed in a container filled with $SF_6$ and a two-electrode air spark gap with an operating threshold voltage of 0.8–0.9 MV were used as pulse sharpening switches. Referring to Fig. 8.19, it can be seen that the cathode (3) of the microwave generator is connected through the spark gap switch (1) in parallel with the opening switch made of electrically exploding conductors.

The output of the MCGs transformer was also connected in series with the solenoid (6) and was used to create the magnetic field within the microwave source. A tubular electron beam with a diameter of 350 mm was generated by the diode and was transported along the corrugated surface of the electrodynamic structure of the microwave source. The diameter of the output window of the source was approximately 1.5 m.

When tested experimentally, this system delivered 100 MW of power into the atmosphere. The radiated signal had a wavelength of 3 cm and a pulse length of 0.8 $\mu$s.

### *8.2.4 Magnetically Insulated Linear Oscillators*

The magnetically insulated linear oscillator (MILO) (see Fig.8.21) is a crossed-field device, like the magnetron, the key feature of which is the use of the self-magnetic field of the current flowing along the cathode to cut off electron flow to the anode [8.25]. Thus the MILO is self-insulating, in that it inherently prevents short circuiting of the anode–cathode gap. As a result, the electron flow takes place along the surface of the cathode. The existence of this electron flow prompted the idea that the MILO might be used to generate and amplify microwave energy. To achieve this, it is nec-

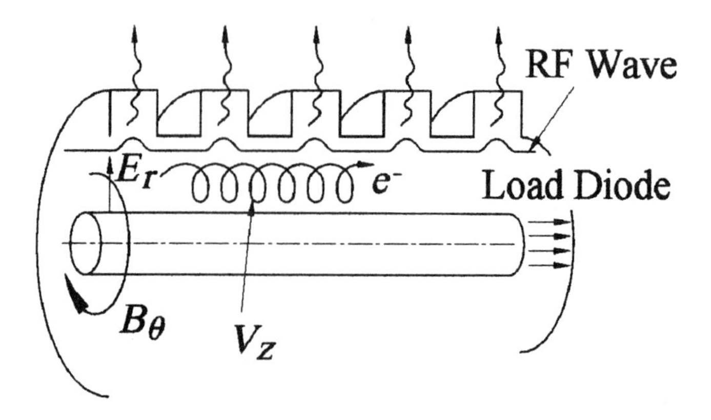


FIGURE 8.21. Diagram of a MILO. The MILO is self-insulating in that its electron beam generates a self-magnetic field that prevents breakdown between the anode and cathode. This eliminates the need for external magnets and reduces the mass and size of the system.

essary that a slow wave propagates along the cathode with a phase velocity less than that of the speed of light in vacuum. This slow-wave mode can be achieved by using one of several techniques, including corrugating one or both of the electrodes or using a series of resonators or comb-type slow-wave structures. Once the geometry of the generator has been determined, the parameters of the electron flow, i.e., beam current, beam energy, and beam dimensions, are fixed. The voltage is used to determine the output frequency and power of the generator, as well as its efficiency. Both coaxial and planar versions of the MILO have been built [8.25].

Two of the greatest difficulties in using the MCG to power high-power microwave sources are the inductive nature of the load, which varies rapidly in time, and the comparatively low output voltage, i.e., $\leq$ 500 kV, of the MCG at load currents of 500–600 kA. Of all the types of relativistic microwave generators, the two that match up best to the output parameters of the MCG are the vircator and the cylindrical MILO. Both of these generators operate at relative low resistances, i.e., of the order of 1 Ohm, and at high voltages of 500 kV. The MILO has the added advantages that it is simple to operate and generates a wide range of wavelengths. In addition, no stringent requirements are placed on the electron beam geometry, energy spread, or beam velocity. These features are particularly important when the MCG is used as the power source, because the output pulse of the MCG cannot be adequately shaped.

Of the vircator, discussed earlier, and the MILO, the MILO is preferred for the following reasons:

- A microwave generator with magnetic insulation permits more efficient coupling of energy from the electrodynamic structure into the antenna than does the vircator.

- The working frequency, band width, and spectral density of the MILO are dictated to a significant degree by the geometry of the electrodynamic structure, and can therefore be easily controlled.

- The MILO has a narrower emission band and a higher spectral density than does the vircator, which makes it possible to construct an efficient antenna feed system.

- The MILO has an increased breakdown voltage, owing to the magnetic insulation.

- The MILO has a reduced mass and size, due to the elimination of an external magnetic field system.

However, the vircator remains a viable alternative to the MILO for use with the MCG, as demonstrated earlier in this chapter. This is true since the vircator is more forgiving than the MILO with respect to the electrical pulse from the MCG. That is, the MILO requires a flat top pulse.

As has been already discussed, there are several types of MCGs, each with its own unique set of characteristics. MCGs that are used to power relativistic electron beam (REB) devices, such as the MILO, need to have the following parameters:

- output voltage not less than 500 kV,

- current not less than 200 kA,

- current pulse rise time out of the switch of approximately 0.1 $\mu$s.

One class of MCG that approaches these requirements is the high-inductance, high-speed POTOK spiral generator.

As discussed earlier, there are two ways to increase the energy gain of the MCG. The first is to connect two or more MCGs in series, when the generators would have to be interconnected using either air-core transformers or magnetic flux traps. The energy gain of this cascaded system is the product of the energy gains of each individual generator. The disadvantages of this concept are

- increased complexity and cost of the prime power system,

- decreased reliability with an increasing number of stages,

- increased weight by a factor as much as 2. For example, an air core transformer has a weight comparable to that of one of the MCGs.

The second way to increase the energy gain of the MCG is to increase the ratio of its initial inductance to that of the load, which can only be done with the spiral generator. It should be noted that an increased initial inductance results in higher voltages, i.e., tens to hundreds of kilovolts, in the active volume of the MCG. Therefore, special measures must be taken to prevent breakdown within the MCG, which would seriously degrade its efficiency. There are two ways to reduce the possibility of breakdown. The first is to cover the spiral coils with a dielectric material. It has been demonstrated experimentally that the electric field strength of the dielectrics must exceed 100 kV/m, owing to the insulation-induced flux losses and the maximum electric field requirements of the generator circuit. The second method is to use different geometries in designing the coil, so as not to exceed the dielectric breakdown voltage.

As noted above, another critical parameter is the risetime of the MCG. The load, i.e., the REB accelerator, requires an input pulse with a fast risetime. There are two ways to achieve this, by using simultaneous axial initiation and/or various types of switches. The disadvantage of the first method is the loss of uniform compression of the magnetic field by the liner due to time variations in firing the detonators. The disadvantages of the switches are that a large amount of energy is lost in the switch and their cost. The cost of the switches required to form the pulses having the requisite parameters is comparable to that of the MCG. In the next section, a modified spiral generator that has potential as a power source for the MILO is introduced.

### Modified Spiral Generator

Studies have shown that it is desirable to separate the magnetic compression and output phases during the operation of the MCG. The diagram of a modified spiral MCG that was designed for this purpose is presented in Fig. 8.22. The MCG consists of a coil (1), liner (2), high explosive (3), detonator (4), and switch (5). The liner is electrically connected to the switch with a metal piston (6). The liner has a dielectric insulator (7) and conducting outlets (8) and (9) situated in the switch.

When the high-explosive charge is detonated, the liner begins to move, compressing the magnetic field created in the generator by a small capacitor bank. This compression creates a current which flows through the "coil–fuse–piston–liner" circuit. As the detonation wave created by the explosive charge moves along the axis of the generator, it pushes the explosive products, which, in turn, exerts a force on the piston. The piston then begins to move and breaks the "coil–fuse–piston–liner" circuit, and the current pulse from the opened circuit is transferred to the load through the outlets (8) and (9). To prevent the current from flowing into the load during the operation of the generator and to sharpen the rise time of the pulse, the generator is connected to the load through a discharge switch.

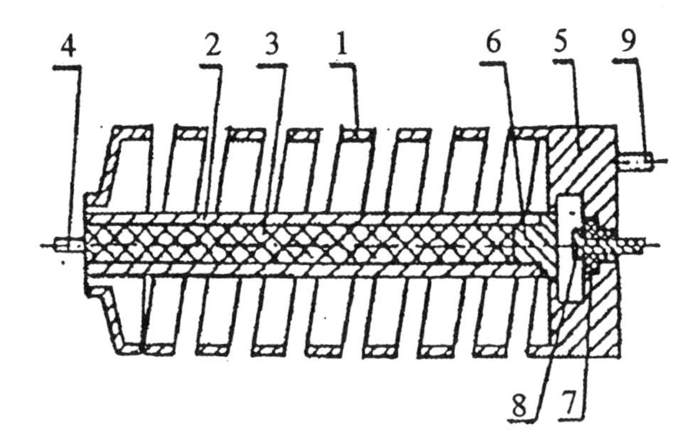


FIGURE 8.22. Modified spiral MCG: 1 – helical coil, 2 – liner, 3 – explosive, 4 – detonator, 5 – switch, 6 – metal piston, 7 – insulator, and 8,9 – electrical contacts.

The processes that take place in this modified spiral generator can be considered by using the equivalent electrical circuit in Fig. 8.23. Let $L_{oc}$ be the initial inductance, $L_r$ the generator inductance at time $t$, and $L_L$ the load inductance. The energy loss in the MCG is characterized by the active resistance $R_1$. The circuit disruption process and the switching of the MCG into the load is represented by the closing of switch $K$. It is assumed that the moment at which the piston begins to move there is an instantaneous closing of the discharge switch $K$. At the onset of compression of the magnetic field in the generator, which is taken to be at $t = 0$, a current $I_0$ flows through the generator, which has an initial inductance of $L_0$. The processes that take place in the circuit are described by the equation

$$\frac{d(L_r I_1)}{dt} + L_{0c}\frac{dI_1}{dt} + R_1 I_1 = 0. \tag{8.1}$$

The solution of this equation is

$$I_1 = I_0 \frac{L_0}{L(t)} \exp\left[-\int \frac{R_1}{L(t)} dt\right], \tag{8.2}$$

where $L(t) = L_r(t) + L_{0c}$.

An important characteristic in the operation of this generator is the decrease in voltage, $V_{0c} = L_{0c} dI_1/dt$, which is due to the residual inductance $L_{0c}$. This residual inductance determines the current flow from the generator to the load. At the onset of movement by the piston, the "liner–piston–fuse–coil" circuit does not open, because of the formation of a current carrying channel (electric arc) between the liner and the piston. The

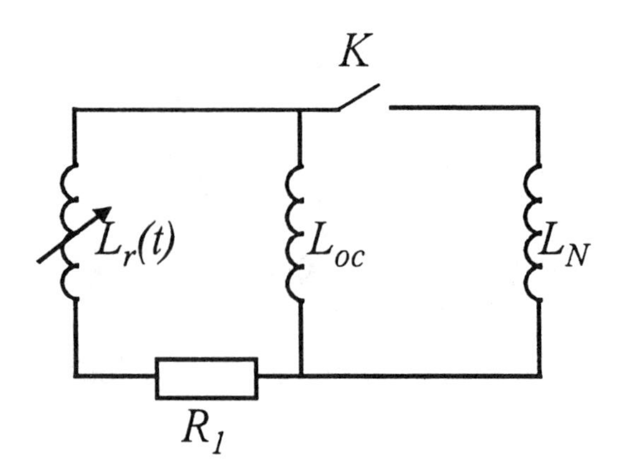


FIGURE 8.23. Equivalent circuit diagram for a modified spiral MCG.

time between the initial motion of the piston and the time the channel extinguishes itself determines the delay time for the transfer of current to the load. Solving Eq. 8.1 for $dI_1/dt$ and substituting the result into Eq.8.2, yields

$$V_{0c} = -I_1 \frac{L_{0c}}{L(t)} \left[ \frac{dL_r}{dt} + R_1 \right]. \tag{8.3}$$

If the inductance of a generator with a coil having a variable pitch is given by $L_r = L_0 \exp(-\alpha t)$, Eq. 8.3 becomes

$$V_{0c} = I_1 \frac{L_{0c}}{L(t)} \left[ \alpha L_0 e^{-\alpha t} - R_1 \right]. \tag{8.4}$$

Substituting Eq. 8.2 into Eq. 8.4 leads to

$$V_{0c} = I_0 \frac{L_0 L_{0c}}{L^2(t)} \left[ \alpha L_0 e^{-\alpha t} - R_1 \right] \varphi. \tag{8.5}$$

At the time when $L(t) = L_{0c}$, Eq. 8.5 reduces to

$$V_{0c} = I_0 L_0 \alpha \left[ 1 - \frac{R_1}{\alpha L_{0c}} \right] \varphi, \tag{8.6}$$

where

$$\varphi = \exp \left[ - \int_0^t \frac{R_1}{L} dt \right] \tag{8.7}$$

is a coefficient associated with the conservation of magnetic current.

In order to find the residual inductance, $L_{0c}$, of the generator, an expression for the expansion angle for copper tubes having a diameter of 0.4–2.0 mm must be found. The expansion angle is a function of a distance that is 12–15 times longer than the distance between the liner and the detonation front, which is no longer a function of time. The expansion angle is

$$\beta = \sqrt{\frac{\gamma+1}{\gamma-1} - 1}\frac{\pi\sqrt{Z_m}}{2}\left[\sqrt{Z_m} + 1.86 + \frac{0.1}{\Delta}\right], \tag{8.8}$$

where

$$Z_m = \frac{\rho_0 z^2}{\rho_1\left(z_2^2 - z^2\right)} \tag{8.9}$$

is the ratio of the mass of the high explosive to that of the liner per unit area, $z_2$ is the outer radius of the liner, $z$ is the inner radius of the liner, $z_1$ is the inner radius of the liner as it expands, $\Delta = (z_1 - z)/z$ is the increase in the radius of the liner, $\rho_0$ is the density of the high explosive, and $\rho_1$ is the density of the liner material.

Having defined the expansion angle, an expression can be written for the residual inductance as

$$L_{0c} = \frac{\mu_0\pi(z_2^2 - z_3^2)w^2}{15(z_1 - z) + (z_3 - z_2)\sin\beta}, \tag{8.10}$$

where $z_3$ is the inner radius of the coil, $\beta$ is the average angle of expansion, $w$ is the number of windings in the segment $l_{0c}$, and $\mu_0$ is the magnetic permeability.

The expression for $\alpha$ depends on the ratio of $L_r(t_{0c})$ to $L_{0c}$, that is,

$$\alpha = -\frac{D}{l_x}\ln\frac{L_{0c}}{L_0}, \tag{8.11}$$

where $t_{0c} = l_r/D$ is the duration of the pulse in the generator, which can be found from the parameters of the generator, $l_r$ is the length of the generator, and $l_x$ is the distance along the generator that the liner has expanded and made contact with the coil. If Eqs. 8.10 and 8.11 are substituted into Eq. 8.5, an expression for the voltage due to the residual inductance of the generator can be derived as

$$V_{0c} = -I_0 L_0 \varphi\frac{D}{l_x}\ln\left(\frac{L_{0c}}{L_0}\right)\left[1 + \frac{R_1 l_x}{D\ln\left(\frac{L_{0c}}{L_0}\right)}\right]. \tag{8.12}$$

Assuming that $R_1 = 0$, Eq. 8.12 reduces to

$$V_{0c} = -I_0 L_0 \frac{D}{l_x} \ln \frac{\mu_0 \pi (z_2^2 - z_3^2) w^2}{[15(z_2 - z) + (z_3 - z_2)\sin\beta]\, L_0}. \quad (8.13)$$

The process of opening the "coil–fuse–piston–liner" circuit by the motion of the piston and, thus, switching the power into the load is represented by the closing of switch $K$. To find the current $I_{0L}$ flowing into the load after closing the switch, the law of conservation of magnetic flux is used:

$$I_1 L_{0c} = I_{0L} L_L + I_{0L} L_{0c}, \quad (8.14)$$

where $I_1$ is the current that flows from the generator at the moment the switch is activated. Rewriting Eq. 8.14, one obtains:

$$I_{0L} = I_1 \frac{L_{0c}}{L_L + L_{0c}}. \quad (8.15)$$

From Eq. 8.15, it can be seen that, when $L_L = L_{0c}$, a maximum current of $I_{0L} = 0.5 I_1$ is delivered to the load. Because the current $I_L$ in the load at time $t$ increases as $L_{0c}$ goes to zero, the dependence of the current on time is described by the equation, which is analogous to Eq. 8.2:

$$I_L = I_1 \frac{L_{0c} + L_L}{L_L + L_{0c}(t)} \exp\left[\int_{t_{0c}+t_k}^{t} \frac{R_2}{L_L + L_{0c}} dt\right]. \quad (8.16)$$

As mentioned above, instantaneous opening of the generator circuit does not take place because of the formation of a current-carrying channel. This delay in transferring the current from the generator into the load is designated $t_k$. To find this delay time, the breakdown voltage of the "piston–liner" gap must be known, since the extinction of the current channel depends on the following condition:

$$\frac{V_{0c}}{l_k} \leq E_{np}, \quad (8.17)$$

where $E_{np}$ is the breakdown voltage and $l_k$ is the length of the channel. If it is assumed that the channel is extinguished at the moment that $V_{0c}/l_k = E_{np}$, then it has been found that the dielectric strength of the expanding detonation products is greater than 70 kV/cm.

To find the velocity of the piston, the following semiempirical expression is used:

$$v_n = \left(0.2093\sqrt{\frac{0.02}{2z}}\sqrt{\frac{m_3}{M}} - 0.038\right) D, \tag{8.18}$$

where $m_3$ is the mass of the high explosive and $M$ is the mass of the piston. This expression holds in the velocity range between 1 and 5 km/s. To calculate $v_n$, the mass of the high explosive, having a length of $l_3$, is substituted into the above equation, since it is assumed that the remaining mass of high explosive is used to accelerate the liner.

Taking into account Eq. 8.18, the expression for the delay time becomes

$$t_k = \frac{v_{0c}}{\left(0.209^3\sqrt[3]{\frac{10}{z}}\sqrt{\frac{\rho_0 \pi z^2 l_{0c}}{M}} - 0.038\right) DE_{np}}. \tag{8.19}$$

The presence of the current-carrying channels leads to additional energy losses. To evaluate these, it is necessary that the inductance, $L_K$, of the current-carrying channels be determined. This inductance can be approximated by using an expression that describes an arc discharge:

$$L_K \approx 2 l_K l_n \frac{b}{a} \approx 14 l_K, \tag{8.20}$$

where $a$ is the radius of the channel. By using Eq. 8.15, it has been found that the energy losses per centimeter due to the channel inductance is small, since:

$$W = W_H + W_{0c} + W_K = 0.5 I_H^2 (L_H + L_{0c} + L_K) \tag{8.21}$$

and

$$\frac{L_K}{L_H + L_{0c}} \ll 1. \tag{8.22}$$

In order to verify the above analysis, a series of MCGs was constructed and tested. Their coils were wound with wire in one, two, and three turns. Copper tubes were used as the liner. The piston was made from copper and had a mass of $M = 5.5$ g. A capacitor bank was used to create the initial magnetic field in the MCG. It was connected to the MCG through a vacuum spark gap switch, with a switching time of the order of tens of nanoseconds. The discharge of the capacitor bank was oscillatory in nature. The maximum current delivered to the MCG was $I_0 = 1.9$ kA. The time taken for the current to reach its peak value was $t = 200$ $\mu$s. The best results obtained in these experiments are presented in Table 8.2.

| | Quantities | Value |
|---|---|---|
| Measured | $l_r$ | 50 cm |
| | $z_3$ | 1.4 cm |
| | $z_2$ | 0.7 cm |
| | $d_n$ | 0.2 cm |
| | $L_0$ | 98 $\mu$H |
| | $R_1$ | 0.1 $\Omega$ |
| | $M$ | 70 g |
| | $\beta$ | $43^o$ |
| | $l_{0c}$ | 2.42 cm |
| | $\alpha$ | 42761/s |
| | $\gamma$ | 23 |
| | $V_{0c}$ | 1.3 kV |
| | $t_K$ | 0.14 $\mu$s |
| | $v_n$ | 1328 m/s |
| | $\phi$ | 0.15 |
| Experimental | $I_0$ | 1.9 kA |
| | $I_1$ | 6.6 kA |
| | $I_H$ | 3.0 kA |
| | $t_u$ | 1.8 $\mu$s |

TABLE 8.2. Parameters of the modified spiral generator.

Analysis of the above data demonstrates that the MCG design proposed in this section will permit the time taken to deliver energy to the load to be shortened from tens of microseconds to microseconds, without using additional peaking switches.

*Powering Relativistic Electron Beam Devices with MCGs*

In this section, an analysis is presented of the conditions required for powering an REB device, such as the MILO, with an MCG through a two-stage peaking switch. The first stage of the switch consists of an explosive switch, while the second stage is an electroexplosive switch.

Powering REB devices with high-inductance, high-speed spiral MCGs with explosive switches requires that the current risetimes are of the order of $\sim 0.1$ $\mu$s, which means that a peaking switch must also be used. Explosive current switches can only generate current pulses with a risetime of 1 $\mu$s. To further sharpen the rise time, a second switch, such as a plasma erosion opening switch (PEOS) or an electroexplosive switch, must be used.

The equivalent circuit diagram of an MCG with a two-stage switch is presented in Fig. 8.24. Time is counted from the moment the explosive switch is activated, i.e., it is assumed that switch $K_1$ is activated at $t = 0$. Switch $K_2$ is activated at time $t = t'$, which is the moment at which the voltage in the electroexplosive switch has reached the assigned value of $U_p$. The system of equations that describes this equivalent circuit is

$$J_K = J_B + J_P, \tag{8.23}$$

$$J_K(0) = J_0, \tag{8.24}$$

$$J_B(0) = J_0, \tag{8.25}$$

$$J_P(0) = J_\phi(0) = 0, \tag{8.26}$$

$$L_K \dot{J}_K + \dot{L}_K J_K + L_P \dot{J}_P + R_K J_K + R_\phi J_\phi = 0, \tag{8.27}$$

$$L_P \dot{J}_P + R_\phi J_\phi - J_B R_B = 0, \tag{8.28}$$

$$\dot{e} = \frac{J_\phi^2 R_\phi}{m}, \tag{8.29}$$

$$e(0) = 0, \tag{8.30}$$

for $0 < t < t'$, and

$$J_K = J_B + J_P, \tag{8.31}$$

$$J_K(t') = J'_K, \tag{8.32}$$

$$J_B(t') = J'_B, \tag{8.33}$$

$$J_P(t') = J'_P, \tag{8.34}$$

$$J_P = J_\phi + J_H, \tag{8.35}$$

$$J_\phi(t') = J'_\phi, \tag{8.36}$$

$$J_H(t') = 0, \tag{8.37}$$

$$L_K \dot{J}_K + \dot{L}_K J_K + L_P \dot{J}_P + L_H \dot{J}_H + R_K J_K + R_H J_H = 0, \tag{8.38}$$

$$L_P \dot{J}_P + L_H \dot{J}_H + R_H J_H - R_B J_B = 0, \tag{8.39}$$

$$L_H \dot{J}_H + R_H J_H - R_\phi J_\phi = 0, \tag{8.40}$$

$$\dot{e} = \frac{J_\phi^2 R_\phi}{m}, \tag{8.41}$$

$$e(t') = e', \tag{8.42}$$

for $t > t'$. The various symbols used in the above equations correspond to those shown in Fig. 8.24. The quantities $m$ and $e$ are the mass and internal energy of a unit mass of high explosive, respectively.

Because the time it takes to switch the current into the load is much less than the operating time of the MCG, the change in inductance can be approximated by the linear relationship

$$L_K(t) = L_0 - at, \tag{8.43}$$

where the rate of change of the load inductance, $a$, is a function of $L_0$. It is also assumed that the active resistance of the MCG, $R_K$, during the switching process does not change. The value of this resistance $R_K$ has been determined experimentally by switching an MCG into an inductive load, and it was found to be 10 mOhm. The dependence of the resistance of

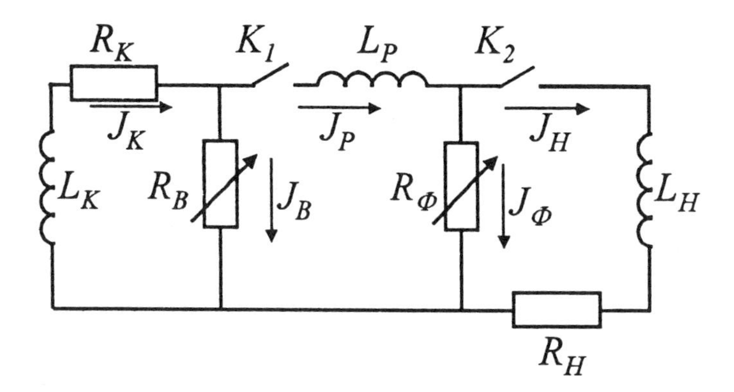


FIGURE 8.24. Equivalent circuit diagram for an MCG with a two-stage switch: $L_k$ and $R_k$ – inductance and resistance of MCG; $L_p$ – inductance of switch, $L_H$ and $R_H$ – inductance and resistance of the load; $R_b$ and $R_\phi$ – resistance of 1st and 2nd cascades; $K_1$ and $K_2$ – closing switches, and $J_k, J_B, J_p, J_\phi$, and $J_H$ – currents in corresponding branches of the circuit.

the explosive switch, $R_B$, on various parameters must be determined experimentally. The resistance of the electroexplosive switch can be calculated by using

$$R_\phi(t) = \frac{\rho l e(t)}{S}, \tag{8.44}$$

where $\rho$, $l$, and $S$ are the specific resistivity, length, and cross-sectional area of the electroexplosive conductor. If it is assumed that the load parameters are $L_H = 50$ nH and $R_H = 1$ Ohm and that the voltage at which switch $K_2$ activates corresponds to the breakdown voltage of the REB device, i.e., $U_P = 100$ kV, then the following parameters must be determined: $L_0$; $a$; the switch inductance $L_P$; the length $l$ and cross-sectional area $S$ of the electroexplosive conductor; the maximum input voltage to the load, i.e., $U_H = J_\phi R_\phi$, which is assumed to exceed 500 kV; the maximum input current to the load, which is assumed to exceed 200 kA; and the risetime of the current pulse into the load, which is approximately 0.1 $\mu$s. To ensure that the switch is operating at peak efficiency, the initial value of the current, $J_0$, at the moment the switch is opened should be kept as low as possible.

Before solving the governing equations, the range of values over which the parameters of the MCG can be varied must be defined. In this analysis, it will be assumed that the MCG has an inductance of 35 nH and that the initial current $J_0$ is limited to a value of 6 MA. The initial inductance, $L_0$, and the rate of change in inductance, $a$, at the moment the switches open, are interrelated, i.e., only one of them is an independent variable. It is also assumed that the initial inductance varies between 20 and 40 nH. These limits are based on the efficiency and electric field strength of typical MCGs, and the value of $a$ ranges from 0 to 7 nH/$\mu$s. The value of the switch inductance, $L_P$, ranges from 10 to 50 nH, with the lower limit of $L_P$ being determined by the dimensions of the explosive current switch and the upper limit by the dimensions of the energy source. The minimum length of the electroexplosive switch conductors is determined by the electric field strength of the explosion products, which is $\sim$ 10 kV/cm. Thus, the length of the conductor must be no more than 500 mm. No limits are placed on the cross-sectional area, $S$, of the conductors.

These calculations were based on the following assumptions:

- The length $l$ of the electroexplosive conductor does not change during the calculations, which is assumed to be 500 mm.
- The inductance of the MCG at the moment the switch opens the circuit is $L_0$.
- The MCG current $J_0$ is at its peak value at the moment the switch opens the circuit.

Based on these assumptions, and assuming that the MCG output voltage is $U_H \geq 500$ kV and that the output current is $J_H \geq 200$ kA, the optimal values of $L_P$ and $S$ are calculated. Using these values, the current $J_0$ is minimized and the optimal values of $L_P$ and $S$ are recalculated. If these values cannot be optimized, a new initial value of $L_0$ is selected and the optimization process repeated. The thickness $b$ of the electroexplosive switch determines the cross-sectional area $S$, and is found by using the expression $b = S/\pi D$, where $D$ is the diameter and has a value of 120 mm.

For example, the values of $L_P$, $S$, and $J_0$, given that $L_0 = 20$ nH and $a = 0$, can be optimized. Several test cases show that given a maximum current of $J_0 = 6$ MA, it is possible to achieve the required output characteristics of the MCG. However, at $J_0 = 6$ MA, the explosive current switch is operating at its peak value, which may lead to unstable operation of the MCG. It has been shown that slight deviations in the current $J_0$ from the required value may lead to noticeable reductions in the output voltage of the MCG. It has also been shown that the current pulse in the load circuit is less sensitive to deviations of $J_0$ from the required value. For this reason, the value of $J_0$ was first reduced to 5 MA, and then to 4 MA. From these new calculations, it was found that the required values for the output parameters of the MCG can be achieved for the given values of $L_0$ and $a$, if the current does not fall below 5 MA. Furthermore, the maximum voltage amplitude in the load is achieved when $J_0 = 5$ MA, if $L_P = 25$ nH and $b = 3$ $\mu$m, or when $J_0 = 4$ MA, if $L_P = 33$ nH and $b = 2$ $\mu$m. These are the optimal values of $L_P$ and $b$, for different values of $J_0$. The current amplitude in the load behaves similarly.

In order to determine the required output characteristics of the MCG, given that $J_0 = 4$ MA, a series of calculations were made for the case when $L_0 = 25$ nH and $a = 0$. It was found that the values of the required output parameters of the MCG could be attained, with the exception that the peak voltage amplitude in the load slightly exceeds its assigned value. It was also shown that the optimal values for $L_P$ and $b$ are 37 nH and 2 $\mu$m, respectively. However, it was also found that an $L_P$ of 37 nH for a foil length of 500 mm would lead to an unacceptable increase in the dimensions of the MCG. Therefore, a series of calculations was made to optimize the parameters of the MCG and the second stage of the switch when $J_0 = 4.5$ MA. It was found that the required output parameters of the second stage switch should lie within a significantly wider range of values of $L_P$ and $b$, than when $J_0 = 4$ MA. The optimal values of $L_P$ and $b$ were found to be 30 nH and 3 $\mu$m, respectively.

In summary, it was found that the output characteristics of the MCG with a two-stage switch required to drive a REB device can be achieved given that the following conditions are met: $L_P$ = 20–30 nH and $b$ = 3–4 $\mu$m. It can be expected that MCGs that meet these criteria will be able to generate output voltages of 700 kV and currents of 360 kA with rise times of 0.1 $\mu$s.

In addition, calculations were performed to determine the effects of load resistance on the output characteristics of the MCG. The analysis was based on an MCG that met the conditions described above for two types of loads: one with an active inductance ($L_H = 50$ nH and $R_H = 1$ Ohm) and one that is purely inductive ($L_H = 50$ nH and $R_H = 0$ Ohm). The calculations show that the presence of an active component in the load resistance leads to an increase in the input current into the load, while the load which was purely inductive results in a decrease in the input current to the load.

Incorporating all these results into the analysis, it has been shown that an MCG with a two-stage switch can deliver to REB devices, such as a MILO, voltage amplitudes no less than 500 kV and current amplitudes no less than 200 kA, where the load inductance is 50 nH and the active resistance is 1 Ohm. In this case, the generator–switch system must meet the following conditions: $J_0$ = 4.5–5.0 MA, $L_0$ = 20–25 nH, $L_P$ = 20–30 nH, $S$ = 1.0–1.5 mm$^2$, and $l = 500$ mm.

### *8.2.5 Transition Radiation Generators*

At present, efficient compact sources of electromagnetic energy capable of generating gigawatt power levels with pulse lengths of 1 to 20 ns in the wavelength band of 1 $\mu$m to 1 mm do not exist. In other words, there are no high power sources of electromagnetic energy that operate in the frequency regime between that of high-power microwave sources and high-power lasers. It has been proposed that the *transition radiation generator* (TRG) may be capable of generating high powers in this spectral band. The TRG emits transition radiation produced by high-current relativistic electron beams.

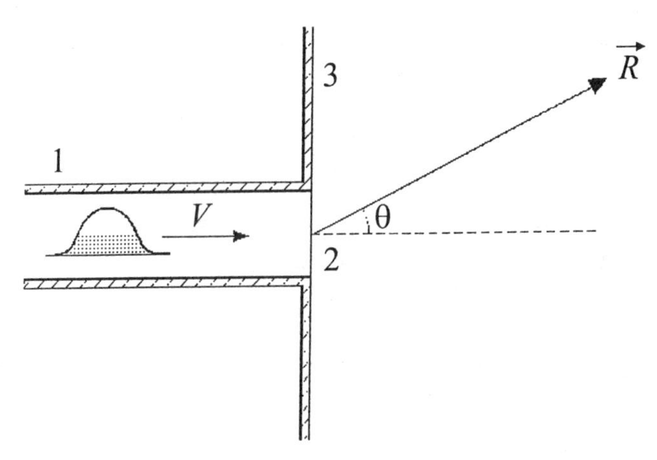


FIGURE 8.25. Waveguide transition radiation generator. When the electron bunch passes from the waveguide into free space, transition radiation is generated due changes in its electromagnetic environment.

In general, certain parameters of the TRG are probably worse than those of conventional electromagnetic sources, but they may be able to generate relative high powers in the spectral range 1 $\mu$m to 1 mm with relative rich spectral densities. While sources of electromagnetic energy have a number of applications including radar, communications, plasma chemistry, and ecology, communications seems to be the most near term application for TRGs, since the parameters of the radiated pulse precisely copy the parameters of the modulated electron beam used to generate the pulse.

*Transition radiation* is generated when charged particles moving in a straight line at a constant speed pass through or near an electrodynamic nonlinearity. These electrodynamic nonlinearities arise because of either inhomogeneities or time variations in the properties of a medium.

There are several mechanisms by which transition radiation may be generated, and all are based on the use of REBs. The TRG that is considered in this section consists of a high-voltage MCG, REB accelerator, and electromagnetic pulse (EMP) generator. The MCG needs to generate a high-voltage pulse; i.e., 0.6–0.8 MV, with a pulse length of 0.1–1.0 $\mu$s. Two EMP generators are considered. The first is simply a waveguide from which the electron beam passes into free space (vacuum or air) as shown in Fig. 8.25. Transition radiation is generated when the beam leaves the waveguide and enters free space. The electron beam also serves as the EMP antenna. The second type of EMP generator has a *converter* that converts the electron beam energy into EMP, which is then fed into a wide band antenna as shown in Fig. 8.26.

The TRG differs from traditional electromagnetic generators in several ways:

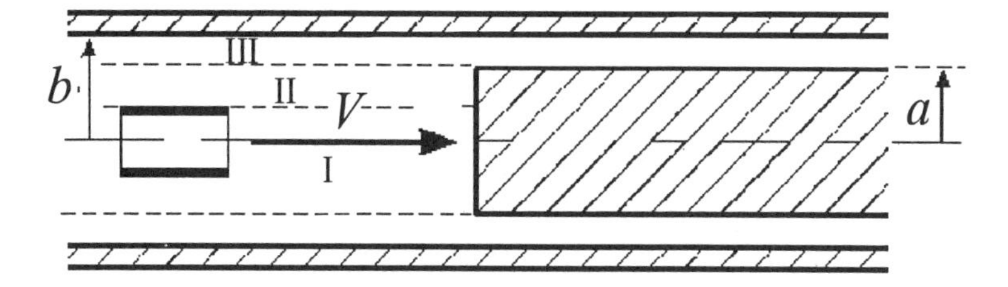


FIGURE 8.26. Transition radiation generator with converter. The converter transitions the electron beam energy into an EMP pulse, which is than radiated by a wide band antenna.

- The TRG does not require a pulse forming line, which is replaced with a modulated electron beam. Such a system will probably be more compact than the pulse forming line in traditional EMP generators.

- The shape of the radiated pulse is determined by the shape of the electron pulses formed by beam modulation. Therefore, by adjusting the shape of the electron pulse, the shape of the radiated pulse can be adjusted. The pulse length of the EMP depends on the pulse length of the electron bunch and the power in the EMP depends on the beam current and energy. Estimates show that the efficiency of TRGs having low beam energies and low currents or high beam energies and low currents is only a fraction of a percent. However, if short electron pulses with a pulse length of 0.1–10 ns, energy of 0.5–1.0 MeV, and peak current of 10–100 kA are used, the efficiency is tens of percent.

- The Fourier spectrum of the radiated pulse shows that the radiated energy lies primarily in the low-frequency portion of the spectral band 1 $\mu$m–1 mm. The characteristic frequencies in the spectrum of the TRG are determined by the pulse length of the electron bunch. There are spectral lines in the higher frequency regime of the spectrum, but the energy content falls off as the frequency increases.

- The energy can be shifted from the low frequencies of the spectrum towards the higher frequencies by decreasing the pulse length of the electron bunch. This has been done in linear resonant accelerators, which generate pulses with lengths that are a fraction of a millisecond.

The radiated output powers of these generators in the short wavelength domain, i.e., $< 1$ mm, may not exceed 1 GW. Realistically, the power levels may only be 1–10 MW for pulse lengths of 10–100 ns, but if the pulse length is decreased to 0.1 ns, the power level increases.

## 8.3 Direct-Drive Devices

In June, 1994, A.B. Prishchepenko at the Central Scientific and Research Institute for Chemistry and Mechanics published a paper on a class of devices, called direct-drive devices, in which MCGs connected to a small capacitive load are used to generate wide-band electromagnetic signals with a frequency content ranging from tens of megahertz to tens of gigahertz [8.28]. Prishchepenko called these devices *electromagnetic ammunition* (EMA). By *direct-drive devices* is meant that there is no source of electromagnetic waves, such as a vircator or MILO. That is, the energy from the MCG is coupled directly into the antenna. These devices range in size from that of a baseball to that of a 105 mm artillery shell. Subsequent papers [8.28–8.38] have revealed that there are several versions of EMA, some of which do not use MCGs as their power source, but rather MHDGs, PEGs, and FMGs, which were all briefly discussed in Chapter 1. In this section, some of these devices are examined in more detail. It should be pointed out that there is much that is not understood about how they work, such as how they generate frequencies greater than 1 GHz; i.e., 1–150 GHz, and high powers, i.e., megawatts to gigawatts.

### *8.3.1 Types of EMAs*

As noted above, there are several different types of EMAs. Some things that bond them together into a family are that they all rely on explosive power sources and employ small capacitive loads to generate electromagnetic energy, as well as their intended application. In identifying these devices (Fig. 8.27) , the names given to them by Prishchepenko are used, and they are

- explosive magnetic generator of frequency (EMGF),
- implosive magnetic generator of frequency (IMGF),
- cylindrical shock wave source (CSWS),
- spherical shock wave source (SSWS),
- magnetohydrodynamic source (MHDS),
- magnetohydrodynamic generator of frequency (MHDGF),
- piezoelectric generator of frequency (PEGF),
- ferromagnetic generator of frequency (FMGF),
- superconductive magnetic field shock wave former (SMFSWF).

EMAs that are combinations of those listed above have also been built. For example, the ferromagnetic generator (FMG) has been used as the

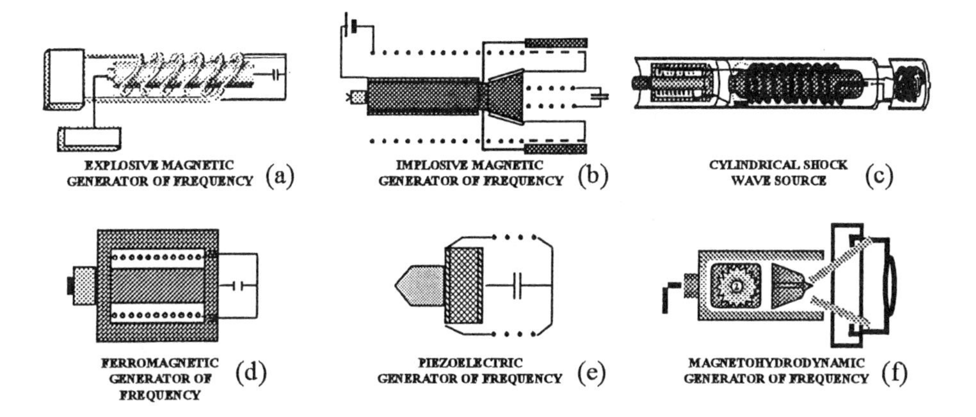


FIGURE 8.27. There are several versions of the EMA: (a) EMGF, (b) IMGF, (c) CSWS, (d) FMGF, (e) MHDGF, and (f) PEGF.

source of seed current for the EMGF, IMGF, and CSWS. The FMGF and PEGF have been joined to form the "combined generator of frequency" (CGF).

The CSWS was the first of this class of device to be tested, on March 2, 1983 [8.36]. One version of CSWS consists of an FMG that provides the seed current to a helical MCG, which drives a solenoid, that contains a working body placed within the solenoid as shown in Fig. 8.28. The current from the helical MCG is fed to the inner coil (1) to create a magnetic field in the working body. This field freely penetrates the working body (2), which is a material such as powdered silicon or monocrystalline cesium iodide. When the seed magnetic field in the solenoid reaches its peak value, the annular explosive charge (3) is detonated. The implosion converts the coil into a monolithic liner, which impacts on the working body, thus forming a converging and electrically conducting shock wave in the working body that compresses the magnetic field. The high pressures within the shock wave in the working body transitions the dielectric material within the shock front into a conducting material. It is this metalized shock front that completes the compression of the magnetic field and the conversion of chemical energy from the high explosive into electromagnetic radiation. The key to making this device work is that the magnetic pressure during the compression process does not exceed the hydrodynamic pressure until the radius of the shock wave is one one-thousandth that of the working body. One of the main advantages of this device is that Rayleigh–Taylor instabilities do not occur in monocrystals. The spherical version of the shock wave sources, i.e., the SSWS, is shown in Fig. 3.18, which is the most difficult and expensive to build.

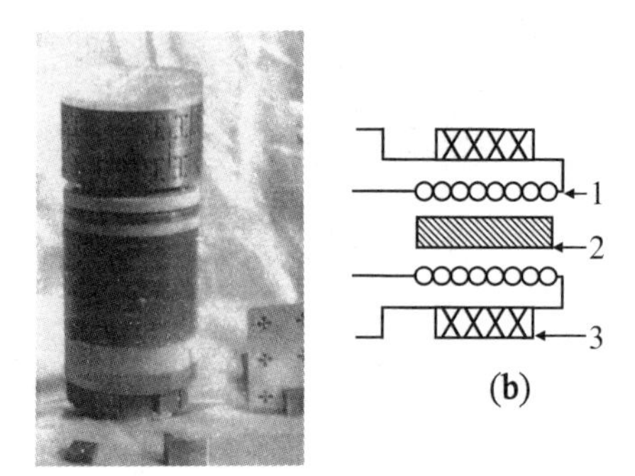


FIGURE 8.28. Photograph (a) and diagram (b) of a cylindrical shock wave source: 1 – coil, 2 – monocrystal, and 3 – explosives.

The EMGF shown in Fig. 8.29 consists of either permanent or electromagnets (1) and a circuit that includes an armature (2), helical coil (3), and capacitor (4). The EMGF is basically a parametric amplifier, which means that it operates only at those moments when the currents are substantial, since only under these conditions is work done by the armature against the magnetic field. Since the equivalent circuit of the device is basically an *LRC* circuit, the induced currents oscillate, and radiates electromagnetic energy. The key to making this device work is the small capacitor. The coil of the MCG also serves as the antenna for the device. This device will be discussed in more detail later.

The diagram of the IMGF is presented in Fig. 8.30. An initial current from either an FMG or capacitor is delivered the feed coil (1), which creates a magnetic field within the coil. When the feed current reaches its peak value, the annular explosive charge (2) is detonated. The feed coil is converted into a monolithic liner, which is used to compress the magnetic field. The radiating coil (3) is connected to the capacitor (4). During the initial stages of compression, there is no substantial diffusion of the magnetic field into the radiating coil. However, the converging liner forces the magnetic field into the radiating coil, which decreases its inductance rapidly, i.e., by a factor of 2 within 100 ns. Against the background of the carrier current pulse, which is of submicrosecond duration, oscillations with increasing frequency are formed. These oscillations radiate electromagnetic energy.

### 8.3.2 *Explosive Magnetic Generator of Frequency*

The explosive magnetic generator of frequency (EMGF) is based on the MCG. According to the literature, there are several variations of the EMGF.

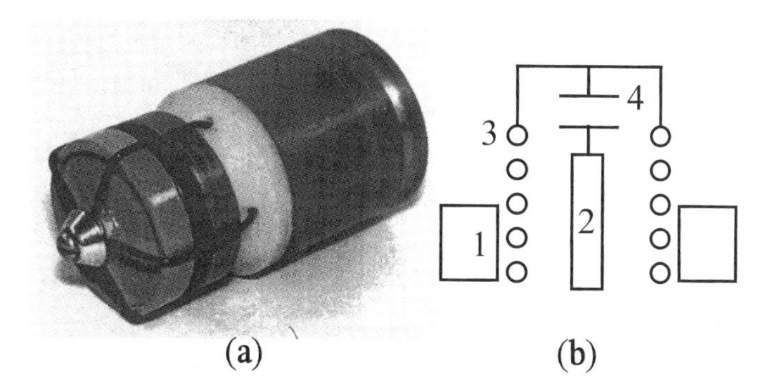


FIGURE 8.29. Photograph (a) and diagram (b) of an explosive magnetic generator of frequency: 1 – permanent magnets, 2 – armature, 3 – helical coil, and 4 – capacitor.

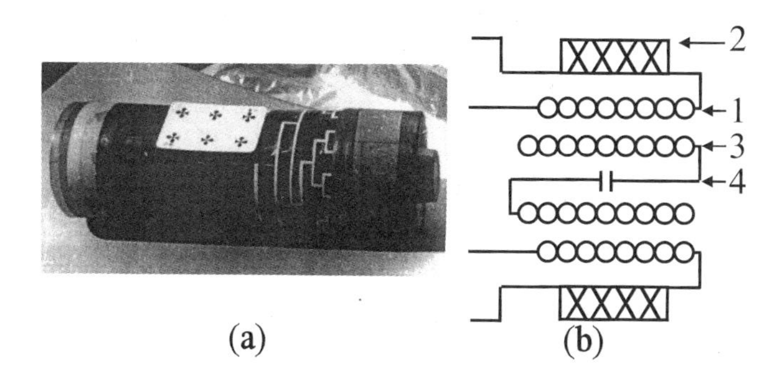


FIGURE 8.30. Photograph (a) and diagram (b) of an implosive magnetic generator of frequency: 1 – feed coil, 2 – explosives, 3 – radiating coil, and 4 – capacitor.

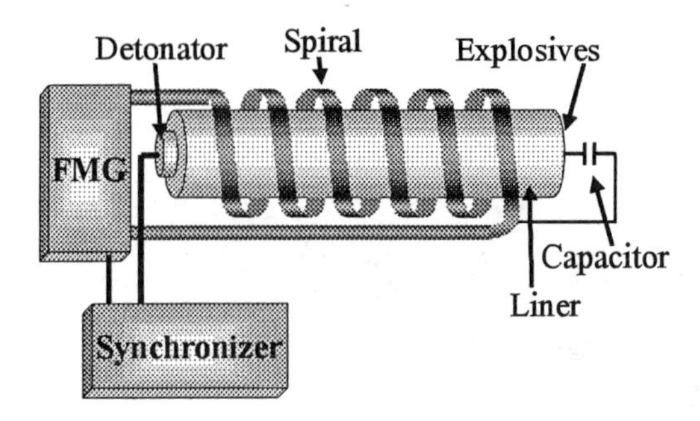


FIGURE 8.31. Single-stage explosive magnetic generator of frequency (EMGF).

Some of these are single stage devices, in which a single MK-1 or MK-2 type MCG is used to drive a capacitor. Others are multistage and consist of two or more MCGs. In the two stage devices, a low energy MCG serves as the seed source for a more powerful energy amplifier. Both versions will be briefly discussed in this section.

### EMGFs Based on a Single MCG

Consider the device in Fig. 8.31. It consists of a seed source, a synchronization circuit, a spiral MCG, and a low value capacitor. The seed source could either be a battery, capacitor, or an FMG. The critical component is the capacitor, which has a capacitance of between 100 to 1000 pF. By adding the capacitor, an LRC circuit is created which rings as the MCG generates electric current. It is this ringing that generates the microwaves, which according to [8.36] has a frequency content ranging from tens of megahertz to 150 GHz. To make these devices compact and light-weight, the spiral winding of the MCG also doubles as the antenna for the EMGF.

#### *Circuit Analysis*

The single-stage EMGF can be represented by the series $LRC$ circuit shown in Fig. 8.32. It is assumed that the inductance and capacitance are functions of time. Summing the potentials around the circuit, yields

$$\frac{d}{dt}(IL) + IR + \frac{1}{C}\int_0^t I\ dt = 0, \tag{8.45}$$

where $I$ is the current. At $t = 0$, $I = I_0$ and Eq. 8.45 can be rewritten as

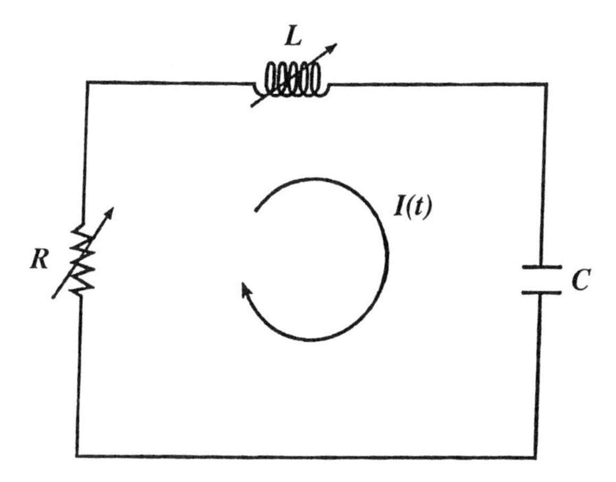


FIGURE 8.32. Equivalent circuit diagram of a single-stage EMGF.

$$L\frac{d^2I}{dt^2}+\left(2\frac{dL}{dt}+R\right)\frac{dI}{dt}+\left(\frac{d^2L}{dt^2}+\frac{dR}{dt}+\frac{1}{C}\right)I=0. \tag{8.46}$$

The inductance of a solenoid is given by

$$L=\frac{\mu N^2 A}{l}, \tag{8.47}$$

where it is assumed that the length of the solenoid is much greater than its radius. In Eq. 8.47, $\mu$ is the permeability, $N$ is the number of turns in the solenoid, $A$ is the cross-sectional area of the solenoid, and $l$ is the length of the solenoid.

When the explosive inside the liner is detonated, the area between the liner and the solenoid decreases, thus compressing the magnetic field. The area between the liner and the solenoid is

$$A=\pi\left(r_0^2-r_i^2\right)=\pi\left(r_0+r_i\right)\left(r_0-r_i\right). \tag{8.48}$$

If it is assumed that the annular region between the liner and solenoid is relatively small, then Eq. 8.48 becomes

$$A\simeq 2\pi r_0^2\left(1-\frac{r_i}{r_0}\right)\simeq 2\pi r_0^2 e^{-\frac{r_i}{r_0}}. \tag{8.49}$$

The radius $r_i$ increases at a rate of $r_i = vt$, where $v$ is the velocity of the liner following the detonation of the high explosive. Therefore, the inductance is

$$L = \frac{2\pi\mu(Nr_0)^2}{l} e^{-\frac{vt}{r_0}}, \tag{8.50}$$

where $r_0/v$ is the characteristic time $\tau_L$ and

$$L_0 = \frac{2\pi\mu(Nr_0)^2}{l}. \tag{8.51}$$

The velocity of the liner is assumed to be the Gurney velocity, which is defined by the expression

$$v = \sqrt{-2\Delta e}\left(\frac{\frac{C}{M}}{1+\frac{C}{2M}}\right)^{\frac{1}{2}}, \tag{8.52}$$

where $\sqrt{-2\Delta e}$ is the Gurney constant, $C$ is the mass of the high explosive, and $M$ is the mass of the liner. The Gurney constant varies with the type of explosive used, and for TNT, it is equal to $2.315 \times 10^3$ m/s. Thus, the time-dependent change in the solenoid current depends on the type of explosive used and the dimensions of the device.

It is assumed that the time-dependent behavior of the inductance and the resistance are the same and are given by

$$L = L_0 e^{-\frac{t}{\tau_L}} \tag{8.53}$$

and

$$R = R_0 e^{-\frac{t}{\tau_L}}, \tag{8.54}$$

where $\tau_L$ is a characteristic time which depends on the dimensions of the device, the type of explosive used, and other physical characteristics of the device. Substituting Eqs. 8.53 and 8.54 into Eq. 8.46 and differentiating yields

$$\frac{d^2I}{dt^2} + \left(\frac{R_0}{L_0} - \frac{2}{\tau_L}\right)\frac{dI}{dt} + \left[\left(\frac{1}{\tau^2} - \frac{R_0}{\tau_L L_0}\right) + \frac{e^{-\frac{t}{\tau_L}}}{L_0 C}\right] I = 0. \tag{8.55}$$

To solve this equation, let

$$z^2 = e^{\frac{t}{\tau_L}}. \tag{8.56}$$

Differentiating yields

$$\frac{dz}{dt} = \frac{z}{2\tau_L}, \tag{8.57}$$

so that the following identities can rewritten as

$$\frac{dI}{dt} = \frac{dI}{dz}\frac{dz}{dt} = \frac{z}{2\tau_L}\frac{dI}{dz} \tag{8.58}$$

and

$$\begin{aligned} \frac{d^2I}{dt^2} &= \frac{d}{dt}\left(\frac{dI}{dt}\right) = \frac{d}{dz}\left(\frac{dI}{dt}\right)\frac{dz}{dt} \\ &= \left(\frac{z}{2\tau_L}\right)^2 \frac{d^2I}{dt^2} + \frac{z}{4\tau_L^2}\frac{dI}{dz}. \end{aligned} \tag{8.59}$$

Substituting Eqs. 8.58 and 8.59 into Eq. 8.55 yields

$$\frac{d^2I}{dz^2} + \frac{1}{z}\left[1 + 2\tau_L\left(\frac{R_0}{L_0} - \frac{2}{\tau_L}\right)\right]\frac{dI}{dz} \tag{8.60}$$

$$+ \left[\frac{4\tau_L^2}{L_0C} + \frac{4}{z^2}\left(1 - \frac{R_0\tau_L}{L_0}\right)\right] I = 0.$$

The general solution of this equation is

$$I = z^{2-\nu}\left[AJ_\nu(\lambda z) + BY_\nu(\lambda z)\right], \tag{8.61}$$

where $\nu = R_0\tau_L/L_0$, $\lambda = 2\tau_L/\sqrt{L_0C}$, $A$ and $B$ are constants that must be determined, and $J(x)$ and $Y(x)$ are Bessel functions of the first and second kind, respectively, of order $\nu$ and argument $x$.

At $t = 0$, the initial conditions are

$$I(0) = I_0 \tag{8.62}$$

and

$$\left.\frac{dI}{dt}\right|_{t=o} \frac{I_0}{\tau_L}\left(1 - \frac{R_0\tau_L}{L_0}\right). \tag{8.63}$$

Using the identities

$$\frac{dJ_\nu(\lambda z)}{dz} = \lambda \left[ J_{\nu-1}(\lambda z) - \frac{\nu}{\lambda z} J_\nu(\lambda z) \right], \tag{8.64}$$

$$\frac{dY_\nu(\lambda z)}{dz} = \lambda \left[ Y_{\nu-1}(\lambda z) - \frac{\nu}{\lambda z} Y_\nu(\lambda z) \right], \tag{8.65}$$

and after significant computation, expressions for $A$ and $B$ are

$$A = \frac{I_0 Y_{\nu-1}(\lambda)}{J_\nu(\lambda)Y_{\nu-1}(\lambda) - Y_\nu(\lambda)J_{\nu-1}(\lambda)}, \tag{8.66}$$

$$B = -\frac{I_0 J_{\nu-1}(\lambda)}{J_\nu(\lambda)Y_{\nu-1}(\lambda) - Y_\nu(\lambda)J_{\nu-1}(\lambda)}. \tag{8.67}$$

Using the identity

$$J_\nu(\lambda)Y_{\nu-1}(\lambda) - Y_\nu(\lambda)J_{\nu-1}(\lambda) = \frac{2}{\pi\lambda}, \tag{8.68}$$

Eqs. 8.66 and 8.67 become

$$A = \frac{\pi\lambda I_0}{2} Y_{\nu-1}(\lambda), \tag{8.69}$$

$$B = -\frac{\pi\lambda I_0}{2} J_{\nu-1}(\lambda). \tag{8.70}$$

Therefore, the solution of Eq. 8.60 is

$$I = \frac{\pi\lambda I_0}{2} z^{2-\nu} \left[ Y_{\nu-1}(\lambda)J(\lambda z) - J_{\nu-1}(\lambda)Y_\nu(\lambda z) \right]. \tag{8.71}$$

When the arguments of the Bessel functions are large ($\lambda \geq 10$), then the identities

$$J_\nu(x) = \sqrt{\frac{2}{\pi x}} \cos\left[ x - \frac{\pi}{4}(1+2\nu) \right], \tag{8.72}$$

$$Y_\nu(x) = \sqrt{\frac{2}{\pi x}} \sin\left[ x - \frac{\pi}{4}(1+2\nu) \right]. \tag{8.73}$$

can be substituted into Eq. 8.71 to obtain

$$I = I_0 z^{\frac{3}{2}-\nu} \cos\left[\lambda(1-z)\right]. \tag{8.74}$$

Examining Eq. 8.74, shows that the induced current is oscillatory. At very early times in the compression process, the following approximations hold:

$$z \approx 1 + \frac{t}{2\tau_L}, \tag{8.75}$$

$$I \approx I_0 \cos\left(\frac{t}{\sqrt{L_0 C}}\right), \tag{8.76}$$

so that the ringing frequency of the current in the EMGF is

$$\omega = 2\pi f = \frac{\pi}{2\tau \ln\left[1 + \frac{\pi}{\tau_L}\sqrt{L_0 C} e^{-\frac{t}{\tau_l}}\right]}, \tag{8.77}$$

which increases as time increases.

*Numerical Example*

The EMGF has been tested, and the results are presented in [8.32]. The reported values of the various parameters are: $I_0 = 20$ A, $L_0 = 50$ $\mu$H, $C = 1$ pF, $\tau_L = 3.5$ $\mu$s, $\epsilon = 1.14 \times 10^{-2}$, $S = 2$, $\nu = 0.72$. No value was given for $R_0$, however using the definition of $\nu$ and the values of the above quantities, $R_0$ can be calculated to be 10.3 $\Omega$.

It should be noted that for $\nu = 0.72$, the amplitude of the current, defined by Eq. 8.74, increases continuously with time. Thus, Eq. 8.74 does not represent a physically acceptable solution throughout the operation of the MCG. The experimental results, Figs. 8.33 and 8.34, indicate that $\nu$ should be a function of time. Using semiempirical arguments, it is found that there is very good agreement between Eq. 8.74 and experimental results, if the value of $\nu$ is taken as

$$\nu = \begin{cases} \frac{R_0 \tau_L}{L_0}, & t \leq t_0 \\ \frac{R_0 \tau}{L_0} e^{T_A(t-t_0)}, & t > t_0 \end{cases}, \tag{8.78}$$

where $t_0 = 6$ $\mu$s and $T_A = 1.8 \times 10^5$. The factor $T_A$ was obtained by fitting the model for the resistance to the published curves for the EMGF.

In Eq. 8.74, when $\nu > 3/2$ for $t > 6$ $\mu$s, the current begins to decay exponentially with time, with the results of these calculations being in

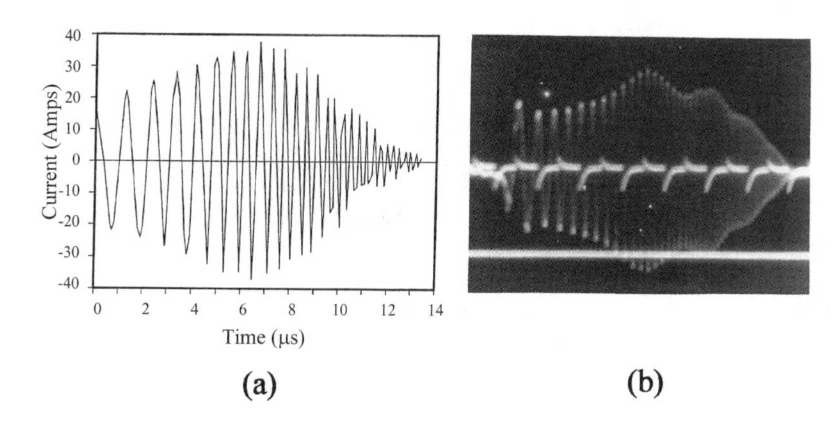


FIGURE 8.33. Plotting Eq. 8.74 using Eq. 8.78, the output current waveform of the EMGF has a "fish" shape structure.

good agreement with the experimental results in [8.32]. The significance of this result is that Eq. 8.74 has only a single physical parameter that needs to be studied in order to understand the operation of the inductive EMGF. Once the behavior of this parameter is understood, this model can be used to optimize the design of the EMGF.

Figure 8.33 shows an oscillogram of the Rogowski signal from an EMGF and the plot of Eq. 8.74. In both the oscillogram and the plot, it can be seen that a ringing is superimposed on the initial exponential growth, and then decay, of the current. It can also be seen that the ringing frequency changes with time.

Figure 8.34 compares plots of the envelope of the current output of the EMGF calculated using Eq. 8.74 to that in [8.32]. The experimental and Prishchepenko curves are taken from [8.32] and the semiempirical model curve calculated by using Eq. 8.74. As can be seen, there is good agreement between the curve calculated using the model proposed in this section and the experimental curve.

In June 1997, two sets of experiments were conducted on the EMGF. The first test was a joint test of Prishchepenko's devices with the High Mountain Geophysical Institute in Nalchik, Russia. The second test was of U.S. version of the EMGF and this test took place at the Energetic Research Center in Socorro, NM. In the Nalchik experiments, the current waveforms in Figs. 8.35 and 8.36 were measured with a Rogowski coil. A photograph of the EMGF tested at Nalchik is presented in Fig. 8.37 and photographs of two versions of the EMGF tested at Socorro are presented in Fig. 8.38. The "fish" waveform for the electrical current was observed in both sets of tests. As can seen in the Rogowski measurements from the Nalchik test, the ringing decays to zero in 13 $\mu s$, but the generator continues to function for 32 $\mu s$. It is unclear why the resistance apparently increases

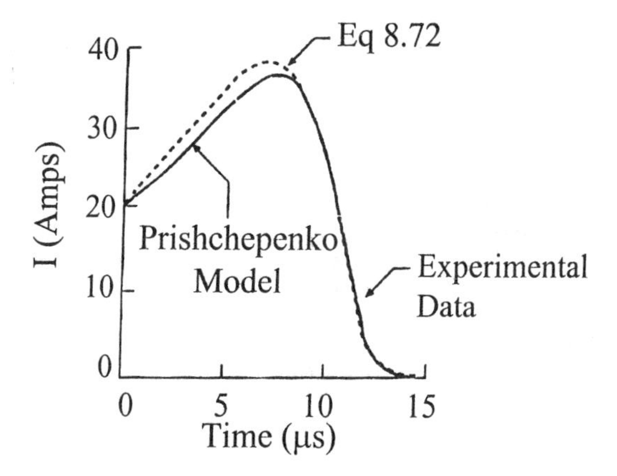


FIGURE 8.34. Plots of envelope of EMGF coil current versus time.

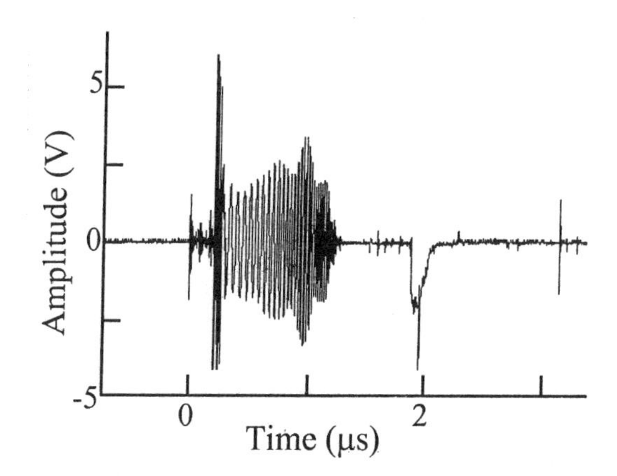


FIGURE 8.35. Coil current measured in an EMGF—shot 1.

from milliohms (the typical resistance of helical MCGs) to values as high as 10 Ohms. Without knowing more about the design of the EMGFs tested, two theories have evolved. One is that, initially, the resistance of the EMGF decreases due to increased numbers of electrons in the plasma formed in the flux compressor, but as the electron density increases, their mobility decreases, thus increasing the resistance. The other is that there may be an opening switch in the circuit. When the current reaches a certain value, the switch opens and the ringing dampens out. This would lead to the formation of high voltages within the EMGF that could account for the generation of a wide band RF signal.

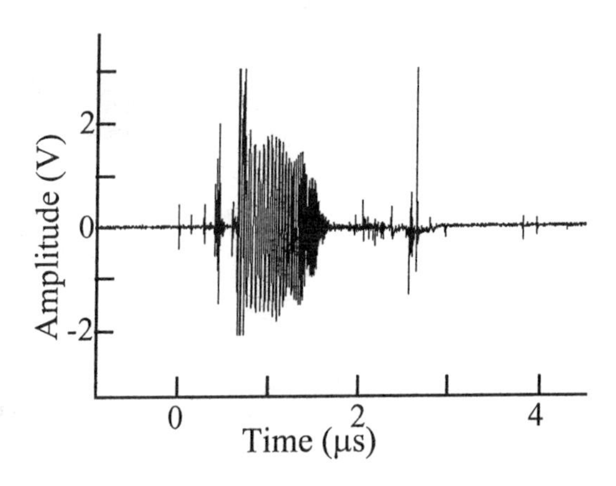


FIGURE 8.36. Coil current measured in an EMGF—shot 2.

FIGURE 8.37. EMGF under test at a test range belonging to the High Mountain Geophysical Institute in Nalchik, Russia in June, 1997.

FIGURE 8.38. A U.S. version of EMGF under test at the Energetic Material Research Center in Socorro, NM. (Courtesy of Explosive Pulsed Power, Inc.)

*Verification of Electrical Resistance Model*

Based on experimental data, the resistance of the EMGF first decreases and then at some point increases. A semiempirical model of these changes was derived from the circuit equations for the EMGF. To verify this model, it was applied to the EF-3, MARK IX, and FLEXY I generators.

Unlike the EMGF, these three devices were tested without a capacitive load; nevertheless, the basic analysis and the semiempirical model developed for the resistance of the EMGF should be applicable. The basic circuit equation for all three generators is

$$\frac{d(LI)}{dt} + RI = 0. \tag{8.79}$$

If it is assumed that the initial time behavior of the inductance and the resistance is the same as that for the EMGF, then the resistance model for all three generators is:

$$R = R_0 e^{-\frac{t}{\tau}} \tag{8.80}$$

for $0 \leq t \leq t_0$, and

$$R_0(1 + T_A t) \exp\left[-\left(\frac{t}{\tau} - T_A(t - t_0)\right)\right] \tag{8.81}$$

for $R_0(1 + T_A t) \exp\left[-\left(\frac{t}{\tau} - T_A(t - t_0)\right)\right]$, where $\tau$ is a characteristic time and the integration of Eq. 8.79 is straightforward:

$$I = I_0 e^{\frac{t}{\tau}\left(1-\frac{R_0\tau}{L_0}\right)} \equiv I_0 z^{2(1-\nu)}, \tag{8.82}$$

where $\nu$ and $z$ have the same definition used for the EMGF. Using this semiempirical model for $\nu$, Eq.8.82 becomes

$$I = I_0 \exp\left[\frac{t}{\tau_L}\left(1 - \frac{R_0\tau_L}{L_0}\right)\right] \tag{8.83}$$

for $t \leq t_0$, and

$$I = I_0 \exp\left[\frac{t}{\tau_L}\left(1 - \frac{R_0\tau}{L_0}\exp\left(T_A(t-t_0)\right)\right)\right] \tag{8.84}$$

for $t > t_0$, where $\tau$, $R_0$, and $t_0$ must be determined empirically. Substituting Eqs. 8.83 and 8.84 into Eqs. 8.80 and 8.81 and solving for $R$, provides a generalized model for the resistance.

The parameters of the EMGF, EF-3, MARK IX, and FLEXY I are presented in Table 8.3. Using these parameters and plotting Eqs. 8.83 and 8.84, it can be seen in Fig. 8 .39–Fig. 8.41 that the proposed model is in good agreement with the experimental data. It should be noted that the factor $1.8 \times 10^5$ in Eqs. 8.83 and 8.84 is applicable to all four helical generators. The physical significance of this is not presently understood, but the fact that it retains the same value for all four different generator designs with different loads is interesting. In order to understand the physical significance of this factor, the similarities of these four generators must be examined. These include that they

- are all helical generators,
- have a finite distance between the liner and the coil, and
- are all gas filled (air in the case of the EMGF, EF-3, and FLEXY I and $SF_6$ in the case of the MARK IX).

### Multistage EMGFs

In a multistage EMGF, a low-energy MCG is used as the prime energy source for a second fast spiral MCG that serves as an energy amplifier, as shown in Fig. 8.42. The seed source for the low-energy MCG may be either a capacitor or a FMG. However, a different design, where the magnetic

| Parameter | EMGF | EF-3 | Mark IX | FLEXY I |
|---|---|---|---|---|
| $I_0$ (kA) | 0.020 | 5.0 | 412.0 | 48.0 |
| $L_0$ ($\mu$H) | 50.0 | 650.0 | 6.0 | 35.0 |
| $\tau$ ($\mu$s) | 3.5 | 31.7 | 39.7 | 30.0 |
| $t_0$ ($\mu$s) | 6.0 | 200.0 | 150.0 | 133.0 |
| $R_0$ ($\Omega$) | 10.3 | 0.0285 | 0.000238 | 0.00205 |

TABLE 8.3. Empirical parameters for the EMGF, EF-3, Mark IX, and FLEXY I MCGs.

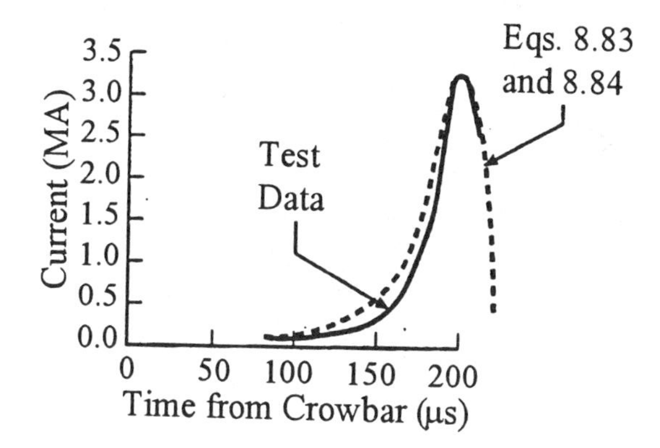


FIGURE 8.39. Comparison of the plot of Eqs. 8.83 and 8.84 using the proposed resistance model with experimental data for the EF-3 generator.

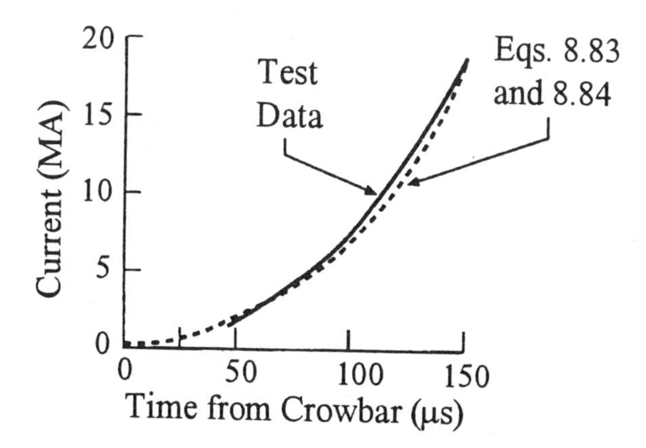


FIGURE 8.40. Comparison of the plot of Eqs. 8.83 amd 8.84 using the proposed resistance model to experimental data for the MARK IX generator.

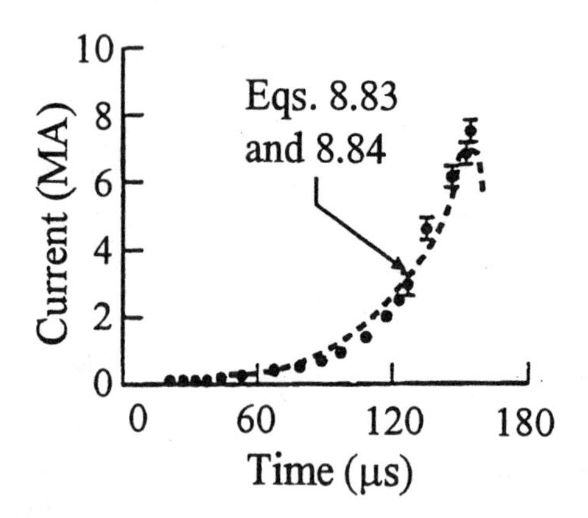


FIGURE 8.41. Comparison of the plot of Eqs. 8.83 and 8.84 using the proposed resistance model to experimental data for the FLEXY I generator.

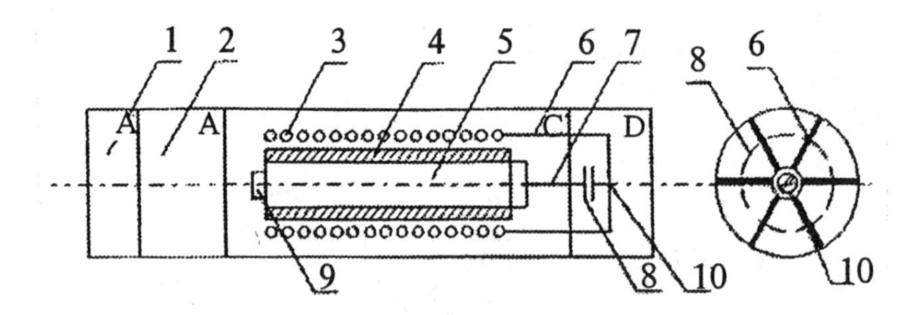


FIGURE 8.42. Multistage EMGF: A – synchronization and control unit, B – low-energy MCG, C - spiral MCG (energy amplifier), and D - small capacitor. The major components are—1 – synchronization and control unit, 2 – prime energy source, 3 – spiral MCG, 4 – metal liner, 5 – explosives, 6 – conductors that connect the sprial coil of the high-energy MCG to the capacitive load, 7 – connectors that connect the liner of the MCG to the capacitive load, 8 – low value capacitor, 9 – detonator for the high-energy MCG, and 10 – cables connecting the capacitive load and high-current conductors.

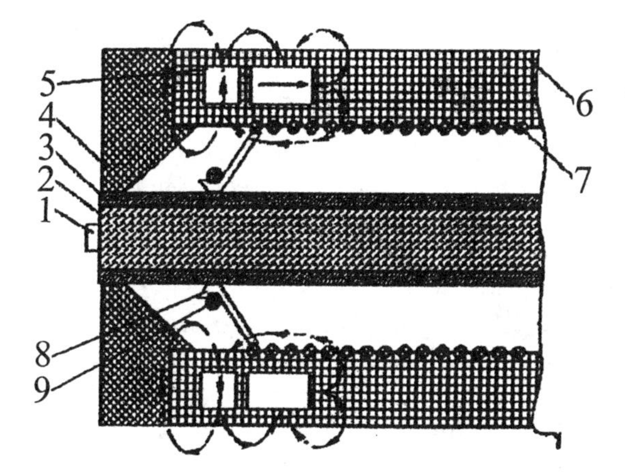


FIGURE 8.43. Low-energy sprial MCG with permanent magnet system and magnetic flux concentrators: 1 – detonator, 2 – explosives, 3 – metal liner, 4 – centering flange, 5 – elements of magnet system, 6 – casing, 7 – sprial winding, 8 – mobile contact, and 9 – holding ring.

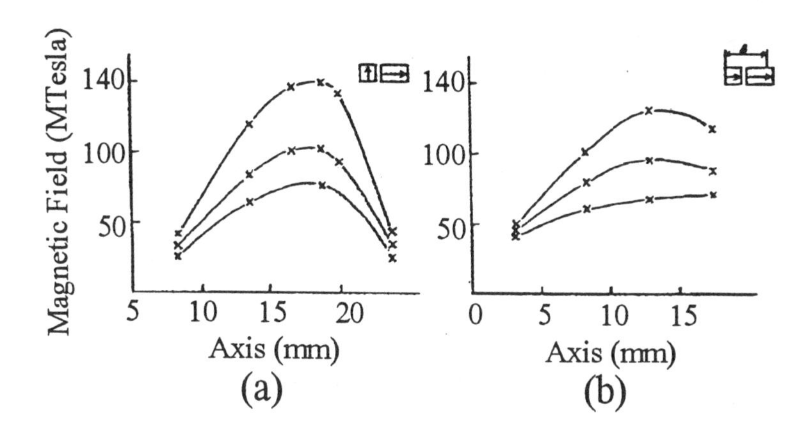


FIGURE 8.44. Distribution of magnetic field for magnetic system with concentration (a) and without concentration (b). The radius is specified in millimeters for each curve, with the axis of the MCG being the $x$-axis and representing the distance from one end of the MCG to the other end.

field in the low energy MCG is created by permanent magnets, will be considered.

The reason for using compact low-energy MCGs as a seed source for energy amplifiers is to generate the highest possible gains in magnetic energy and electrical current. In addition, since compact devices having a diameter of 30 to 50 mm are being considered, it is feasible to employ multisegmented permanent magnets to create the initial magnetic field in the low-yield MCGs. The use of modern magnetic materials, as well as magnetic flux concentrators, permits the generation of magnetic fluxes sufficient for the operation of the low-energy MCG. The schematic of such a low-energy MCG with permanent magnets is presented in Fig. 8.43.

When a low-energy MCG serves as the seed source for an MCG energy amplifier, it does not require optimized parameters. Therefore, the magnetic flux density created within the spiral winding of the low energy MCG is limited by the capabilities of the magnetic materials used and is on the order of a tenth of a Tesla. The magnetic field is usually concentrated in the initial sections of the windings, where the greater density of turns per unit length permits the generation of the maximum magnetic flux from the limited initial magnetic induction. The magnetic flux can be concentrated in the initial section of the spiral winding by superimposing two types of magnetic elements. The distribution of the magnetic field within the magnetic system that creates flux concentration is presented in Fig. 8.44.

The energy amplifier is connected to a small capacitive load and, with the exception of using a low-energy MCG as the seed source, behaves like the single-stage EMGF described in the previous section. The main reason for using a multistage EMGF is to enhance its power output.

### *8.3.3 Cylindrical Shock-Wave Source*

Shock-wave sources were discussed in Section 3.8. In this section, the focus is on the cylindrical version of a shock-wave generator. A photograph and schematic drawing of the CSWS is presented in Fig. 8.45. A feed current from an MCG is used to create a magnetic field of about 1 T within coil 1. This field penetrates into the working body 2, which is a single nonconducting crystal such as CsI. When the current in the coil is at its peak value, the annular explosive charge 3 is detonated and coil 1 is converted into a monolithic liner. The converging liner compresses the magnetic field in the annular region between the liner and the working body, and then impinges on the surface of the working body. This in turn creates a shock wave within the working body (Fig. 8.46). At the high shock pressures, which range from hundreds of kilobars up to 1 Mbar, the nonconducting crystal becomes conducting, creating a converging electrically conducting shock front. This shock front further compresses the magnetic field, but in such a way that the magnetic pressure within the unshocked region remains less than the hydrodynamic pressure of the shock wave. The shock

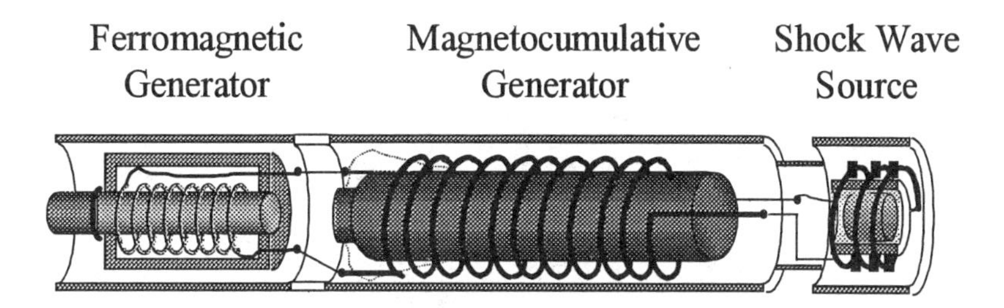


FIGURE 8.45. Cylindrical shock wave source consists of three major components: the seed source which is either a capacitor or a ferromagnetic generator, an MCG, and a solenoid containing a monocrystalline working body such as CsI. When current in the solenoid and, thus, the magnetic field in the monocrystal is at its peak value, the explosive is detonated and drives the solenoid into the surface of the crystal. Owing to the impact of the solenoid, a converging shock wave is generated within the crystal. At shock pressures, the crystal becomes electrically conducting and the converging shock wave compresses the magnetic field.

front converges to a radius that is one-thousandth that of the radius of the working body. Very high magnetic fields are generated within the working body, and at some point the shock wave breaks down and the compressed magnetic energy is reportedly [8.32] radiated as microwaves within a pulse length of a nanosecond. In addition, azimuthal currents flow through the converging shock front, which also radiates electromagnetic energy. It is also possible to use permanent magnets to create the initial magnetic field in the working body, as is done in the spherical version of the SSWS [8.35].

As discussed in Chapter 2, the main effect that limits the peak fields that can be generated in MCGs is Rayleigh–Taylor instabilities. The use of semiconductors, powders, and porous materials in which the conductivity sharply increases under compression allows stable shock-wave fronts to be obtained, which work as a piston to compress magnetic fields. The key parameter in this process is the variable conductivity of the working material. To model this, a 1D cylindrical MHD code was developed to perform computer simulations [8.39]. This code accounts for changes in electrical conductivity.

To account for the 1000-fold radial compression reported in [8.32], an effective rezoning procedure was used with a conservative finite difference scheme to approximate the governing equations. By choosing different models for conductivity growth, the time history of the magnetic field ahead of the shock wave can be calculated. It was found that this time history and the distribution of the magnetic field in the conducting medium is dependent on the model used for conductivity growth. It was also found that

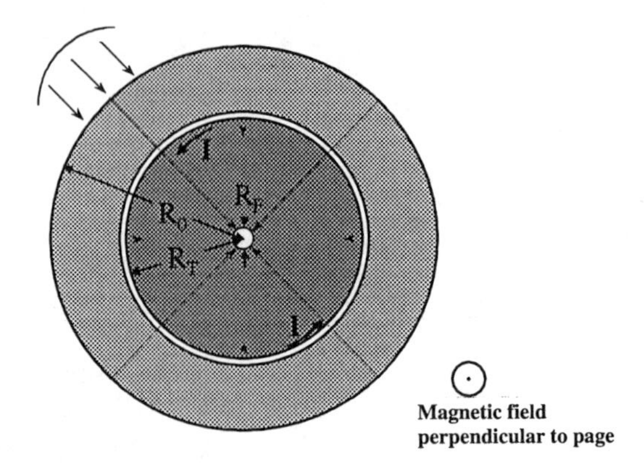


FIGURE 8.46. The shock wave converges to a radius one one-thousandth that of the crystal. This is thought to be due to the magnetic pressure, which retards the motion of the shock wave, remaining less than that of the dynamic shock pressure because of diffusion of the magnetic field through the shock front. As the converging conducting shock wave compresses the magnetic field within the working body, currents are induced within the shock front.

breakdown of the nonconducting medium ahead of the shock wave limits the maximum achievable magnetic fields [8.39].

The working equations for the cylindrical compression of a magnetic field by an imploding shock wave in a medium with variable conductivity are

$$\frac{\partial}{\partial t}\left(\frac{1}{\rho}\right) = \frac{\partial(rv)}{\partial s} \tag{8.85}$$

representing the conversation of mass,

$$\frac{\partial v}{\partial t} = -r\frac{\partial p}{\partial s} - r\frac{\partial}{\partial s}\frac{H^2}{8\pi} \tag{8.86}$$

representing the conservation of momentum,

$$\frac{\partial \varepsilon}{\partial t} = -p\frac{\partial(rv)}{\partial s} + q_j \tag{8.87}$$

representing the conservation of energy,

$$\frac{\partial}{\partial t}\frac{H}{\rho} = -\frac{\partial(rE\}}{\partial s} \tag{8.88}$$

representing magnetic diffusion,

$$\frac{\partial r}{\partial t} = v, \tag{8.89}$$

representing the equation of motion, where $t$ is time in microseconds, $r$ is radial distance in centimeters, $\rho$ is density in g/cm$^3$, $v$ is the velocity in cm/s, $\varepsilon$ is the specific internal energy, $p = p(\varepsilon, \rho)$ is the hydrodynamic pressure in megabars, $H$ is the axial component of the magnetic field in megagauss, $E_\phi = -(\rho r/4\pi\sigma)\,\partial H/\partial s$ is the angular component of the electric field, $q_j = \sigma E^2/\rho$ is the joule heating, $E$ is the electric field strength in $10^4$ V/cm, $\sigma = \sigma(\varepsilon, \rho, E, t)$ is the electrical conductivity of the medium in $10^3\ \Omega^{-1}\text{cm}^{-1}$, and $S$ is the Lagrangian mass coordinate $S = \int_0^r \rho(r')r'dr'$. The pressure $p$ depends on the equation of state of the working material. Neglecting the thermal conductivity in Eq. 8.87 and assuming that $\sigma(\varepsilon, \rho, E, t) \geq \sigma_0 > 0$ everywhere, the boundary conditions are $p - P_{out}(t)$ and $H = H_0$ at $r = R_{out}(t)$ and $v = 0$ and $\partial H/\partial s = 0$ at $r = 0$.

Since it is difficult to find an expression for the rate of change of conductivity, when the increase is from an almost nonconducting state to a highly conductive one, an empirical relationship is used [8.39]:

$$\sigma = \max(\sigma_0, \sigma_d, \sigma_{BD}), \tag{8.90}$$

where $\sigma_0$ is the initial conductivity of the working body, $\sigma_d$ is the conductivity due to the increase in density, and $\sigma_{BD}$ is the conductivity due to breakdown. An empirical relationship for $\sigma_d$ is

$$\sigma_d = \begin{cases} \sigma_{\max}\dfrac{\left(\frac{\rho}{\rho_0} - \rho_\sigma\right)^{N_e}}{\sigma_1 + \left(\frac{\rho}{\rho_0} - \rho_\sigma\right)^{N_e}}, & \text{for } \left(\frac{\rho}{\rho_0} - \rho_\sigma\right) > 0 \\ 0, & \text{for } \left(\frac{\rho}{\rho_0} - \rho_\sigma\right) \leq 0 \end{cases} \tag{8.91}$$

By varying $\rho_\sigma$ in the above equation, it can be seen in Fig. 8.47 that two physically different cases can be observed. The first is when the conductivity only becomes appreciable near the final stage of compression behind the shock wave, and the second is when the conductivity starts to become substantial during the initial stages of compression. If the latter is true, the magnetic field "freezes" in the slightly compressed region, which increases as the density increases. An analogous situation arises if the material in front of the shock wave becomes conducting due to breakdown. Breakdown occurs because of both high electric fields generated in front of the shock wave and the formation of ionizing radiation in the shock wave. A good approximation of the electric field near the shock wave is $E = D_{SW}H$, where $D_{SW}$ is the velocity of the shock wave. A simple model for avalanche breakdown is

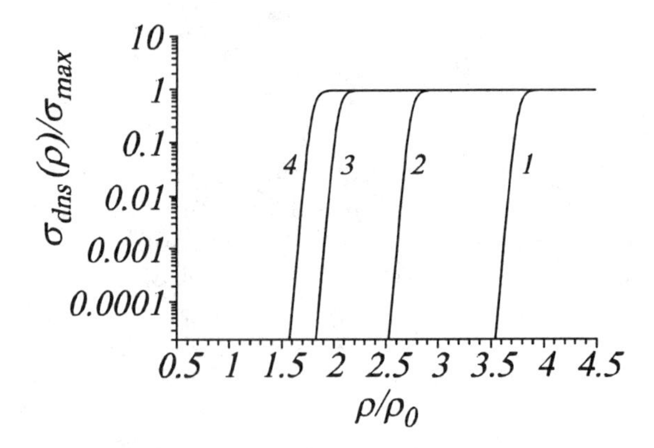


FIGURE 8.47. Changes in conductivity due to changes in density [8.39]. Two physically different cases can be observed: The first is substantial changes in conductivity during the initial stages of compression, which results in the magnetic field being frozen in the compressed region and increasing as the density increases. The second is substantial changes during the final stages of compression.

$$\frac{d\sigma_{BD}}{dt} = \frac{\sigma_{BD}}{\tau_{BD}}\left[1 - \left(\frac{\sigma_{BD}}{\sigma_{\max}}\right)^2\right], \tag{8.92}$$

where the breakdown time $\tau_{BD}$ depends on the electric field:

$$\frac{1}{\tau_{BD}} = \begin{cases} \frac{1}{\tau_{BD}(0)}\left[\left(\frac{E}{E_{BD}}\right)^2 - 1\right], & \text{if } E > E_{BD} \\ 0, & \text{if } E \leq E_{BD} \end{cases} \tag{8.93}$$

and $\sigma_{BD}(0) = \sigma_0$.

The effects of breakdown on magnetic flux compression can be seen in Fig. 8.48. It can be seen that if breakdown does occur, the magnetic field ahead of the shock wave is defined by the breakdown electric field and depends weakly on the initial magnetic field. It has been estimated that the final magnetic field ahead of the shock wave, when breakdown occurs, is two to three times the ratio $E_{BD}/D_{SW}$.

In order to account for the electromagnetic energy radiated by the CSWS, it is assumed that the converging shock wave acts as a magnetic dipole. It was found that the magnetic field is compressed according to the relationship

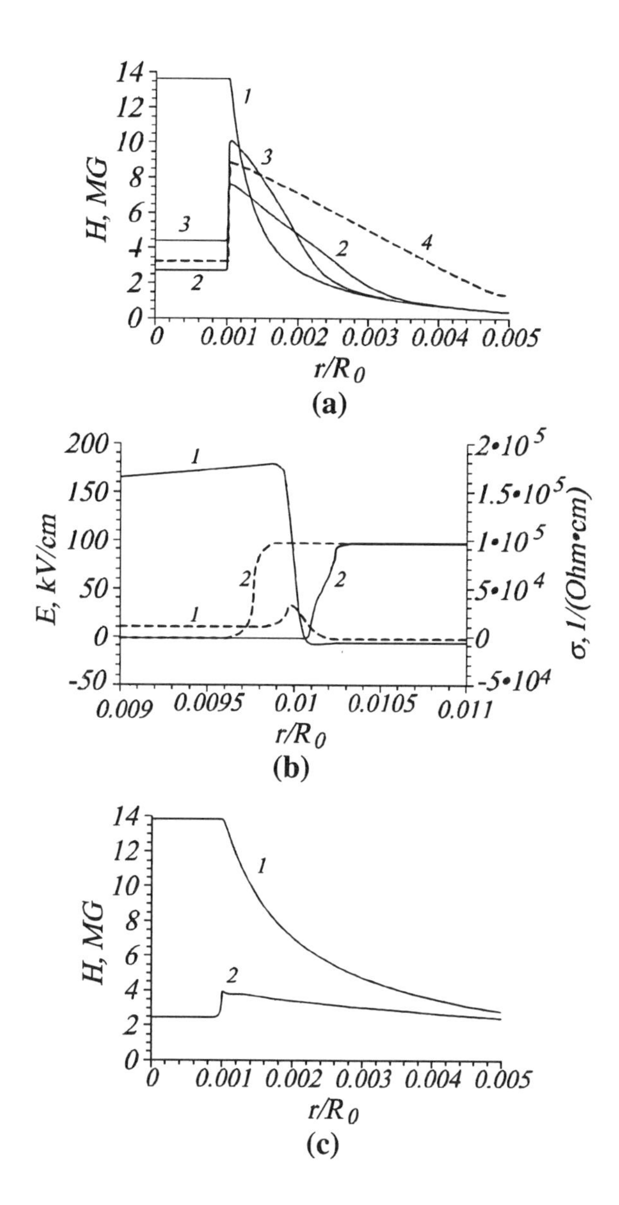


FIGURE 8.48. Effects of breakdown on magnetic flux compression [8.39]. When breakdown occurs, it has been estimated that the final magnetic field ahead of the shock wave is two to three times the ratio of the breakdown electric field to the shock wave velocity.

$$B = B_0 \left[ 1 + \left( \frac{1}{R_{SW} + \frac{1}{R_0}\sqrt{\frac{t}{\mu_0 \sigma_0}}} \right)^2 \right], \tag{8.94}$$

where $R_{SW}$ is the radius of the shock wave and is represented by

$$R_{SW} = 1 - 3.29 t_{SW}^{1.14}, \tag{8.95}$$

if the working body is CsI. The conductivity is found to be

$$\frac{\sigma}{\sigma_0} = \begin{cases} 1.31 \times 10^{-3} E^{4.7} \sqrt{\frac{T}{T_0}}, \text{ for } \frac{T}{T_0} \leq 2 \\ \sigma_0, \text{ for } \frac{T}{T_0} > 2 \end{cases}, \tag{8.96}$$

where $\sigma_0 = 8.89 \times 10^6$ mho/m. The shock temperature is related to the shock pressure by

$$T(^oK) = 1.10 \times 10^4 (1 - e^{-1.28 \times 10^{-3} P}), \tag{8.97}$$

if $P \leq 500$ kbar, and

$$\begin{aligned} T(^oK) &= 1.10 \times 10^4 (1 - e^{-1.28 \times 10^{-3} P}) \\ &\quad + 3.50 \times 10^3 (1 - e^{-7.50 \times 10^{-3} (P - 500)}), \end{aligned} \tag{8.98}$$

if $P > 500$ kbar. There is good agreement between this equation and experimental data as is shown in Fig. 8.49 [8.40]. The relationship between the shock pressure and density is

$$P(\text{Mbar}) = 143 e^{-10.6 \frac{\rho}{\rho_0}}. \tag{8.99}$$

Again, there is good agreement between this equation and experimental data as shown in Fig. 8.50.

Using the above relationships, the magnetic flux density of the magnetic dipole can be found by using:

$$\frac{B_d}{B_0} = 1 + 78.8 (1 - R_{SW})^{11} e^{-(1 - R_{SW})^{11}}, \tag{8.100}$$

which can be used to calculate the radiated field.

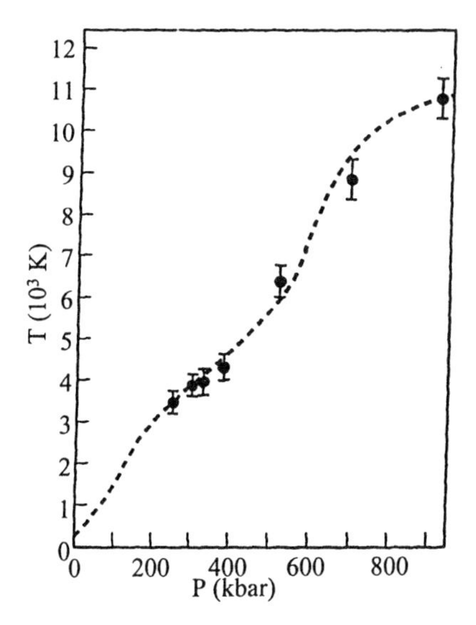


FIGURE 8.49. Comparison of the plot of the pressure versus temperature equation (Eq. 8.98) of state for CsI to experimental data.

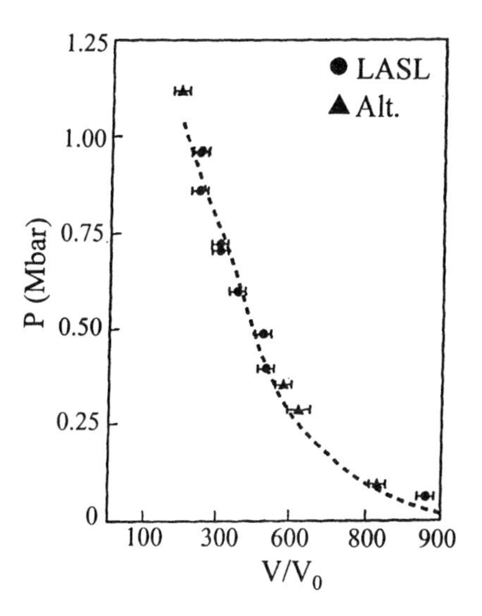


FIGURE 8.50. Comparison of the plot of the volume versus pressure equation of state (Eq. 8.99) for CsI to experimental data.

## 8.4 Summary

In summary, several specific examples in which MCGs have been as the power source have been examined in this chapter. The focus was on directed energy devices, i.e., lasers and microwave sources; however, MCGs can be used in basic physics experiments. For example, in recent articles by Spielman and by Carl Ekdahl, Max Fowler, and Johndale Solem [8.41], they describe six experiments that use the high magnetic fields created by MCGs: 1) a search for the quantum limit phenomena in a two-dimensional organic metal, 2) an attempt to observe high-magnetic-field-induced superconductivity, 3) an attempt to observe a field-induced transition to conductivity in several Kondo insulators, 4) use of the Zeeman effect to break one bond of a quadruply bonded transition metal complex, 5) exploration of the Faraday rotation in samples containing $Eu^{+3}$ and $Sm^{+3}$ ions for calibrating ultrahigh-magnetic-field probes, and 6) exploration of nonlinear Faraday rotation in cadmium manganese tellurium at low temperatures and ultrahigh fields. While these science experiments have both scientific and practical value, the main application for MCGs is as power sources, especially with the creation of the nonexplosive long-pulse 60-tesla magnet at Los Alamos. MCGs provide high power, high current pulses and can be made compact for use in specific applications. Until new pulsed power technologies are developed that can meet these criteria, the MCG will remain the source of choice..

# References


[8.1] *Proceedings of Conference on Megagauss Magnetic Field Generation by Explosives and Related Experiments* (eds. F. Herlach and H. Knoepfel), Euratom EUR 2750.e, Bruxelle, July (1966).

[8.2] *Megagauss Physics and Technology, Proceedings of the 2nd International Conference on Megagauss Magnetic Field Generation* (ed. P.J. Turchi), Plenum Press, New York (1980).

[8.3] *Ultrahigh Magnetic Fields: Physics, Techniques, Applications, Proceedings of the 3rd International Conference on Megagauss Magnetic Field Generation and Related Topics* (eds. V.M. Titov and G.A. Shvetsov), Nauka, (1983).

[8.4] *Megagauss Technology and Pulsed Power Applications*, Megagauss IV (eds. C.M. Fowler, R.S. Caird, and P.J. Erickson), Plenum Press, New York (1987).

[8.5] *Megagauss Fields and Pulsed Power Systems,* Megagauss V (eds. V.M. Titov and G.A. Shvetsov), Nova Science Publisher, Inc., New York (1990).

[8.6] *Megagauss and Magnetic Field Generation and Pulsed Power Applications*, Megagauss VI, Parts 1 and 2 (eds. M. Cowan and R.B. Spielman), Nova Science Publishers, Inc., New York (1994).

[8.7] A.I. Pavlovskii, R.Z. Lyudaev, V.N. Plyashkevich, N.B. Romanenko, G.M. Spirov, and L.B. Sukhanov, "MCG Application for Powered

Channeling Neodim Laser," *Megagauss Magnetic Field Generation and Pulsed Power Applications* (eds. M. Cowan and R.B. Spielman), Nova Science Publishers, Inc., pp. 969–976 (1994).

[8.8] A.I. Pavlovskii, A. Ya. Brodskii, B.P. Giterman, V. Ye. Gurin, A.S. Kravchenko, B.V. Lazhintsev, R.Z. Lyudaev, N.N. Petrov, L.N. Plyashkevich, G.M. Spirov, and L.V. Sukhanov, "Magnetic Cumulation Generator Applications for Photodissociation Laser Powering," *Megagauss Magnetic Field Generation and Pulsed Power Applications* (eds. M. Cowan and R.B. Spielman), Nova Science Publishers, Inc., pp. 977–984 (1994).

[8.9] B.M. Novac and I.R. Smith, "Autonomous Compact Sources for High Power Microwave Applications," to be published in the Proceedings of the Megagauss VIII Conference, Tallahassee, FL, Oct. 1998.

[8.10] R.W. Lemke and M.C. Clark, "Theory and Simulation of High-Power Microwave Generation in a Magnetically Insulated Transmission Line Oscillator," J. Appl. Phys., **62**, pp. 3436–3470 (1987).

[8.11] B.M. Novac, I.R. Smith, H.R. Stewardson, P. Senior, V.V. Vadher, and M.C. Enache, "Design, Construction, and Testing of Explosive-Driven Helical Generators", J. Phys. D Appl. Phys., **28**, pp. 807 – 823 (1985).

[8.12] R.S. Caird and C.M. Fowler, "Conceptual Design for a Short-Pulse Explosive-Driven Generator," in *Megagauss Technology and Pulsed Power Applications* (eds. C.M. Fowler, R.S. Caird, and D.J. Erikson), Plenum Press, New York, pp. 425–431 (1987).

[8.13] C.M. Fowler, R.S. Caird, B.L. Freeman, and S.P. March, "Design of the Marx 101 Magnetic Flux Compression Generator," in *Megagauss Technology and Pulsed Power Applications* (eds. C.M. Fowler, R.S. Caird, and D.J. Erikson), Plenum Press, New York, pp. 433–439 (1987).

[8.14] G.A. Shvetsov and A.D. Matrosov, "Explosive MK–Generator with External Excitation," *Ultrahigh Magnetic Fields: Physics, Techniques, Applications* (eds. V.M. Titov and G.A. Shvetsov), Nauka, Moscow, pp. 263–264 (1984).

[8.15] A.I. Pavlovski, P.E. Lyudaev, L.N. Plyaschkevich, and V.E. Gurin, "Magneto-Explosive Generator," Patent No. 266100, USSR (Russia), (1969).

[8.16] B.M. Novac, I.R. Smith, M.C. Enache, and H.R. Stewardson, "Two-Dimensional Modelling of Inductively Coupled Helical Flux-Compression Generator—the FLUXAR System," *Lasers and Particle Beams*, to be published (1997).

[8.17] J.H. Goforth, A.H. Williams, and S.P. Marsh, "Multi-Megampere Current Interruption from Explosive Deformation of Conductors," *5th IEEE International Pulsed Power Conference* (eds. P.J. Turchi and M.F. Rose), pp. 200–203 (1985).

[8.18] H.R. Stewardson, S.M. Miran, I.R. Smith, B.M. Novac, and V.V. Vadher, "Fuse Conditioning of the Output of a Capacitor Bank to Drive a PEOS," *10th International Pulsed Power Conference* (eds. W. Baker and G. Cooperstein), **2**, Albuquerque, pp. 1115–1120 (1995).

[8.19] J.C. Martin and I.D. Smith, "Improvements in or Relating to High-Voltage Pulse-Generating Transformers and Circuits," UK Patent No. 1, 114713 (1968).

[8.20] V. M. Bistritsky, G.A. Mesyats, A.A. Kim, B.H. Kovalchuk, and Ya.E. Krasik, "Microsecond Plasma Opening Switches," *Atomic, Nuclear and Particle Physics*, **23**(1), pp. 19–57 (1992).

[8.21] H.R. Stewardson, B.M.Novac, S.M. Miran, I.R. Smith, and M.C. Enache, "A Two-Stage Exploding-Foil/Plasma Erosion Opening Switch Conditioning System," *J. Phys. D: Appl. Phys.,* **30**, pp. 1011–1016 (1997).

[8.22] J.E. Vorthman, C.M. Fowler, R.F. Hoeberling, and M.V. Fazio, "Battery-Powered Flux Compression Generator System," *Megagauss Fields and Pulsed Power Systems* (eds. C.M. Fowler, R.S. Caird, and D.J. Erikson), Plenum Press, New York, pp. 437–440 (1990).

[8.23] V.A. Demidov, et al., "Helical Cascade FCG Powered by Piezogenerator," *11th International Pulsed Power Conference*, Baltimore, **II**, pp. 1476–1489 (1997).

[8.24] D.J. Erikson, R.S. Caird, C.M. Fowler, B.L. Freeman, W.B. Garn, and J.H. Goforth, "A Megavolt Transformer Powered by a Fast Plate Generator," *Ultrahigh Magnetic Fields: Physics, Techniques, Applications* (eds. V.M. Titov and G.A. Shvetsov), Nauka, Moscow, pp. 333–340 (1984).

[8.25] J. Benford and J. Swegle, *High-Power Microwaves*, Artech House, Boston (1992).

[8.26] A.E. Sheindlin and V.E. Fortov, *Pulsed MHD-Converters of Chemical Energy into Electrical Energy*, Enrgoatomizdat, Moscow (1997).

[8.27] A.I. Pavlovskii, A.S. Kravchenko, V.D. Selemir, A. Ya. Brodskii, Yu. B. Bragin, V.V. Ivanov, I.V. Konovalov, V.G. Suvorov, K.V. Cherepenin, V.A. Vdovin, A.V. Korzhenevskii, and S.A.

Sokolov, "EMG Magnetic Energy for Superpower Electromagnetic Microwave Pulse Generation," *Megagauss Magnetic Field Generation and Pulsed Power Applications* (eds. M. Cowan and R.B. Spielman), Nova Science Publishers, Inc., pp. 961–968 (1994).

[8.28] A.B. Prishchepenko and V.V. Kiseljov, "Radio Frequency Weapon at the Future Battlefield," Proceedings of EUROEM (1994).

[8.29] A.B. Prishchepenko, "Electromagnetic Weapons in Future Battle," Morskoy Sbornik, No. 3, pp. 71–72 (1995).

[8.30] A.B. Prishchepenko, "Invisible Death of Electronics," Soldat Ukachi, No. 3, pp. 45–46 (1996).

[8.31] A.B. Prishchepenko and M.V. Shchelkachev, "Dissipation and Diffusion Losses in a Spiral Explosive Magnetic Generator," Electricity, No. 8, pp. 31–36 (1993).

[8.32] A.B. Prishchepenko, "Regimes of Work for Magnetic Generator with Capacitive Load and Magnetic Losses," Applied Mechanics and Technical Physics, pp. 31–37 (1991).

[8.33] A.B. Prishchepenko, "A Device for Measuring the Inductance of the Winding of a Coil Explosion Magnetic Generator," Instruments and Experimental Techniques, **37**(4), Part 1, pp. 429–430 (1994).

[8.34] A.B. Prishchepenko, "Device Built Around Permanent Magnet for Generating an Initial Current in Helical Explosive Magnetic Generator," Instrument and Experimental Techniques, **38**(4), Part 2, pp. 515–520.

[8.35] A.B. Prishchepenko and V.P. Zhitnikov, "Microwave Ammunitions: SUUM CUIQUE," Proceedings of AMREM (1996).

[8.36] A. B. Prishchepenko, A.A. Barmin, and O.E. Melnik, "Microwave Emissions by Overcritical Magnetic Field Penetration into the Superconductive Cylinder," Preprint (1996).

[8.37] A.B. Prishchepenko, D.V. Treejakov, and M.V. Shchelkachev, "Energy Balance by Explosive Piezoelectric Generator of Frequency Work," Preprint (1996).

[8.38] A.B. Prishchepenko and M.V. Shchelkavchev, "The Work of the Implosive Type Generator with Capacitive Load," Preprint (1996).

[8.39] H.G. Almstrom, G. Bjarnholt, S.M. Golberg, and M.A. Liberman, "Numerical Modeling of Magnetic Flux Compression by Cylindrical Imploding Ionizing Shock Wave," to be published in the Proceedings of the 7th Megagauss Conference (1996).

[8.40] R.R. Radousky, A.C. Mitchell, and W.J. Nellis, "Shock Temperatures and Melting in CsI," Phys. Rev. B, **31**(3), pp. 1457–1462 (1985).

[8.41] R.B. Spielman, C.A. Ekdahl, C.M. Fowler, and J. Solem, Comments on "Magnetic Fields are Getting Higher on Research Lab Agendas," Physics Today, 11, pp. 11–12 (1998).

# Index